The Anthropology of the Body

Association of Social Anthropologists

A Series of Monographs

13. J. B. Loudon, Social Anthropology and Medicine - 1976
14. I. Hamnett, Social Anthropology and Law - 1977
15. J. Blacking, The Anthropology of the Body - 1977
16. R. P. Werbner, Regional Cults - 1977

A.S.A. MONOGRAPH 15

The Anthropology of the Body

Edited by

JOHN BLACKING

Department of Social Anthropology
The Queen's University of Belfast
Northern Ireland

1977

ACADEMIC PRESS

London · New York · San Francisco

A Subsidiary of Harcourt Brace Jovanovich, Publishers

ACADEMIC PRESS INC. (LONDON) LTD.
24/28 Oval Road,
London NW1

United States Edition published by
ACADEMIC PRESS INC.
111 Fifth Avenue,
New York, New York 10003

This volume derives mainly from material presented at a conference on the anthropology of the body sponsored by the Association of Social Anthropologists of the Commonwealth held at The Queen's University of Belfast, April 3rd—5th 1975

Library of Congress Catalog Card Number: 0-12-103250-7
ISBN: 77-71793

Printed in Great Britain

PREFACE

The ASA Conference on The Anthropology of the Body was intended to focus on the human body as the link between the nature and the culture present in all human activities. It was not motivated by a desire to emphasize physical anthropology, ethology, or the problems of inter-species similarities, but concerned with the development of social and cultural forms and hence indirectly with changes in the quality of life of the members of a single species, *homo sapiens sapiens*.

Many anthropologists since Tylor have been interested in the evolution of culture through the development of the power of mind, given the same human organisms in a variety of historical and geographical environments. We now know a great deal more about the organism and its cognitive capacities, and we know that language can no longer be regarded as the most distinctly human characteristic or the prerequisite of conceptual thought. We are beginning to know more about the workings of the different hemispheres of the brain, and about the links between the development and exploration of the body and the use and growth of intelligence, and between social experience and creative thought. We have a better idea of the species-specific capacities of the human organism, and know how much man depends on interaction with fellowman to develop them. Thus the processes and products of the human mind can usefully be seen as extensions of human bodies in community, in which the basic structures of relationships between parts are probably limited by the structures of the body, but the range of parts that can be related is infinite. Since human bodies are the instruments that both discover and make decisions about self, others, and the worlds of nature and cultural tradition, the anthropology of the body is concerned with the interface between the body and society, the ways in which the physical organism constrains and inspires patterns of social interaction and the invention of culture.

The call for papers began by proclaiming that the conference would be

> concerned with the structure of the body as a source of cultural creativity. Its chief aim will be to examine, in different cultural contexts, uses of the body as a nonverbal medium of expression (eg, in dance, gesture, posture, games, etc.), and extensions of the body in skills, crafts and corporate ritual; to review available techniques of research and notation; and to consider areas of social interaction that ought to be more carefully recorded in the field. Attention will be given to the problem of distinguishing different somatic states, such as those loosely described as religious conversion, ecstasy, trance and possession; and to the notation of changes in posture, expression and movement brought about by sickness or variations of social situation. I hope that we may make some progress towards understanding better the distinctions between body and mind, intuition and reason, manual and mental labour, that are probably derived from experience in culture rather than basic biological structures.
>
> One theory of behaviour that will be explored in the conference is that each human being is born with a latent repertoire of capabilities peculiar to the species. The development of most of these capabilities is subject to the conditions of social interaction of the organism, and human cultures are based on experiences that can be shared and transmitted, and often conceptualised in language.
>
> Thus there may be a moral element to the most 'physical' of bodily activities or experiences. For example, athletic prowess may depend on having a body of appropriate proportions, but social factors may help a less adequate body to perform better than expected.

If the idea of an anthropology of the body adds anything to the programme of research originally advocated by Tylor and his contemporaries, it is in its emphasis on the role of subjective experience in the scientific study of man.

Firstly, there is the subjective experience in the social interaction that anthropologists observe, and secondly there is the subjective experience inherent in the process of observation.

The significance of 'affectually orientated actions' has been recognized by Max Weber and other social theorists and it is often invoked by anthropologists in a rather vague way, but it has not been analysed in detail in studies that concentrate on the minutiae of social interaction. More attention should be given to states of the body and to how people *feel*, both in themselves and about others, when they make the decisions that affect patterns of social interaction, and when they relate to existing ideas or create new ones. This is not simply a matter of investigating the social consequences of sore heads and empty stomachs, or even of altered states of consciousness, but of considering the problem of feeling and emotion in human behaviour and action. Some somatic states can really 'blow the mind' and precipitate important new activities, and whether or not people 'take to' each other may be the most important factor in their decision-making.

The subjectivity of the observer is generally acknowledged even in the most objective of scientific experiments. The social sciences are inevitably more subjective than the natural sciences: no matter how much anthropologists may seek objectivity, or claim to analyse only the actors' views and explanations of their world, anthropological research may reveal more about anthropologists and their own societies than it does about the societies they study. For those who agree with Tylor that anthropology is 'the reformer's science' and hold that an important aim of research is to help us to change our own societies more effectively, this subjectivity is not a serious disadvantage. Anthropology can be seen as a discipline whose development was motivated by collective doubt about values and ways of living in anthropologists' own societies, so that its ultimate purpose is to re-examine the foundations, form, and function of all institutions in those societies, in much the same way that the most rarified and disinterested medical research is intended to improve health and the quality of life. For those anthropologists who consider that their business is not to associate themselves specifically with current problems of their own or any other society, but to study social interaction as objectively as possible and seek to understand social and cultural organization, the very nature of their investigation still renders its subjectivity problematic. One solution is to use subjectivity rather than try to push it aside. In studying man, the observer can relate to other people's ways of thinking and acting in a way that no physicist can relate to the particles he accelerates: and so subjectivity might well be included in analyses and used consciously as a research tool. If I understood him correctly, this is an aspect of what David Pocock described at the 1973 ASA Conference as Personal Anthropology. Unfortunately, there was no paper addressed specifically to this problem at the 1975 meeting.

I have allowed considerable diversity of topic and purpose into this volume, because I wish it to reflect the variety of responses to the original call for papers. If the book conveys what was described as 'the chaotic structure of the conference' it is possibly because the conference was too ambitious. It was intended to take the argument further than Jonathan Benthall's ICA Conference on The Body as a Medium of Expression, which was concerned with expressive qualities and uses of the body, and particularly with nonverbal communication, and to break down the dichotomies of body and mind, emotion and reason, nonverbal and verbal, in our analyses of social and cultural organization. Paul Ekman quotes in his paper Robert Hinde's suggestion 'that we should expect similarities among cultures in signals that concern the personal or the emotional and we should expect differences in movements which take the place of words, which symbolize aspects of culture or which depend on languages' (see p. 76 of this volume). It is also necessary to distinguish analytically between human behaviour and human action, between motor events that happen to individuals and events that are intended to have consequences, as Ekman distinguishes between nonverbal phenomena that are less or more cultural, more 'in the body' or more 'in the mind'. Above all, I am convinced that we should look more carefully at language, the intellect, the mind, reason, and the other characteristics that are considered peculiarly human, as extensions of nonverbal communication, the senses, the body, and emotion, even when they appear to be wholly cultural. While conscious action may seem to be all-important at each moment of our lives, perhaps it is our adaptive behaviour, such as our feeling-responses, that is more important in the long run, and especially in the creation of the structures that outlive our momentary experiences.

The response of participants was, perhaps, not as diffuse as I have suggested, and the reader must judge this from the papers that follow. Not all of them were given at the conference, and some that were presented could not be included in this volume. Brenda Beck, whose excellent paper on 'The symbolic merger of body, space and cosmos in Hindu Tamilnad' had, unfortunately to be published elsewhere (Beck 1976), was one participant who discerned 'the existence of a common field of problems':

> Of these, three issues deserve special mention. First, those present appeared to agree that social anthropologists must give the facts of shifting somatic states and of their expression in bodily form increased importance in their efforts to understand the functioning of social structures

> and codes. Second, it was generally felt that more work must be focused on the qualitative processes of interaction, as these serve both as a foundation and as a constant reinforcement for collective constructs. Finally, there was apparent agreement about the fact that symbolic and social systems become more intelligible when it is recognized that they continually utilize the restricted, yet extremely rich, set of metaphorical possibilities which the body provides (Beck 1975:486).

My own paper is essentially programmatic, and Francis Huxley's raises issues that I overlooked. Paul Ekman reviews in detail the interplay of biological and social factors in body movements and facial expression, and so distinguishes between behaviour and action.

Ivan Polunin discusses in general nonverbal manifestations of health and sickness, and Andrew Strathern describes how the Melpa of Papua New Guinea, expect 'shame' to show on the skin, but only skin deep, so that it cannot produce sickness in people. The problem of fear and the possibility of psychogenic death is taken up by Gilbert Lewis in an analysis of more evidence from New Guinea, and Vieda Skultans looks at nineteenth century ideas on the behavioural symptoms of mental states. Joe Loudon examines attitudes towards the body's katabolic functions.

There follow two essays on ideas and behaviour surrounding male and female sexuality. Katherine Arnold explores the relationship between changing images of men and women in Peruvian society and changing positions in sexual intercourse, and Leora Rosenthal reports an exception to the generalization that female sexuality is uniformly defined as passive when their status is inferior.

Judith Hanna's study of the psychobiological bases of dance introduces a group of papers on two basic modes of institutionalized nonverbal expression: dance and music. Gilbert Rouget briefly disposes of the claim that certain kinds of music have the intrinsic power to precipitate trance. Roderyk Lange demonstrates how different performances of Polish dances reflect different social conditions, and Gerhard Kubik discusses some of the forms and functions of body movement in Angolan initiation schools. Roderyk Lange uses the standard Labanotation for recording movements, while Kubik has devised his own notation. In a detailed study of movement patterns and musical performance on four plucked Afghan lutes, John Baily relates recent developments in the form of the instruments and their music both to extensions of the body and to general social and technological changes. The essay on the Benesh movement notation by the late Rudolf Benesh and his wife introduces a third system of movement notation which, though designed for ballet, can be adapted for a variety of other purposes.

Brian Street's comment on the Benesh system of notation provides the bridge back to the mental uses of the body as the source for systems of classification, as he stresses that the observer has to make critical distinctions between kinds of action and sources of movement. Roy Ellen's analysis of Nuaulu symbolism illustrates how the parts of the body can be divided and sub-divided to provide a rich system of classification, and Anne Sutherland's paper on the use of the body as a social symbol among the Rom introduces an upper-lower distinction which provides an interesting contrast to the more familiar left-right divisions of the body.

Finally, Lorna McDougall brings us back to the body as a source of cultural forms by suggesting that the experience of phosphenes may be an important factor in the creation of symbols.

I would like to express my thanks to all who have contributed to this volume, and to those who participated in the conference: to the British Academy, the Wenner Gren Foundation, and the British Council, for grants to defray conference expenses; to my students and our departmental secretary, Elspeth Larkin, for assistance in preparing and distributing the papers; to David Riches, who administered the conference most efficiently; and to the staff of the Queen's Elms Halls of Residence, who were always, helpful, co-operative, and courteous. Finally, I should like to thank Gerd Baumann for compiling the index.

REFERENCES

Beck, B.E.F. (1975). The Anthropology of the Body, *Current Anthropology*, 16 (3): 486.

Beck, B.E.F. (1976). The symbolic merger of body, space and cosmos in Hindu Tamilnad. *Contributions to Indian Sociology*, 10.

CONTENTS

TOWARDS AN ANTHROPOLOGY OF THE BODY

John Blacking

> It is through my body that I understand other people; just as it is through my body that I perceive 'things'. The meaning of a gesture thus 'understood' is not behind it, it is intermingled with the structure of the world, outlined by the gesture, and which I take up on my own account. It is arrayed all over the gesture itself (Merleau-Ponty 1962:186).

> It is, however, quite clear that constituted speech, as it operates in daily life, assumes that the decisive step of expression has been taken. Our view of man will remain superficial so long as we fail to go back to that origin, so long as we fail to find, beneath the chatter of words, the primordial silence, and as long as we do not describe the action which breaks this silence. The spoken word is a gesture, and its meaning, a world' (Merleau-Ponty 1962: 184. Quotations taken from Poole in Benthall and Polhemus 1975:88 and 89).

I

What I mean by the anthropology of the body probably differs considerably from the meanings assigned by other contributors and by those who came to participate. But although I begin with quotations from Merleau-Ponty that sound phenomenological, and although ambiguous and divisive chatter may reinforce rather than clarify our differences, I hope there can be some consensus about the need to study the biological and affective foundations of our social constructions of reality, even if we cannot all agree that they are the proper concern of social anthropologists.

It may seem premature to try to go further than the contributors in the excellent volumes edited by Robert Hinde on

Non-Verbal Communication (1972) and by Benthall and Polhemus on *The Body as a Medium of Expression* (1975), and presumptuous to think that we can attain a degree of detachment from and objectivity about forces to which we are all subject. But with the disappearance of the nature/culture divide as a relevant boundary of our discipline, it is legitimate and desirable to see the body's means of expression as products of more general environmental and biological processes. Also, by studying cross-culturally the somatic states involved in human behaviour and action, we may be able to describe better the latent repertoire of the human body from which cultural transformations are ultimately derived. The topic merits further discussion if only because it has something to offer the growing number of scholars concerned with the study of interpersonal behaviour: anthropology provides a crucial testing-ground for their theories.

The area of discourse that I call *The Anthropology of the Body* is not a new name for physical or biosocial anthropology, though it must rest on the assumption that human behaviour and action are functions of organisms with a unique set of species characteristics in particular social and physical environments, and that they are adaptive in the context of biological evolution. In claiming that explanations of human behaviour must begin with a full understanding of the body as a total physical system, I am not suggesting that anthropologists should become psychologists or physiologists. Our chief concern is with the cultural processes and products that are externalizations and extensions of the body in varying contexts of social interaction. The division of physical and cultural (or social) anthropology is no longer useful, since it has become clear that the shape of the human body has been influenced by culture, and an apparently cultural phenomenon such as language is biologically based (Lenneberg 1967).

Leslie White claimed that Tylor's 'discovery' of the science of culture may 'one day rank in importance in the history of science with the helicentric theory of Copernicus or the discovery of the cellular basis of all living forms' (White 1949: 4-5). We may agree that patterns of culture, once created by man, can have a force and logic of their own, as one system of controlling energy leads to another, and that they are more pervasive than the forces of production in their influence on human action. But White's view that 'human behaviour is a function of culture.... As the culture varies so will the behaviour' (White 1949:403) ignores biological factors in the invention of culture, even if it is allowed that White is referring to the conscious, goal-directed aspects of behaviour, which I prefer to call 'action'. It implicitly sets man apart from other animals in a way that is not warranted by the facts. Man is not necessarily tied to the products of cultural adaptation as are giraffes to their long necks or were mammoths to woolly coats; but nor is human behaviour separated from that of other animals by a simple culture/nature dichotomy.

Human behaviour and action are only in part a function of culture, and there are no grounds for supposing that at some point in history the development of culture, reason, language, or technological expertise released *homo sapiens sapiens* from the constraints, forces and proclivities imposed by the characteristics of his species and by the evolutionary process that brought the species into being. The ability to transmit culture does not in itself set man apart from other species, because many other animals have some sort of culture. And although the development of culture seems to have been a critical factor in the biological evolution of modern man, we must posit other kinds of behavioural and biological changes to account for the peculiar qualities of human culture.

That is, although it has been argued that 'the large size of the brain of certain hominids was a relatively late development due to new selection pressures *after* bipedalism and consequent upon the use of tools' (Washburn and Howell 1960: 49ff.) and 'the uniqueness of modern man is... the result of a technical-social life which tripled the size of the brain, reduced the face, and modified many other structures of the body' (*ibid*), it can also be argued that the technical-social life was founded on refinements of sensori-motor capacities as much as on the development of a hunting way of life. Special developments in the communication and interaction of organisms may have been more crucial for the evolution of modern man and human culture than the extension of animal 'culture' into a mode of production: cooperation in production presupposes a peculiarly human form of cooperation. Similarly, the faculty of speech is a unique characteristic of the human species that is biologically determined; but it requires specific social conditions for normal development (Lenneberg 1967:373-4). Since a biologically based human faculty requires social interaction, we can say that the social conditions are also biologically determined.

Again, anthropologists claim that man has a special genetic capacity for culture, an attribute of the species without which an organism is not normal, not truly human. If culture is to a certain extent biological, its realization depends on social interaction of the right kind at the right time in the ontogenetic development of the organism. Sense-deprivations such as congenital blindness or deafness may inhibit or change the course of development, but they do not impair the humanity of the organism as does social deprivation. Serious sense-deprivation may interfere with the development of basic features of human cognition, such as symbolic thinking and the ability to differentiate and interrelate categories, as in the case of Helen Keller; but it does not destroy man's more fundamental attributes of educability, plasticity and fellow-feeling.

Our understanding of what may be called the sociology of the body has been advanced by Marcel Mauss, Robert Hertz, and most recently Mary Douglas. I hope that an anthropology of the body may be able to incorporate their insights and techniques of analysis with those of ethologists and biosocial anthropologists. Thus Mauss argues that, 'Dans toute société, tout le monde sait et doit savoir et apprendre ce qu'il doit dans toutes conditions' (Mauss 1936:384), and Douglas writes:

> The social body constrains the way the physical body is perceived. The physical experience of the body, always modified by the social categories through which it is known, sustains a particular view of society. There is a continual exchange of meanings between the two kinds of bodily experience so that each reinforces the categories of the other. As a result of this interaction the body itself is a highly restricted medium of expression (Douglas 1973:93).

There is a further problem, though. Techniques of the body are not entirely learnt *from* others so much as discovered through others. The cognitive consensus that makes both the social and the socio-physical bodies possible is not always fully perceived or cognized. Many things happen to us for which society has no labels. Since I maintain that the body is never infinitely malleable, it follows that there can be tensions between people's training and education in the techniques of the body and their bodies' own inner forces and proclivities. I would even go so far as to say that this tension between 'inner' and 'outer' reality is a crucial factor motivating human behaviour and action, and that the measure of its influence is to be found in the meaning of feeling in any given situation. It is to this kind of problem that Merleau-Ponty addressed himself; but when Roger Poole criticises phenomenology for failing to examine bodily expressivity in a historical context, and rightly argues that 'it is from a specific *historical and ethical* context that the individual derives the expressive possibilities of his body' (Poole 1975:95), it must also be remembered that each body in history is also an event in the history of the body, and that its genetic constitution and its place in biological evolution exercises some constraints on its freedom to act.

II

Study of the *quality* of human feelings and of the structure of affect becomes a priority in an anthropology of the body. After extensive research into affect, Silvan Tomkins summarised the position in 1964:

> (1) *Affects are the primary motives of man;* (2) affects are not private obscure internal intestinal responses but facial responses that communicate and motivate... (3) these communications are more often *understood* than misunderstood; (4) affects have profound effects on cognition and action, and conversely both cognition and action have profound effects on affect. Therefore the problem of measurement involves the measurement of both the *awareness* of affect by the *observer* and by the *self* and also the measurement of the impact of affect on cognition and action, and the measurement of the impact of cognition and action on affect (Tomkins and Izard 1966:vii. Italics in original).

Social scientists frequently describe relationships in affective terminology, or tend to assume that the more often people associate the more they care for each other, but in all the literature on kinship, marriage and religion, there is very little precise information on falling in love in different cultures, on the feelings of spouses towards each other, or on people's experiences of ecstasy, transcendence, or joy in dancing. Feeling is the catalyst that transforms acquired knowledge into understanding, and so adds the dimension of commitment to action. It is the mediator between the body and what is generally called mind, because it provides the value that selects what shall be shared and conceptualized from what remains private sensation. The human mind is basically an expression of the feelings of the social body (social, in the biological rather than nominal sense). To say that such somatic states cannot be described, if only by analogy, is a council of defeat.

Firstly, techniques for investigating emotional states are becoming increasingly sophisticated, and can no longer be glossed over or ignored in detailed studies of human interaction (see, for instance, Ekman, Friesen and Ellsworth 1972, Ekman 1975), and secondly, observation of the nonverbal communication of others may reveal patterns of behaviour that are important in sequences of interaction, but which neither observer nor actors had cognized. Thirdly, since it is possible for an observer to relate to the human objects of his investigation in a way that is not possible in the natural sciences, the observer's body may serve as a diagnostic tool. My knowledge is both generated and restricted by the perceptions and cognitive processes of my society, but through my body I can sometimes understand more than I know through my or another's society, because I have more experience than society labels. It should therefore be our task, as anthropologists, to experience others' bodies through our own bodies and to learn more about some of the somatic states that we can understand but about which we know little beyond the inadequate verbal descriptions of our society. I suggest we can and ought to be rational about feelings, and that conscious subjectivity can be scientifically productive.

It cannot be assumed that somatic states necessarily remain neutral until some social interpretation or value is assigned to them. Some somatic states may have intrinsic qualities that command attention, expand consciousness, and actually suggest their own interpretation. Altered states of consciousness are common experiences for many oppressed peoples of the world, and very often these states seem to postpone rather than promote a positive response to the political situation. But some somatic states may also help a generalized collective consciousness to emerge, and from these feelings, rather than rational assessments of the perceived situation, a more specific class consciousness can develop and serve as the basis for political action. Similar situations may occur when people are affected by non-referential musical experiences: because music can create a world of virtual time in which things are no longer subject to time and space, it can make people more aware of feelings they have experienced, or partly experienced, and so restore the conditions of fellow-feeling, body awareness, educability and plasticity that are basic to the survival of the species. What needs to be known is whether it is the same kind of somatic state that brings about these different results and to what extent its effectiveness depends on its cultural connotations and on variations in social situation. Or are there different kinds of somatic state? There is evidence that some states involve both hemispheres of the brain, and particularly the right hemisphere. Perhaps when the left is cut out we have a state in which the individual becomes like an automaton, but where *both* hemispheres are working with equal energy we have a more enlightened, creative state.

The most dramatic of such somatic states are trance and ecstasy, and they seem to depend very much on the actual or imagined presence of others. Abraham Maslow has argued that such experiences are an integral part of human bodies, and he calls one type a peak-experience. 'Man has a higher and transcendent nature, and this is part of his essence, that is, his biological nature as a member of a species which has evolved' (Maslow 1970:xvi). 'Spiritual values have naturalistic meaning... they are not the exclusive possession of organized churches' (*op.cit.*:4). In other words, people may be 'surprised by joy' and receive divine grace in all its fullness without any of the theological preparation and explanation that is generally considered necessary for the experience to be defined as such.

It is not necessary to recite ethnographic evidence to support the view that throughout the world societies attach supreme value to transcendental somatic states and experiences of bodily resonance, or true fellow-feeling. Nature affects us when we 'belong' to the sky or grass and can identify with trees and flowers. A successful party is one in which the conversation flows effortlessly because people are sensitive to each other's moods and interests. In a good seminar, all present become very physically involved in the discussion and are uplifted and often surprised by the bright remarks that come out of their mouths. Music-making is most exciting when performers and instruments are one, and being played; and this is something that is infectious and felt by audiences. Paintings and sculptures are enjoyed when they cease to be detached objects in space, when the viewer is 'lost' in the rhythms of their colours or shapes and surfaces. Love is most rapturous when it transcends the obviously sexual, when genital arousal subsides and gives way to a more general awareness of the whole body, and bodies are carried away in a counterpoint of movement, of which actual coitus is but one phase. Similar somatic states seem to be induced by drink or drugs or fasting; but I think it would be more correct to say that such devices do not induce the state so much as help to suppress the cultural rules that have inhibited their natural expression.

Given the malleability of the human organism, it is perhaps surprising that people are not more different and that altered states of consciousness do not generate more individualistic, self-centred action. For this reason, amongst others, it is necessary to investigate how somatic states differ, and to what extent, if any, they are really shared. That is, both by measurements of the body and by observing behaviour and recording verbal accounts of it or related to it, we must analyse similar states in different societies and by cross-cultural comparison, distinguish between them. We need to know both how the brain and nervous system are working and what people say and do in relation to these states. How is it that some fire-walkers burn their feet and others do not, that some devotees draw blood with knives and pins and others do not, that some'trance-states' are creative and others turn people into automatons? Are music and dance therapy effective because certain patterns of sound and movement may automatically affect the human body, or because the intensity of the associated social situation generates energy and/or releases inhibitory mechanisms? We need to know which states are or can be consciously shared, so that two or more bodies may be aware of each other in experiencing a similar state. Such states are common in 'religious experiences, and in music and dance, and their accuracy is often validated by the precise diagnosis of pain and sickness derived from corresponding feelings in the body of the diagnostician. What constrains people from developing limitless differences is not simply social convention, but also the bonding of the sensory and communication system of the species (cf. Eibl-Eibesfeldt 1971).

III

The first premise of an anthropology of the body is therefore that society is 'not a nominal being created by reason, but a system of active forces' (Durkheim 1961:495). Human society is not merely *like* a single organism: it *is* a biological phenomenon, a product of the evolutionary process. Thus we ought to talk not of the species man, but of the species man and fellowman. Co-operation and social interaction are not the consequence of rational contract or of habits learned during a long period of infant-mother dependency: they are biologically programmed and a necessary condition for the growth of distinctly human organisms. Durkheim's reference to 'a system of active forces' is surely more than a statement about the nature of society as a social fact which both constrains and creates behaviour; it implies powers of sensory awareness, of resonance and of communication between the individual parts of the social organism. It provides the basis of a science of society by suggesting that human beings are meant to share somatic states and that this species-specific characteristic is necessary for the development of their cognitive processes and capacities.

In spite of overwhelming evidence that people in all societies share somatic states in pairs and often in considerably larger numbers, and in spite of all the music, literature and daily experience concerned with love (or perhaps because of this data), Durkheim's successors have generally avoided the seemingly mystical aspects of his theory. Instead of studying societies as they really are, as systems of active forces, anthropologists and sociologists have studies them as actors and social scientists *say* they are: the fundamental biological solidarity of groups has not been distinguished analytically from their nominal solidarity. Indeed, certain kinds of social behaviour have been referred to as the expression of 'herd instincts', and so deprived of human significance and purpose, as if biologically induced solidarity can only be a negative factor in truly human affairs. Analytical distinctions have been made between nominal and real structures of social interaction, or between manifest and latent function, but they are derived from the observation of behaviour that is expressed largely in spoken language.

Language is only one form of communication, and it can be extraordinarily ambiguous and inaccurate, especially when describing feelings. In general 'Words tag the processes by which the species deals cognitively with its environment' (Lenneberg 1967:334), though sometimes they may in themselves generate ideas. Language, too, appears to be a comparatively late development in human cultural history, though admittedly it is a crucial factor in the behaviour of the species that is the concern of an anthropology of the body. The earlier beings who made Acheulean 'handaxes' clearly had a common material culture, and therefore presumably conceptual thought, and they are regarded as human; but it is now widely accepted that they did not speak a language as we know it. Thus, although language is almost certainly a basic ingredient in the entire cultural history of *homo sapiens sapiens*, the roots of that history, and perhaps even of language itself, were established with other modes of communication (see, for instance, Hewes 1973 and Livingstone 1973).

Crucial factors in the development of cultural forms are the possibility of shared somatic states, the structures of the bodies that share them, and the rhythms of interaction that transform commonly experienced internal sensations into externally visible and transmissible forms. This process does not in theory require spoken language or even linguistic concepts, though in practice language use and its cognitive consequences are an integral part of the experience of normal human beings. The shared states of different bodies can generate different sets of rules for the construction of behaviour and action by means of repeated movements in space and time that can be transmitted from one generation to another. Thus an infinite number of cultural variations are possible, based on the variables of shared somatic states, movements of the body, numbers and shapes of organisms involved, dimensions of space and durations of time. The effective operation of these species-specific capabilities may be restricted by external constraints such as the dulling of the senses or the restricted definitions of fellowman imprinted by some socio-cultural systems. The human proclivities for relating to others in terms of feeling-response and for active co-operation with available people are blunted from the earliest age. 'Feelings' towards others are channelled into appropriate domains on the basis of prescribed statuses, roles, and social situations, rather than allowed to develop according to inner direction; and attempts to co-operate are restricted by the overriding demands of others who have the physical maturity to control social interaction in their own interest. As children become fluent in language, so the barrier between biosocial proclivity and cultural constraint becomes even more acute. Children who attempt, but are not allowed, to call their parents by the personal names that the parents use in speaking to each other, may find it harder to love and react spontaneously even when encouraged to do so, because they thereby learn that affection towards individuals must be tempered by the required relationships of authority. This does not estrange parents from their children, however, because others to whom children might feel equally affectionate are defined as distant, and this nominal classification reinforces the real physical presence of the close group and concentrates affective impulse.

Shared somatic states are consequences of the sensory and communication system of the human species which is the basic condition for social interaction. Telepathy and bodily empathy, for instance, are not paranormal but normal, although they are often suppressed or allowed to atrophy in cultures in which excessive importance is attached to verbal communication. As soon as people begin to analyse certain kinds of experience, they begin to note that they may happen too often to be coincidences. For instance, one may feel a desire to ring someone; as one approaches the telephone it rings, and there at the other end is the voice of the person whom one was about to ring. I would argue that this is a special expression of the general bond between man and fellowman, in which a powerful feeling is translated into social action. A good example of the opposite situation, the effect of social interaction on somatic states, is discussed by Tiger (1975), who quotes Martha McClintock's (1971) report of 'a significant increase in synchronization' of menstrual onset among room-mates and close friends.

IV

The second premise of an anthropology of the body is that every normal member of the species possesses not only a common repertoire of somatic states and a common potential for the altered states of consciousness that have been classed, especially by humanistic psychologists, as part of normal human behaviour (see, for instance, Rowan 1976), but also the same specific properties of cognitive function. We must distinguish between performance and competence, between what people do and what they are capable of doing. If some human beings can perform certain skills, it should be possible for any member of the species to do so, given the appropriate social and cultural environment, and similar opportunities and incentives. This is a logical consequence of the widely accepted principle of the psychic unity of mankind, but its implications are not generally pursued. Even Lévi-Strauss succumbs to the elitist views of his own culture when he argues that whereas any reasonably educated person could write poetry, the composition of music requires special gifts that are not common to all members of the species (Lévi-Strauss 1964:26).

The biological foundations and the cultural forms of language provide a model that might be useful in analysing other fields of human action. All normal human beings can speak a natural language, and there is evidence to suggest that musical competence is likewise based on species-specific capabilities, even though the incidence of musical performance varies considerably from one society to another (Blacking 1971 and 1973). Human behaviour is what humans do, and most things that one human feels can be felt by another. The chief problem is to identify significant domains of social behaviour, and it is in this respect that biological boundaries (as in language) may be usefully compared with different societies' and anthropologists' categorizations of areas of behaviour. That is, there is clearly a domain of language that can be partially isolated biologically and culturally. Even if other domains such as politics, religion, music, kinship or economic life are recognized in all societies' cultural categories, we need to know how, if at all, such categories relate to different or related sets of biological functions, and above all to kinds of shared experience that can exist among human beings.

The first and second premises can be linked to make two related propositions. In the shared world of culture and society, and allowing for certain limitations imposed by sexual differentiation, *what one human organism can do, any organism can do; but no human can do anything, or even become human, without fellow humans*. Human behaviour and action are extensions of capabilities that are already in the body, and the forms and content of these extensions are generated by patterns of interaction between bodies in the context of different social and physical environments. The question immediately arises: if there are basic human skills derived from species-specific capabilities, why do they not emerge ontogenetically in all members of the species, like language? I suggest that, as with language, such skills may be programmed to a maturational schedule, and that if they are not utilized in society at the appropriate time they may become atrophied.

We ought to observe the precise social environment of the emergence of so-called child prodigies, for instance, as if their development were normal rather than exceptional. Thus we might learn more about the optimum conditions and stages for learning certain skills. Some systems of education have been based on such theories, and most recently the Suzuki method has demonstrated that socialized learning of the violin at a very early age can produce remarkable results.

This begs the question 'what is a normal human being?' It is sometimes argued that significant progress in the evolution of cultures, especially in activities called 'the arts', has frequently been made by people who are not only deviant but who have been or should be described as abnormal. Within anthropology, there has been debate about supposed mental illness that finds institutional expression in spirit-mediumship or certain professions. It was even claimed by a psychiatrist that the 'antic disposition' (*Ukuthwasa*) of the Xhosa which precedes training and often a successful career in traditional medicine, is evidence of fundamental abnormality. Thus, in discussing a number of cases examined, he wrote, for example: 'Evidence of madness in the family: his father and grandfather were doctors'! Although there are different definitions of normalcy both within the medical profession and in different cultures, I take as normal any human being who has not the misfortune to be born with a serious physical impediment that would incur medical classifications such as mentally retarded, deaf-mute, or palsied, and require special care.

A further problem, which I am not qualified to discuss in detail, is the distinction between normal individual genetic variation and an abnormal mutation. I would not challenge the notion of mutation as a possible explanation of certain exceptional human beings, but such exceptions are not of much interest to anthropology, as the science of culture, unless the same mutation occurs many times and is a necessary condition for a particular cultural trait. For instance, it has been suggested (Smith and Worthington 1967) that Nicolo Paganini suffered from a heritable disorder of connective tissue, such as the Ehlers-Danlos or Marfan's syndromes, which may have contributed to the innovations that he made in violin technique. Once made by this man with an extraordinary body, they became part of the technical repertoire of virtuosos, and indeed no really new device for the violin has been invented since his time. Even if this diagnosis is correct and the disorder was caused by a gene, it is only of passing interest, since hundreds of fiddlers have subsequently been able to play Paganini's music without suffering from this physical disorder. The 'uniqueness' Paganini's body is significant in the development of culture only in the way that a dream or flight of imagination spawns a poem or a new machine. The same innovations *could* have been made by any imaginative fiddler, and the fact that they were not, does not necessarily prove that their inventor was significantly abnormal. In fact, one could argue the exact opposite: that he stretched his normal human capabilities to the maximum in violin-playing, as subsequent fiddlers have been compelled to do, so that what was extraordinary for Paganini's time is now expected of any first-rate fiddler.

Since all individuals are biologically unique, what is most interesting in the analysis of culture is the experiences and ideas people can and do share. Mozart was undoubtedly a remarkable individual, but what is most interesting about him is that he was so extra-*ordinary*, so much in command of his human capabilities, that he could resonate with millions of other humans of different epochs and cultures.

Clearly some social (or more often anti-social!) behaviour is idiosyncratic and may be brought on by somatic states ranging from a stomach-ache to an epileptic fit. Somatic idiosyncracies are the concern of an anthropology of the body only if they need be invoked to explain a person's action in a particular social situation, or if they are taken up by a social group as a part of its culture. An anthropology of the body must be concerned primarily with somatic states and capabilities that normal organisms can share, either by varieties of phylic communion or by social experience in a common culture.

V

The third premise of an anthropology arises from the first and second premises. If the basic condition of human society is a generalized state of fellow-feeling that can be perceived through the sensations of individual organisms, nonverbal forms of interaction are fundamental. Kinesics and proxemics are concerned with microscopic aspects of human movement (see, for instance, Birdwhistell 1970 and Hall 1968), and demography and population genetics with the patterns of adaptation of larger social groups. An anthropology of the body is the meeting point of the micro and the macro: wherever possible, it must be able both to detach body movements from verbal associations, and to specify the levels at which bodies respond consciously or unconsciously to the evolutionary relationship of the species and its environment at any given time or place.

Bateson has stressed that 'the discourse of nonverbal communication is precisely concerned with matters of relationship - love, hate, respect, fear, dependency, etc. - between self and vis-à-vis or between self and environment and... the nature of human society is such that falsification of this discourse rapidly becomes pathogenic' (Bateson 1973:388). Culturally prescribed, word-based categories and systems of thought may conflict with information that comes from within the body, and so build up tensions in feeling-states that find expression in behaviour and social action.

A useful distinction can be drawn between the particular ritual forms of different cultures and a universal biological model that might be called proto-ritual, a shared somatic state of the social body that generates special kinds of feelings and apparently spontaneous movements and interaction between bodies in space and time. (The situation is probably *not* spontaneous biologically, though it may appear to be so). States of proto-ritual may be compared to the ritualization of some species of animals, but in man they are distinguished by being situational rather than maturational. That is, there is a 'species-specific, and hence adaptive, ritual behaviour of man, as distinct from the ritualistic behaviour of some men' that co-ordinates in time all organisms who

> happen to be within an immediately observable and/or ultimately conceivable space.... Experiences in time and space are therefore externalized in an *externally* given situation, so that theoretically any number of members of the human species can interact ritually at any time of life. For most animal rituals, the choice of time and place seems to depend on *internal* factors such as the stages of maturation of those involved' (Blacking 1976:10).

Observation of the cultural *forms* of such states shows transformations of individual facial expression and body movement, and of the corporate movements in space and time of the bodies involved. 'Waves' of feeling are generated in the body and between bodies, not unlike fits of sneezing or hiccoughs, and discrete sequences of tempo and patterns of movement can be discerned, analogous to the ebb and flow of a piece of music. Obviously, the forms that the movements take are much affected by their cultural framework; but at the same time (and especially in situations of communal worship when people submit to the power of the spirit) a general pattern of interaction and movement emerges, which, though often related to cultural experience, is shaped from within the body and monitored by patterns of energy flow that transcend the actors' conscious attempts to manipulate the situation. In *Natural Symbols*, Mary Douglas has stated that 'communication depends on the use of symbols' (Douglas 1973:29), and particularly the symbols of language, and that 'there can be' no 'organization without symbolic expression' (*op.cit.*:73). She has also argued that emotion is not susceptible to discrete division. In the proto-ritual situations to which I have referred, there is a level of communication and organization that takes place without language and without symbols, and emotion is often expressed in contrasting patterns of movement that *are* susceptible to discrete division. That is, not all of the events are directly related to the shared cultural categories, and it ought to be possible to distinguish the particular forms from the universal by careful analyses of body movement.

Clifford Geertz also rejects the notion of an innate, pre-symbolic tendency to order in human interaction, and extends the argument beyond the field of intentional communication to feeling:

> Undirected by culture patterns - organized systems of significant symbols - man's behaviour would be virtually ungovernable, a mere chaos of pointless acts and exploding emotions, his experience virtually shapeless (Geertz 1975: 46).

> Without the guiding patterns of human culture... man would, quite literally, not know how to feel (Geertz 1964:47).

It is perhaps unfair to quote Geertz without reference to Susanne Langer's argument: 'by virtue of our thought and imagination we have not only feelings, but *a life of feeling*' (Langer 1953:372. Italics in original). Geertz's statements are more acceptable if he is referring to a life of feeling rather than feelings, and in another context he suggests that this is so:

> In order to make up our minds we must know how to feel about things; and to know how we feel about things we need the public images of sentiment that only ritual, myth, and art can provide (Geertz 1975:82).

Reverting to Tomkins's argument quoted above (Tomkins and Izard 1966:vii), it is necessary to distinguish between feelings and a life of feeling, and in particular to assess the impact of feelings on thought and imagination and to recognize that even 'exploding emotions' are not pointless, haphazard responses. Karl Pribram (1967) has also suggested that emotions should be seen as neural processes that structure input as an alternative to instrumental action. It is not the business of an anthropologist of the body to investigate how these states operate physiologically, though it is necessary to recognize that because they are situational their expressive forms will vary according to time, place and number of people involved, and that although they arise from acceptance of the automatic complexity of the body, they are not 'abandoned'. It is their very repetition and concentration that makes them so distinctively human: for an unspecified time, the world of the actors stands still as they react to each other and the perceived permanence of their social life, rather than to the impermanence of their anatomical existence. The really astonishing quality about human beings is not change but *non*-change, the ability to stop the world and repeat actions, to become detached from the constantly changing environment and cycle of life.

It is this process that breaks 'the primordial silence' of which Merleau-Ponty wrote, and amongst its features that must be measured by anthropologists are its tempi and patterns of movement, for these are keys to its meanings. As Morton Marks has effectively demonstrated in a study of Afro-American music (Marks 1974), ritual is most emphatically *not* communication without information. All that is known about the minds of non-human animals is derived from observation of their movements, gestures and non-verbal communcation: much can be learnt about the human animal when it is studied in the same way.

Many of the body's functions depend on the sequence and timing of events, and proper timing is a sign of good health. This is particularly true of language, where most of the 'CNS disorders of speech and language may be characterized as disorders of timing mechanisms' (Lenneberg 1967:218. See also p. 116ff.). Metre and rhythmic coordination are important in speech therapy, and more generally in the treatment of mentally handicapped children (Nordoff and Robbins 1971:30-32, 103 etc.). Research has not yet shown how the various natural rhythms of the body intermesh, but there is little doubt that the circadian cycle is an important organizing principle in human physiology (Luce 1973). These phenomena are part of the automatic complexity of the body, and in so far as they are only dimly (if at all) perceived through changes of feeling, they are part of 'the primordial silence'.

Biological rhythms may influence feelings and performance, but they do not necessarily determine *experience* of time: this is a consequence of the coding of events, in which the duration of time is perceived in relation to the amount of information processed or recalled (Ornstein 1969). Now the coding of events is essentially an activity of the body, and particularly of the brain, that breaks the primordial silence. It is made possible by recurrent somatic states in which time appears to stand still.

Rationalizations of transcendental somatic states and experiences of bodily resonance vary from one culture to another as much as the titles that Rodin gave to a sculpture before he found one that satisfied him. Probably because of this, the forms of these highly valued 'peak-experiences' have received less attention than 'the chatter of words' surrounding them. Social systems are rightly conceived and studied as systems of values; but systems of values and levels of agreement between members of a society cannot be understood adequately at the verbal level only, either through actors' rationalization of events or an observer's cumulative observation of them. People may not only deceive others; they may deceive themselves, in their reification of their symbolic and social systems. For instance, a system of body symbolism, though appearing to spring from the real needs and nature of the body, may subtly express the interests of a dominant group, and become a strait-jacket that inhibits self-expression and distorts posture. Again, the values of a social system that might seem adaptive in the context of a particular culture could be biologically non-adaptive, owing to faulty rationalizations of biosocial forces; the force of this contrast between real and rationalized values would generate conflict. As Maslow has argued, 'dichotomizing pathologizes (and pathology dichotomizes). Isolating two interrelated parts of a whole from each other, parts that need each other, parts that are truly "parts" and not wholes, distorts them both, sickens and contaminates them' (Maslow 1970:13). When the 'mind' is artificially separated from the body, or thought is separated from the feelings and movements of the body that generate it, both parts suffer. Cultures are extensions and adaptations of feelings, and especially of fellow-feeling, that are universal to the species, and if the biological base of behaviour is ignored or misrepresented, laws of nature must inevitable override man's attempts to escape them. Many conflicts arise from inconsistencies between the social forces that bind human societies and man's inadequate cultural descriptions of social situations.

How can we measure the apparently invisible, and how can we presume to say that something not recognized by the actors is real? It is as real for purposes of explanation as atoms, genes, or successive differential counts of leucocytes: that is, it may be inferred from certain kinds of observable behaviour. It is *not* invisible: it can be observed in the breakdown and transformation of organisms, such as the morbidity and epidemiology of a community or changes in gesture and posture, especially inconsistency of posture in contrasting roleplay. Ekman and Friesen, in studies of facial expression and hand movements, have shown how people's movement patterns vary in sickness and in health, and how it is possible to detect deception from the body or face (Ekman and Friesen 1973 and 1974).

We need to contrast the movements of people subjected to authoritarian situations with the movements that they make in a familiar environment. Although the movement patterns of Jamaicans, Pakistanis and other immigrants may vary in their homes according to certain cultural traits, in the official situations of passport and employment offices and police stations they may respond with common sets of movements. We need to study the movements, gestures and body contact that babies and children experience, as these experiences probably exercise the most profound influence on conceptual thought, social behaviour, health and the development of skills.

At the same time, we have to recognize the possibility, argued by Cicourel (Cicourel in Benthall and Polhemus 1975: 220ff.), that oral and gestural modes of communication differ cognitively, because there are two alternative coding systems for verbal symbolic processes and visual imagery. In this respect, the thinking processes of the deaf might provide crucial evidence. Cicourel argues that the thinking of the deaf involves language, but, as I have suggested, I would expect both oral and gestural thought to be aspects of a deeper movement system of the body.

VI

The fourth premise of an anthropology of the body is that the mind cannot be separated from the body. This is linked to the third premise, for it is in the areas of nonverbal communication, especially dancing and music, that we may observe mind at work through movements of bodies in space and time. Moreover, it is in the verbalisation and mentalisation of the essentially nonverbal, manual 'arts' that we see the most insidious and dehumanizing attack on the further evolution of man and culture. The mind/body dichotomy can be partly explained as a cultural phenomenon, as an artefact of certain modes of production and social formations, reflecting the division of manual and mental labour in production; but there is also evidence that it can be the product of a lopsided use of the body, which could therefore energe even in a truly socialist society where mental and manual labour are not separated or falsely evaluated. Dance, music, painting and sculpture are really forms of nonverbal manual labour, but they have been invested with the 'dignity' of quasi-verbal mental labour: the artists 'speaks' to lesser mortals of 'higher' things, and so his consciousness is distinguished from his practice of a craft. This props up the bourgeois argument for class difference, namely, that one class must do the manual labour so that another can look after the mental needs of society. Thus even where the mode of production is reflected in an ideological separation of body and mind, society admits to the importance of a form of activity that is at once manual and mental and mediates between the world of production and the world of thought and mental constructs. The question then arises to what extent such 'arts' are part of the ideological superstructure of social life necessary to resolve the contradictions created by the division of manual and mental labour, and to what extent they are a special form of manual labour, and, if so, why they exist at all in their present forms or why they ever came into being.

If, as products of mental labour, they are instruments of self-definition in man's development of his senses, they are only part of a phase of human history that will pass when 'the non-labour of the few' has ceased to be a condition 'for the development of the general powers of the human head' (Marx 1973:705). When man has attained ownership of the senses and 'overcome the alienation from his true essence.... Art as such will be superseded, as Art and Life will have become one' (Hess 1973:313). If 'ideas do not exist separate from language', then indeed the Arts as such will disappear when the ideas that they express have been worked out through the resolution of social contradictions. But if the purpose of the Arts is to reflect ideas through different kinds of 'artistic' manual labour, why should they use nonverbal modes of communication? Why try to 'say' in dance or music what can equally or better be said in words? Music, for example, is not as much an immediately understood language that can be expected to produce specific responses as it is a metaphorical expression of feeling. It cannot communicate anything new to its listeners except unfamiliar patterns of sound which, through tonal and rhythmic contrasts, may elicit nervous tension and motor impulse - provided that they are presented in a culturally familiar setting and as behaviour that is perceived as musical. Why should music have survival value if it is merely a *reflection* of the contradictions inherent in the division of manual and mental labour?

It is precisely because dance and music are special, nonverbal modes of communication that they have value in human society.

Recent research has suggested that in the lateralization of brain function

> the left hemisphere is predominantly involved with analytic thinking, especially language and logic. This hemisphere seems to process information sequentially.... The right hemisphere, by contrast, appears to be primarily responsible for our orientation in space, artistic talents, body awareness, and recognition of faces. It processes

> information more diffusely than the left hemisphere does, and integrates material in a simultaneous, rather than linear, fashion' (Ornstein 1973:87).
>
> The specialization of the two hemispheres seems to be unique to human beings and to be related to the evolution of language.... The complementary workings of our two thought processes permit our highest achievements, but most occupations value one mode over the other... *a complete human consciousness should include both modes of thought* (op.cit.:92. Italics mine).

The mind/body dichotomy itself can therefore be seen as an extension of the body, expressing symbolically the separate functions of the left and right hemispheres of the brain. Because much social life is dominated by language and linear thinking, the arts are necessary not merely at an ideological level as mediators between mental and manual labour: they have a biological function in activating both hemispheres of the brain and so contributing to a more complete human consciousness. Some musical capabilities appear to be located in the right hemispheres, whilst other segmental aspects of musical structures must surely derive from left hemisphere operations. If this is so, then music is a particularly suitable co-ordinator of brain activity.

The consequences of lateralization of brain function are that unequal values attached to mental and manual labour may be reinforced or reduced according to societies' emphases on different bodily activities. However, since under ideal (?or normal?) circumstances both hemispheres should be fully operational, there is always the possibility that such 'open' somatic states may occur spontaneously and affect a hitherto closed mind-body, even in a society whose patterns of mental constructs and bodily activities are heavily biased against it.

The dismal mind/body dichotomy provides a convenient escape route from many tough problems, such as explaining concepts like mercy, pity, truth and justice in evolutionary terms. How could such concepts have evoked from people really deep feelings, as distinct from lip-service, unless they had some adaptive value for the species? How could such concepts be detached from feelings, or have come into being, without biological foundations? Could they ever have been invented, if they had not first been felt? If we say that these concepts are largely autonomous creations of the mind, we do not have to ask such awkward questions: they can all be related to actual social situations, and can even be seen as ideas that arise from strategic, self-interested responses to social problems. (For instance, I invent and invest with value the idea of mercy when a prisoner-of-war, who should have been killed, saves me from death.) The important common denominator is how people's *bodies* respond to such situations, rather than the supposed mental gymnastics by which the same concepts may have been arrived at in different societies and cultures. If society is seen as a system of active forces in which every body is highly sensitive to other people's feelings, then mercy, pity, truth and justice can be understood as distinct feelings, with corresponding conceptualizations, that arise when the overriding and biologically necessary experience of fellow-feeling is outraged in some way.

The pathological mind/body dichotomy is also reflected in contradictory cultural conceptualizations of physical experience. For instance, 'falling in love' and 'being in love' are usually distinguished from 'having sex' as higher forms of an essentially animal activity. 'Falling in love' generally refers to an association of bodies that is supposed to include more than obviously sexual attraction, and can be described as an activity of the body that depends on feelings in relation to another body; 'having sex' refers to a situation in which a partner could be performing sexual functions adequately and pleasurably while thinking of something or someone else, and can be described as a learnt activity of the mind that depends on concepts of women, manliness, sexuality, and so on. Curiously, for a society that believes in a mind superior to the body, we extol behaviour in which the mind is said to be absorbed in the body, in contrast to a sexual act in which the mind may be detached from the body.

My concern for an anthropology of the body rests on a conviction that feelings, and particularly fellow-feeling, expressed as movements of bodies in space and time and often without verbal connotations, are the basis of mental life. If the early Acheulean tools are evidence of a new way of thinking, it is because they are primarily the products of new ways of feeling and moving the body. The most striking feature of these early artefacts is not their function, as instruments of production, but their form, as products of a being who exhibits a sense of order that may be called 'aesthetic' or 'artistic'. They reflect concentration, a systematic pattern of execution, repeated body movements in space and time and a continuity of style that implies a common culture, but above all, a feeling for form.

J.Z. Young has argued that the creation of new aesthetic forms 'has been the most fundamentally productive of all forms of human activity. Whoever creates new artistic conventions has found methods of interchange between people about matters that were incommunicable before. The capacity to do this has been the basis of the whole of human history' (Young 1971:519). Gregory Bateson has pointed out that 'unaided by art' consciousness 'can never appreciate... the *systemic* nature of mind' (Bateson 1973:118). The essential function of the artistic process is to mediate between the impermanent and permanent in man to capture the force of feeling with forms that must always begin as extensions of the body. If the creation of a new aesthetic form may seem to be a product of conscious thought. I think that most creative artists will agree that it begins as an 'explosion' of the body which produces apparently spontaneous results.

The use of mirror forms in music and dance provides the most interesting example of an 'explosion' of the body, because they are forms that man could not possibly discover except through his body. There is a logical sequence from the observation of mirror forms in nature to their use in design, but not to their use in music and dance. If, however, mirror transformations are part of the cognitive functions of the body, they could be applied creatively in any field without prior observations or experience, as in dance and music. They could be a basic biological structure, responsible for mirror forms in the human body and in nature in general. It is also possible that the perception and creation of mirror transformations emerges only through experience of social interaction. What is most significant is that they can be both sensed and used at an unconscious level (Schoenberg 1951: Ch. 5), and that they can be an important factor in musical communication. That is, even when composers exercise their conscious minds to produce a unified musical structure, the finishing touches to their work are often done by the body; and because it is a human body that is creating symbolic order with the minimum of 'cultural' interference, the music resonates with a wide variety of people who may share little or nothing of the composer's background (cf. Reti 1961, Walker 1962). It may not obey the rules of the conservatoire, but audiences take it to their hearts, because it reflects what Bateson calls the 'algorithms of the heart' (Bateson 1973:112).

Although music and dance are forms of nonverbal communication, they are always social facts, and so they can never be completely free of the verbal concepts and constructions of reality that predominate in different societies. It is not simply that words may be set to music and so affect its structure, but that the 'languages' of music and dance may be required to express other sets of symbols, and that music and dance may be precisely demarcated from non-music and non-dance. Nevertheless, it can be shown (cf. Ruwet 1972, Blacking 1974) that there are forms intrinsic to music and dance that are not modelled on language. We may look beyond the 'languages' of dancing, for instance, to the dances of language and thought.

As conscious movement is in our thinking, so thinking may come from movement, and especially shared, or conceptual, thought from communal movement (Blacking 1976). And just as the ultimate aim of dancing is to be able to move *without* thinking, to *be* danced, so the ultimate achievement in thinking is to be moved to think, to *be* thought. It is sometimes called inspiration, insight, genius, creativity, and so on. But essentially it is a form of unconscious cerebration, a movement of the body. We are moved into thinking. Body and mind are one.

One immediate objection to the notion that the concept of mind is really a way of talking about man's social *body*, rather than some almost autonomous phenomenon, comes from evidence which suggests that the moral element is a critical factor in many bodily activities and experiences. I have already mentioned fire-walking, in which it is sometimes said that mind conquers body, or matter. In more mundane fields, athletic prowess or pianistic ability may seem to depend on having a body of appropriate proportions, and yet a mysterious quality such as determination or will may help a less adequate body to perform better than expected. Similar explanations of survival and recovery are often given in cases of acute sickness, or even conquest of a common cold.

When the automatic complexity of the body, and particularly of the social body, is better understood, I think it will become apparent that many mind-over-body situations can be explained in biosocial terms. Firstly, in sickness the purely physical effects of concentration of other people's energy flow may do more for a patient than his own 'will', and secondly, the physical activity that is incidental to an effort of 'mind' probably does more than anything else to allow the body's healing process to operate; and if the body is beyond repair, all the will in the world is unlikely to be of much help. In modern hospitals the importance of moving sick bodies, even soon after a strenuous operation, is now generally recognized; and a number of studies by James Prescott and others have shown the importance of movement, touching and sensory experience in mental development (see, Prescott 1971). There are also many physical activities, such as communal dancing and music, that generate more rather than less physical energy. Moreover, the secret of many athletic, gymnastic and musical techniques is to be found in relaxation and rhythm, and leaving the body to do its own work after a period of practice - which is surely the very opposite of mind over body!

As I see it, the anthropology of the body is not seeking for instinctual bases of behaviour, and certainly not opposing instincts with conceptual thought and verbal speech, as Konrad Lorenz seems to do (Lorenz 1967:29). It seeks a better understanding of general principles of human action by studying cultural variety in relation to biological proclivities and constraints and the general framework of evolution. I am particularly concerned with behavioural capabilities that, like speech, combine specificity with plasticity, and with the possibility that many human capabilities are best developed, and perhaps only capable of development, during specific periods in the growth of the organism.

I am concerned with what Alland calls the human imperative, but I do not share his view that the cognitive system on which human behaviour in specific groups is based, 'consists of classifications in which the environmental field (including other human beings) is divided according to an *arbitrary* but traditional system. Out of this classification system develops a set of rules for action' (Alland 1972:152. Italics mine). The case of music, which Lévi-Strauss has called 'the supreme mystery of the science of man' (Lévi-Strauss 1964, see also Blacking 1975) convinces me that at least this area of human action follows rules that are often far from being arbitrary, no matter how arbitrary the division of music and non-music and other aspects of the classification system may seem to be. What is more, the deep structures of musical behaviour can often be measured with a satisfactory degree of objectivity. For instance, if we take a large sample of performances on the same instrument by different musicians within the same tradition, and also find out from performers and audiences from whom and where the melodies were learnt and from where they originate, we can compare the variations in musical performance with folk classifications and perceptions. We may then find that even if a group of Irish fiddlers should claim that they were influenced by a particular individual who comes from County Donegal or County Clare, their continued residence in Country Tyrone has influenced them to develop styles that betray the characteristics of that region. Their perceptions of the music *per se* could be further tested by asking them to identify the provenance of a series of recordings, and to note whether their answers show consistency in recognizing individual or local styles.

The study of feelings, of systems of nonverbal behaviour and of the extensions of the body in culture adds important dimensions to the stock-in-trade of social anthropologists. It can reveal more fundamental, long-term patterns of interaction between social bodies and their environment that transcend what seem to be more immediately pressing socio-economic realities. I have been accused of choosing an irrelevant topic for serious anthropological study, and one that is scandalously inappropriate for a conference in Northern Ireland in 1975. If I felt that there was any action that I could usefully take to relieve famine, economic exploitation, class oppression or racism in any part of the world, the last thing I would do would be to convene an anthropological conference to talk about it. By and large, I accept the criticisms of colonialist bias in our discipline and welcome the current concern for the plight of ethnic minorities, for problems of development, and for the application of the theory of modes of production to anthropological data. All these matters are very important, but I do not find them problematic: they seem to call for political action rather than theoretical discussion. What has become problematic for me in the last few years is the question of how much control man has over his own destiny and to what extent his cultural freedom is restricted by biological and evolutionary constraints. In restructuring our education system and creating a truly equalitarian society in which the full potential of every human being can be developed, it is necessary to know what proclivities, what capacities, what sensitivities, can be released and at what stages in life they should be cultivated. This requires a fuller understanding of the social body because, as everyone has at some time or other learnt by experience, it is only when we bring our cultural acquisitions into harmony with the forces and structure of the body and accept its conditions, that we have the freedom to live creatively. The first step to the ownership of our senses in a truly free society is an understanding of the limitations and possibilities of individual and social bodies. And that is one of the main reasons why I am interested in research into the anthropology of the body.

ACKNOWLEDGEMENTS

I am most grateful to Brenda Beck, Zureena Desai, Hilary Standing, Paul Ekman, Roy Willis and to colleagues in my department, for criticisms of earlier drafts of this paper.

REFERENCES

Alland, A., Jr. (1972). *The Human Imperative*. New York: Columbia University Press.

Bateson, G. (1973). *Steps to an ecology of mind*. Paladin.

Benthall, J. and Polhemus, T. (eds.) (1975). *The Body as a medium of expression*. London: Allen Lane.

Birdwhistell, R.L. (1970). *Kinesics and context: essays on body motion communication*. Philadelphia: University of Pennsylvania Press.

Blacking, J. (1971). Towards a theory of musical competence. In E.J. de Jager (ed.) *Man: Anthropological essays*

presented to O.F. Raum. Capetown: C. Struik, 19-34.
Blacking, J. (1973). *How musical is man?* Seattle: University of Washington Press.
Blacking, J. (1974). Ethnomusicology as a key subject in the social sciences. In *In memoriam Antonio Jorge Dias*, Vol. III, Lisbon, 71-93.
Blacking, J. (1976). Dance, conceptual thought and production in the archaeological record. In G. de G. Sieveking, I.H. Longworth and K.E. Wilson (eds.), *Problems in economic and social archaeology*, London: Duckworth, 3-13.
Douglas, M. (1973). *Natural symbols: explorations in cosmology.* Pelican Books.
Durkheim, E. (1961). *The elementary forms of the religious life* (tr. J.W. Swain). New York: Collier Books.
Eibl-Eibesfeldt, I. (1971). *Love and hate* (tr. Geoffrey Strachan) London: Methuen.
Ekman, P., Friesen, W.V. and Ellsworth, P. (1972). *Emotion in the human face.* Elmsford, New York: Pergamon.
Ekman, P. and Friesen, W.V. (1973). Nonverbal behavior and psychopathology. In R.J. Friedman and M.M. Katz (eds.), *The Psychology of Depression: Contemporary Theory and Research*, 203-231.
Ekman, P. and Friesen, W.V. (1974). Detecting deception from the body or face. *Journal of Personality and Social Psychology, 29* (3): 288-298.
Ekman, P. and Freisen, W.V. (1975). *Unmasking the face: a guide to recognizing emotions from facial cues.* New Jersey: Prentice-Hall.
Fox, R. (ed.) (1975). *Biosocial anthropology.* ASA Studies. London: Malaby Press.
Geertz, C. (1964). The transition to humanity. In Sol Tax (ed.), *Horizons of anthropology*, London: Allen and Unwin, 37-48.
Geertz, C. (1975). *The interpretation of cultures.* London: Hutchinson (c. 1973).
Hall, E.T. (1968). Proxemics. *Current Anthropology, 9* (2-3): 83-108.
Hess, H. (1973). Is there a theory of art in Marx? *Marxism Today*, October 1973.
Hewes, G.W. (1973). Primate communication and the gestural origin of language. *Current Anthropology, 14* (1-2): 5-24.
Hinde, R. (1972). *Non-verbal communication.* Cambridge University Press.
Langer, S. (1953). *Feeling and form.* London: Routledge and Kegan Paul.
Lenneberg, E.H. (1967). *Biological Foundations of Language.* New York: Wiley.

Lévi-Strauss, C. (1964). *Le cru et le cuit.* Paris: Plon.
Livingstone, F. (1973). Did the australopithecines sing? *Current Anthropology, 14* (1-2): 25-29.
Lorenz, K. (1967). *On Aggression.* New York: Bentham Books.
Luce, G.C. (1973). *Body time: the natural rhythms of the body.* Paladin.
Marks, M. (1974). Uncovering ritual structures in Afro-American Music. In Irving I. Zaretsky and Mark P. Leone (eds.) *Religious Movements in Contemporary America,* Princeton University Press.
Marx, K. (1973). *Grundrisse* (tr. Martin Nicolaus). Penguin Books.
Maslow, A.H. (1970). *Religions, values, and peak-experiences.* New-York: Viking Compass Book.
Mauss, M. (1936). Les techniques du corps. *Journal de la psychologie, 32.*
McClintock, M. (1971). Menstrual synchrony and suppression. *Nature, 229,* 244-245.
Merleau-Ponty, M. (1962). *Phenomenology of Perception* (tr. Colin Smith). London: Routledge and Kegan Paul.
Nordoff, P. and Robbins, C. (1973). *Therapy in music for handicapped children.* London: Gollancz.
Ornstein, R.E. (1969). *On the experience of time.* Penguin Books.
Ornstein, R.E. (1973). Right and left thinking. *Psychology Today,* May: 87-92.
Poole, R. (1965). Objective sign and subjective meaning. In Jonathan Benthall and Ted Polhemus (eds.) *The body as a medium of expression:* 74-104.
Prescott, J.W. (1971). Early somatosensory deprivation as an ontogenetic process in the abnormal development of the brain and behaviour. In I.E. Goldsmith and J. Moor-Jankowski (eds.), *Medical Primatology.* 1970, Basel: S. Karger, 356-375.
Pribram, K.H. (1967). The new neurology and the biology of emotion: a structural approach. *American Psychologist, 22:* 830-838. Reprinted in K.H. Pribram (ed.), (Brain and behaviour, 4), Penguin Books (1969): 452-466.
Reti, R. (1961). *The thematic process in music.* London: Faber and Faber.
Rowan, J. (1976). *Ordinary ecstasy: humanistic psychology in action.* London: Routledge and Kegan Paul.
Ruwet, N. (1972). *Langage, musique, poésie.* Paris:Seuil.
Schoenberg, A. (1951). *Style and idea.* London: Williams and Norgate.
Smith, R.D. and Worthington, J.W. (1967). Paganini: The riddle and connective tissue. *Journal of the American Medical*

Association, 199 (11): 820-824.

Tiger, L. (1975). Somatic factors and social behaviour. In R. Fox (ed.), *Biosocial anthropology*.

Tomkins, S.S. and Izard, C.E. (eds.) (1965). *Affect, cognition, and personality*. London: Tavistock.

Walker, A. (1962). *A study in musical analysis*. London: Barrie and Rockliff.

Washburn, S.L. and Howell, F.C. (1960). Human evolution and culture. In S. Tax (ed.), *The evolution of man*, Chicago University Press.

White, L.A. (1949). *The Science of Culture*. New York: Farrar, Straus and Cudahy.

Young, J.Z. (1971). *An introduction to the study of man*. Oxford: at the Clarendon Press.

THE BODY AND THE PLAY WITHIN THE PLAY

Francis Huxley

'Body' is one of those words whose meaning changes according to context. As its relation to 'butt' and 'bottle' implies, its basic meaning is that of a container, and as such it is often used to designate the whole physical organism. It can also be used in a narrower sense to designate the trunk, as does the semantically similar term 'chest', so that we can then differentiate body from limbs, from tail and, of course, from head. On another level it can refer to matter or substance, in contradistinction to such terms as mind, life, soul and spirit, and so on. Such usages tell us that the word is to be understood in terms of a chimerical whole that is composed of different categorical realities - a difference sometimes expressed in the distinction between *having* a body and *being* one.

The distinction entrains many existential and phenomenological niceties which, though of considerable interest in themselves, by-pass what I take to be a fundamental anthropological concern: namely, that of understanding how the social power incarnated by the symbolic mode allows a tradition to perpetuate itself. The body is then not just a vehicle for meaning but the instrumentation of the symbolic awareness *per se*. The distinction between having and being a body can therefore be rephrased in language that does justice to this state of affairs, as Owen Barfield does in the following aphorism:

> The relation between yourself and nature is, not a relation between your body and all else in nature, but the relation between yourself on the one hand and, on the other, your body as at once a part of nature and her epitome (Barfield 1965:114).

This fine turn of phrase is, of course, not complete. For, having formulated a relation between self and nature in such a way that the body of the self is in synechdoche with nature (and, we must add, with society), we still have to determine what is meant by 'self'. This can only refer to the tautologous function that brings the synechdoche to consciousness in the first place, as the etymology of 'self' makes clear: for it comes from a root that is at once reflexive, privative and possessive. It is also worth noting that in structuralist parlance 'self' denotes the epistemic subject and not the existential one, whose role in constructing the dual concept of body cannot be separated from the body's physical performance in that construction. Indeed, as Piaget remarks, (1973:35) every element in this symbolic anatomy is 'always simultaneously form to the content it subsumes and content for some higher form.' As ho goes on to say, this allows one to break into what first appears as a circular argument and turn it into a spiral - a figure which itself has bodily reference, as will be seen.

The various formal peculiarities to be found in this brief examination of what is meant by 'body' are to be seen in a host of every day affairs ranging from dreams, the pathetic fallacy, *participation mystique*, and magic, to personifications: all of these expressing in their different ways what happens when the body acts itself out as a part of nature on the stage provided by it as her epitome. It is then but a step to the common order of things in which the world is thought of as a great man, and the individual human being as a small world.

The classic exposition of this doctrine is to be found in Plato's *Timaeus*, and some points are worth noting. His intention being at once moral and didactic, he first assumes the existence of a Creator who forms the world from an admixture of two different ratios called the Same and the Other. This world is termed an animal, and also a god; it has stereomorphic properties, is ruled by number, geometry and proportion, and is the measure of Time. Within it, as a secondary creation, is the world of sublunary nature and of Man, which is ruled by Time and has animistic properties. But this world, as Plato confesses, cannot really be accounted for by the science of ratio as number, and he therefore sets up another frame of discourse in which ratio is expressed in various concrete metaphors. His problem, then, is to make these two incommensurate systems cohere or approximate, and he does this at the end of the *Timaeus* by bringing in the doctrine of the immortality of the soul. The soul proves to be the shuttle weaving the two systems together, for at death it lifts to the sphere of the fixed stars, that is, of number, and after judgment descends to animate a new physical vehicle, whose nature depends on the rectitude of its previous life. The stereomorphic and animistic aspects of the Platonic scheme are thus harmonised by means of a quite theatrical device, to express the moral intention he began with; and there is no doubt that the whole construction is best regarded in dramatic terms, complete with plot, actors, roles, scenes and final purpose.

The Platonic model is but one of a large number of similar constructions, and the basic form has lately been given a new lease of life by Lévi-Strauss (1962:201) with his Totemic Operator. This scheme is also synechdochic, in putting the parts of one individual into correspondence with individuals of different species; and coherence between these different levels of category is achieved by a stereomorphic twist at the interface of the two sets, a twist that can also be expressed by a social algebra. The animistic part of totemism thus has its magical attributes redefined in terms of a semantic function or, in Piaget's words, as the epistemic subject of the Operator: which, by virtue of the synchronic method it exemplifies, also rules out of court the dramatic issues that Plato included.

There is no denying the power and clarity that come with structuralism, but we may well ask if it is not time for it to translate its method back into the physicalities that gave it birth, as Plato deemed proper for a departed soul. A first step in this direction is to ask what the semantic function linking two structuralist sets corresponds to in the mind-body economy. The psychoanalytic answer, which is worth hearing, is the 'preconscious', that two-way staging post between the perceptual and the motor systems whose importance was first seen by Bergson and later by Scherner, whose observations gave Freud his key to the interpretation of dreams.

One dream reported by Scherner was of two rows of fair, handsome boys attacking each other on a bridge and then retreating in order to let the dreamer sit on the bridge and draw a long tooth from his jaw (Freud 1930:252). Scherner concluded that the dream had its source in sensations coming from the dreamer's teeth, of which it was the rebus: a conclusion long known in the nursery, as witnessed the old riddle; *Thirty white horses on a white hill, Now they champ, now they stamp, now they stand still.* The fact that physical sensation can thus be spontaneously translated into mental imagery is now a commonplace, for instance, amongst Gestalt therapists, though this does not mean that there is any necessary constancy in an image so produced, at least in its superficial aspects. The following passage from Piaget shows the kind of variations that may be plaed on one theme:

> An adult frequently went to sleep with his hand resting against the angle of his lower jaw, and could feel the blood pumping in the carotid artery. At widely separated times this position gave rise to the following images: (1) A stream of bubbles against the corner of a rock, with a rhythm exactly corresponding to that of the heart beats (as the subject was able to verify when he woke up). (2) A horse galloping with the

same rhythm. (3) The curves described in the water by the *Gordius aquaticus*, a worm more than a yard long and as thin as a piece of string. Here again the intervals between the curves corresponded to the heart beats (Piaget 1962:199).

The use Freud made of these observations was indeed a major advance, since he undertook to discover the reasons for one particular image coming from an organic source rather than another. As a result he could formulate the concept of *cathexis*, or the *enthronement* of a dramatic event as a memory upon that part of the psychophysical economy that experienced it, principally around the mouth, the anus and the genitals. A knowledge of these and other psychoanalytic findings is naturally of great value in laying the groundwork for an anthropology of the body. But the complexity of these processes is now understood to be greater than may at first have been suspected, if only because Freud's preoccupation with the cathectic power of the sphincter muscles obscured the equally important role played by the skeletal musculature in the organisation of perception and the formation of concepts and symbols. Here again Piaget's work is of great importance, with his view of the sensori-motor system and its two functional factors: 'assimilation, the process whereby an action is actively reproduced and comes to incorporate new objects into itself... and accommodation, the process whereby the schemes of assimilation themselves become modified in being applied to a diversity of objects'(1973:63).

To recapitulate: the relation we have with our body is in the symbolic mode, and its instrumentality in making relations possible with others is reflected by means of these relations, when it makes itself known to us. Some of these relationships are crucial in our development, since they reorganise past experience and extend our power over the future: and both Freud and Plato in their different ways use animistic language when noting the dramatic nature of these transformations. Finally, Plato and Lévi-Strauss show how the epitomising function of the body lies at the heart of cosmological thought.

Before we proceed, it is as well to remember that this function is based on certain automatisms, notably that of imitation; and that as a result we must expect it to suffer from a particular kind of infection. One of the simpler forms this infection can take may be seen in body-english, for example when a spectator tries to help a high-jumper over the bar by sketching out the movement in his own body. Its florid and pathological aspects emerge in such hysterical disorders as echolalia and echopraxis, which cause the sufferer to imitate in full what is said or done before him without the possibility of stopping himself. But since the function so affected can work in opposite directions, its suggestibility can well be made use of therapeutically. For an example I cannot do better than quote the following curious passage from Rawcliffe:

> The older clinical cures of functional disease often seemed mysterious even to the psychotherapists themselves. Lys, for example, never understood the principles involved when he cured patients by first hypnotising another person who then, at the word of command, commenced to indulge in mimicry of the patient's symptoms in front of the latter (Rawcliffe 1959:213).

Rawcliffe goes on to remark on the astonishing cures affected by hypnosis, and leaves the reader to work out for himself why the patient was cured while the mimic went unscathed. Since 'suggestion' has always been an easy word to bandy about when dealing with such matters, it is worth seeing in more detail what Lys was about.

For a start, it is plain that his technique, of which I have been unable to find any further information, must be understood as obeying the rules of drama as set out by Aristotle in the *Poetics*. He defines all the human arts here as being imitations of an action, with Tragedy as the main exemplar. This action, or *praxis*, can in turn be understood as an intentional impulse, not necessarily conscious, or what Cyrano de Bergerac (1965:145) called 'the energetic idiom'. It calls for a particular scene in which to act itself out, which reflects it both in its symbolic furniture and in the roles taken up by the action. The principal actor or Agon then suffers his peripeteia, or the Reversal of his situation, which leads him to the stage of Recognition. In a *Grammar of Motives* (1962) Kenneth Burke has described this movement as going through the three stages of poiema, pathema, mathema: that is, doing, suffering and understanding. In the *Poetics*, this last stage is called 'theoria', a looking or view, from which word we also get 'theory' and 'theatre'. The conclusion is obvious, if facile: the theory of body symbolism can only be actualised when it has a theatre for its exposition.

Drama has two principal modes in Europe, the comic and the tragic. The Cambridge school, comprising Frazer, Jane Harrison, Gilbert Murray and their followers, held that these originated in one coherent cyclic ritual. The comic mode has largely to do with sexual and mercantile topics: its actors are allowed to change role in midstream, and the *telos* or end of the drama is marriage. The tragic mode has to do with power, hierarchy and sacrifice, its Agon being destroyed by the role he occupies: its *telos* is death. There is good reason to believe that both had cosmic as well as social relevance, and that their purpose is to harmonise the incommensurable orders of chance and choice.

Comedy does so by restoring good humour with laughter; tragedy, by purging ill humour with pity and terror.

Laughter is of course infectious, which accounts for part of its curative properties: while another part may be accounted for by its diaphragmatic spasms, that affect not only the breathing but the entire postural tonus. But the view laughter gives of the world is not so piercing as that offered by pity and terror, two emotions that anthropologists will recognise to be at the heart of, for instance, the shamanic vocation, of which more anon. Fergusson, in his introduction to the *Poetics*, quotes the following remarkable definition of these two states from Joyce's *Portrait of the Artist as a Young Man:* 'Pity is the feeling which arrests the mind in the presence of whatsoever is grave and constant in human suffering, and unites it with the human sufferer. Terror is the feeling which arrests the mind in whatsoever is grave and constant in human suffering, and unites it with the secret cause' (Fergusson 1964:34).

The secret cause is, of course, the heart of the action imitated by the tragedy, the energetic idiom that is imprisoned in the body or cosmos it makes for itself. This can be seen in terms of sacrifice, which is a crux for body symbolism in being the method by which the material and spiritual aspects of a representation are first put together and then separated, according to the proportion that the sacrificer is to the victim as the victim is to the god. This triadic relation is also found in gift exchanges, and indeed in all cases where the signifier is related to the signified by means of a vehicle that stands for both.

What Aristotle called the imitation of an action is, saving the difference in context, much the same as what Freud took as being the image of one. The Freudian action is cathected upon a bodily function, the Aristotelian upon a dramatic role; they are both seated, as it were, upon their instrumentality, and as if by design. I use this word in a restricted sense, and as an outcome of the structuralist correspondences which the dual nature of the body makes possible. For if a part of the body is designated by a symbol, the symbol can then be used for a purpose: though this purpose properly requires a cosmos for its elaboration.

We can now see some of the conditions under which suggestion must operate if it is to be effective. They place us in that theatre or cosmos where an action is imitated both on the physical and mental levels, both within and without the body, by the same figures. Seeing the outer performance, the self-reflexive nature of consciousness willingly suspends its disbelief and is automatically stirred into a kind of body-english to match the happenings on stage. This is the peripeteia of the audience and the reversal of their situation, which puts them directly into touch with human suffering and its secret cause: in Artaud's phraseology, the theatre of cruelty has lanced the bubos of social plague. The audience is then left with the *theoria* of the action, as also happens with every full ritual, whether it has to do with initiation into adulthood, diagnosis, healing or divination.

In such ceremonies the various orders of experience have to be marshalled and put into play with and against each other, and the client is led through them and pointed towards their secret cause. The cause is secret, of course, because it is dialectic, not only in terms of the actors who are moved by it, but because of what Kenneth Burke terms the scene-act ratio. He uses this phrase to express how, in a unitary drama, an action that moves through the scene in which it is first set requires a second scene, different in aspect but similar in structure, so that it can show forth the different levels of meaning that it cannot but generate. The client or agon is forced to epitomise in himself these different scenic relations set up by the action, and after the Reversal is dissociated from those parts embodying the disease or disorder by being carried to a new setting in which the disease has, so to speak, homeopathically created its own antidote. Burke's 'Symbolic Action in a Poem by Keats', which appears as an appendix to his *Grammar of Motives*, shows what can be achieved by his method, and how the sequence of poetic images embodies what we know of Keats's life at that time and of his physical state, and leads us through their various transformations to a (partially) curative resolution.

The anthropological analysis of symbolic action has hardly reached such rich and precise conclusions, though Lévi-Strauss's paper on Mu-Igala, a medicine song from the Cunas of Panama (1958: ch.x), is a notable essay in this genre. One of the difficulties facing the anthropologist is, of course, that symbolic action in a rite is a collective and generalised affair and not just the singular expression of an individual. Another is that he cannot be expected to pick up the nuances of another language and note the subterranean associations that can be made to resonate through the course of the performance. But it must also be said that the theatrical props used in healing rites have seldom been fully described in terms of the bodily effects they produce. Many of them, of course, are obvious enough: the use of seclusion huts when incubating an illness or a state of mind, the regimen of diet, prayer, song, and so on, that stretches the mind from its present condition to the one that is to be hatched out. We can occasionally match the meaning of a song or a symbolic act to a particular psychic condition, as I have sketched out elsewhere (Huxley 1966). But the details, which are so often enshrined in cadence and rhythm, have so far escaped us, and we have but the crudest appreciation of why in voodoo, again, certain drum-beats are used to invoke certain kinds of deity and how their energetic idiom allows a possessed servitor to express the character of a divinity.

There is of course no doubt that drumming and dance are prime instruments in the production of dissociation and possession. We can trace the physical pathways by which they have their effect without much difficulty: they centre on the inner ear whose automatisms, when thrown into disorder, affect posture, heartbeat, and the use of the eyes, producing nausea, giddiness, fainting and amnesia. The symbolic representatives of giddiness can be found visually in such things as the spirals painted around the centre post of the temple, defined as the course the gods follow down when they are invoked; spirals which can also be read upwards, since the rite is staged in a temple to show these gods in action as the saving counterparts of the disordered nature of their human vehicles. The transition from dissociation to possession is obvious and dramatic: the eyes that were closed are open, the head that was jerking backwards in a hysterial spasm is now erect, the breath that had been laboured is lively, and the voice is either in the chest, in the throat and mouth, or in the nose, according to the nature of the possessing entity. The one possessed is typically amnesiac after the event, though he may remember what he had done later in a dream. This must be because he is a particularly vigorous sleep-walker, and the authenticity of the state is determined by the degree in which he is not himself.

This amnesia can also be accounted for by the fact that the servitor in question is an agon who has not fully undergone the Reversal of his situation, so that his faculty of self-witness is still fragmented. That some servitors have partial knowledge of their possession is, however, of great interest, since we know from Diderot and William Archer (1957) that an actor seized in the throes of inspiration does not forget himself. But this stage is formally reached only by a second initiation into the priesthood and the acquisition of a maraca or rattle, an all-important instrument, with which the priest can conduct the necessary rhythms of the cult and by doing so keep a grip on his own volition. In Haiti this stage is called 'la prise des yeux', which is to say that the initiate can see on two different levels simultaneously. This state goes with a sense of enormous power in the body, especially in the shoulders and at the nape of the neck; by which we may understand that the inner ear, far from being disregarded, is now functioning with a great increase of efficiency and stimulating certain functions that are normally dormant.

Here the nominal distinction made in anthropology between possession states and the shamanic one disappears, for the shaman is in the same control of this remarkable state. His dissociation appears as a crisis of dismemberment, which can be understood as a radical unseating of cathexes throughout the body: a crisis into which he is sometimes drawn as though down a tube, whose bodily referent remains to be diagnosed. He is in touch with spirits, which must be those cathexes transformed from parasites of the body-mind into its instruments. They travel through the various levels and domains of a well-ordered cosmos, or personified image of nature; and they stage their adventures in front of an audience for the benefit of a client, a patient, or an entire congregation. It is not difficult to identify some of the props a shaman uses with the various states he goes through before being fully inspired, especially if we know that he has to cope with the initial spiral of dissociation and giddiness by shifting gear and rhythm as he accelerates through the critical levels of his cosmos; Eliade (1951) describes many of them, though as yet no attempt has been made to find the bodily referents to such figures as Sedna, or Moon, or the bubble which the Eskimo see as representing the spirit. The shaman's art of fascinating an audience, by acting out its preoccupations in symbolic form is more easily understandable, together with his sleight of hand when producing a believable miracle, though even here the bodily effects of all this are usually taken for granted.

These admittedly thin descriptions must suffice to indicate some of the areas in which an anthropology of the body could develop its methodology. The anthropological instances I have chosen exist, of course, at the limits of normal human life: a fact which throws a sharp light on the status of the body as incarnating two different experiential modes simultaneously, and on the nature of some of the disorders that can emerge when these modes contradict one another. Anthropology likewise often finds that its functional theories are at odds with its structuralist ones, and it may perhaps take heart from the ethnographic evidence, which tells us to stage our thought in a theatre of several levels. There, a change of level produces an inversion of the preceding figure according to the rules of drama, through which body symbolism and cosmic imagery are eventually harmonised, the animism of the one finding its consummation in the logic of the other.

REFERENCES

Barfield, O. (1965). *Unancestral Voice*. London: Faber and Faber.

Burke, K. (1962). *A Grammar of Motives*. Cleveland and New York: World Publishing Co.

Cyrano de Bergerac (1965). *Other Worlds*. Translated by Geoffrey

Strachan. London: Oxford University Press.
Diderot, D. and Archer, W. (1964). *The Paradox of Acting; Masks and Faces.* New York: Hill and Wang.
Eliade, M. (1951). *Le Chamanisme.* Paris: Payot.
Fergusson, F. (1964). *Aristotle's Poetics.* New York: Hill and Wang.
Freud, S. (1936). *The Interpretation of Dreams.* In *The Basic Writings of Sigmund Freud.* New York: Random House.
Huxley, F. (1966). *The Invisibles.* London: Hart-Davis.
Lévi-Strauss, C. (1958). *Anthropologie Structurelle.* Paris: Plon.
Lévi-Strauss, C. (1962). *La Pensée Sauvage.* Paris: Plon.
Piaget, J. (1962). *Play Dreams and Imitation in Children.* Translated by C. Gattegno and F.M. Hodgson. London: Routledge and Kegan Paul.
Piaget, J. (1973). *Structuralism.* Translated and edited by C. Maschler. London: Routledge and Kegan Paul.

BIOLOGICAL AND CULTURAL CONTRIBUTIONS TO BODY AND FACIAL MOVEMENT[1]

Paul Ekman

INTRODUCTION

There is no agreed upon single name for the range of facial or bodily activities I will describe. Part of my purpose will be to suggest that there should *not* be, for this is not one phenomenon but many. Terms such as nonverbal behaviour, nonverbal communication, kinesics, kinetics, expressive behaviour, body language, etc. have been used by one or another author to refer to some activity of the face and/or body. It has not always been clear whether actions which seem to have some instrumental purpose, such as pencil sharpening or push-ups, were included. If not then why did they include head scratching? Is the phenomenon restricted to movement or does it include static position of body or face (postures)? If limited to movements, it has not always been clear whether those too fast or too small for visibility under normal circumstances have been included. No matter. While a single suitable term[2] would be a boon when writing about the face and/or body, it could further confuse study of the phenomena, adding to the tendency to ignore or gloss over fundamental differences in what a person does with his face and body.

Some of those who study nonverbal behaviours (kinesics, expressive behaviours, etc.) have argued for an evolutionary perspective. They sought and found universals. Others, arguing for the utility of a linguistic model or the perspective of cultural relativism, sought and found extraordinary differences among cultural groups. Both studied the nonverbal, but it is likely that each examined entirely different activities: activities which differ in origin (how the behaviour became part of the repertoire), in coding (the relationship between movement and what it signifies), and in the circumstances of usage. Perhaps because the phenomena are all the result of muscular action, or because vision is the means for perceiving these activities, or the camera the method of recording, investigators may have expected to find but one type of activity, just one principle of organization, just one type of function, just one major determinant. There are many.

I will describe the four activities we have studied most extensively: emblems, body manipulators, illustrators and emotional expressions.[3] Evidence about each of these activities is summarized, including some findings previously not reported. My emphasis is upon the conceptual framework for interpreting these findings. A central issue will be the extent of variation across cultures. This issue is central not just because this report is given at an anthropological meeting. Such a focus requires discussion of a problem which is the key to understanding and distinguishing among facial and body movements - to explain how biological and social factors have quite different influences upon emblems, body manipulators, illustrators and emotional expressions.

EMBLEMS

We have employed Efron's (1941) term *emblem* to refer to symbolic actions where the movement has a very specific verbal meaning, known to most subscribers to a sub-culture or culture, and typically is employed with the intention of sending a message. The head nods 'yes' or 'no' are examples of emblems. The person performing the emblem takes responsibility for having communicated, for having said something with his face or body. He can be held accountable for his message. The person who sees an emblem considers it was performed for his benefit, to tell him something. Not so with many other actions which nevertheless can be informative. The viewer may infer nervousness when observing a high incidence of body manipulator actions (e.g. scratching, rubbing, picking, etc.), but he will not consider these movements were performed to tell that to him. The knowledge gained from such body manipulator actions is stolen, the message provided by the emblem is given. The meaning of an emblem is not context free. Like words, emblems signify something rather precise, but the exact meaning and interpretation depends upon who is performing the emblem, to whom, in what conversational context, showing which other concomitant behaviours.

Emblems are often used when we could speak but choose not to. For example an emblem may be employed when recognizing a friend seated across the aisle during a theatrical performance, or an emblem may be used to avoid the likelihood of being caught in an extended dialogue when hoping to pass quickly a colleague in a corridor. Of course emblems are also used when verbal discourse is just not possible; for example, because of noise (see Meisner & Philpotts (1975) study of emblems among sawmill workers).

Emblems occur also during conversation, by either speaker or listener. When speaking an emblem may replace a word entirely or repeat the message spoken in a word, or comment or qualify what is being said in words. When emblems repeat what is said in words, they may precede, accompany or follow the verbal statement of the message. Depending upon what is being said, how it is said, the speaker's stance towards what he is saying, there may or may not be latitude about where an emblem must be placed in relation to the word whose message it repeats.

We have been studying the location of a particular emblem, the shrug (Fig. 1) within the flow of speech among Americans. This emblem most often occurs at the start of a speaker's turn as an initial response to the question of the other conversant. The 'I don't know' (or some such verbalization) which the shrug symbolizes may not be spoken at all, or it may be said in words as well as in action. The second most frequent location of the shrug emblem is within the speaker's turn, at the end of a phonemic clause. The least frequent location for this emblem was in a filled pause (*aaahhh*) within a phonmic clause.

Emblems are used in conversation by the person listening and serve as what Dittmann (1972) has called listener responses. He has shown that head nods 'yes' or 'no', and vocal equivalents (*mmhmm*) are employed by the listener in a non-random fashion. Their placement by the listener is related to pauses and to whether the juncture at the end of a phonemic clause in the speaker's conversation is rising or falling.

I have been discussing the usage of emblems, about which all too little is known. Let us turn briefly to the issue of the coding of this form of activity. The coding of emblems can be either iconic (the movement in some way looks like what it signifies) or arbitrary, and often it is difficult to determine which. Is the wink (an emblem signifying collusive agreement or flirtation) arbitrary in coding, or does the closing of the eyes refer to not looking or to future bedtime activities? With some emblems, the iconic base is quite obvious (See examples in Fig. 2).

Most of our research on emblems has attempted to survey the emblem repertoire, the vocabulary if you like, for particular cultures. We have developed a standardized procedure for this purpose, (see Johnson, Ekman & Friesen 1975, for description)[4]. A list of messages is presented to informants who are asked if they know of a commonly used body or facial movement to communicate each message. Their performances are analyzed to sift out those in which a message is performed in a similar way by

Fig. 1. *Shrug Emblem.*

most informants. These are then presented to a new group of informants who 'back translate' from the performance to the message. The informants also evaluate whether each performance is an action they have seen in actual conversations. This procedure is workable with persons in preliterate as well as literate cultures. So far, it has been applied to a study of middle-class white Americans (Johnson, Ekman & Friesen 1975), Iranians (Trupin 1976), Israelis (Broidie 1977), urban Japanese and a preliterate culture and comparisons among the five cultures are reported in Ekman, 1976[5].

The number of emblems found by these procedures has varied, from less than 100 in the U.S. to a few hundred for the Israelis. We have found no universal emblems; that is, instances in which the same movement is employed to signify the same message in each of the five cultures studied. We had not expected universals in this type of nonverbal behaviour. We had hypothesized that emblems are learned much like language, and the

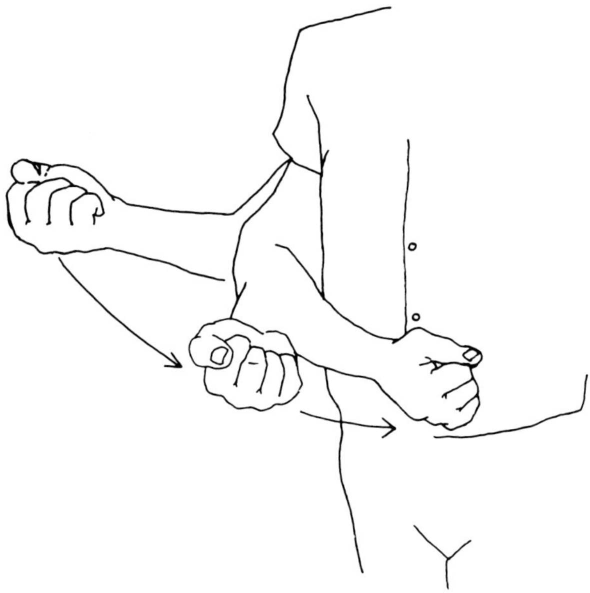

Fig. 2. (A) *Suicide emblems in Japan depicting use of sword.*

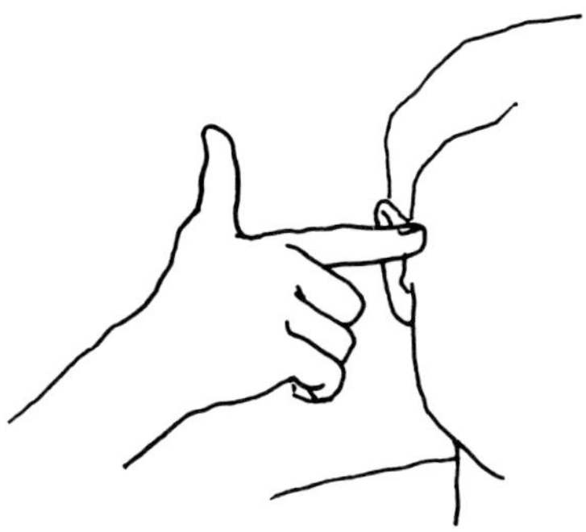

Fig. 2. (B) *Suicide emblems in United States depicting use of a gun.*

emblem vocabulary would vary with culture. We did expect to find some near universal emblems, when the message performed in an emblem refers to bodily functions or use of the body in a way which is anatomically constrained. In such instances, if there is an emblem for the message in more than one culture, it is likely that the performance utilized to symbolize the message will have to be the same in at least some respects. Examples would be the head in a prone position or eyes closed

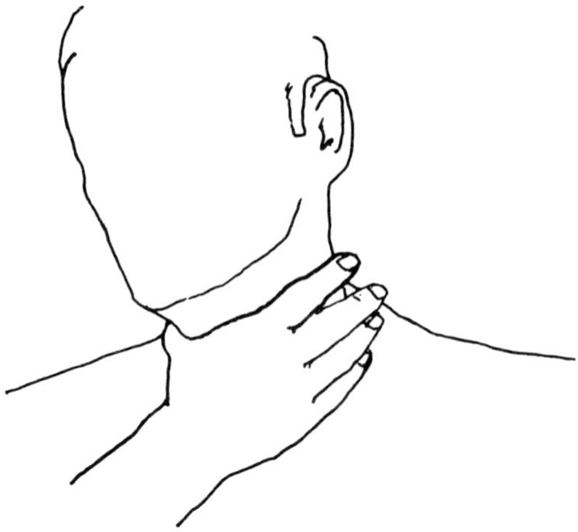

Fig. 2. (C) *Suicide emblems in the South Fore of New Guinea depicting strangling.*

for an emblem referring to sleep; or, reaching to mouth or stomach to refer to eating, hunger or satisfied hunger.

There are interesting similarities and differences across cultures in the types of messages for which there are emblems. Every society we have studied has some emblem (although the specific movement varies), for greeting and departure, for insults, for replies (such as yes, no, uncertainty), for directing locomotion, for referring to the physical state of the person (tiredness, pain, etc.), and to the affective state of the person (angry, surprised, etc.) Often variations in the specific movement pattern employed to symbolize a message reflect the use of different tools. (See Fig. 2 for suicide emblems.) Cultures differ in the number of emblems they have for a class of messages. For example, the Israelis have fewer emblems for affect than the other groups we have studied even though the Israelis have more emblems *in toto*. Before we can make sense of such differences more cultures will have to be studied, picked with some sampling principle in mind, rather than the accident of opportunity which has guided our past selection.

Emblems most often involve the hands and/or arms, but also can utilize head movement, facial movement, postural shifts, and even leg movements (or voice as well). Typically they are not elaborated from smaller to larger units. There are few compound emblems, where each element in the compound has an emblematic meaning, either the same or different from the compound. The shrug is an exception, since there is a shoulders element, and a hand/arm element, each of which means 'I don't know', and together they have that meaning.

Trupin, in her analysis of Iranian emblems, has shown the utility of applying Stokoe's (1974) formulation to understanding the differences among Iranian emblems. She has shown how performances having the same configuration signify different matters when performed at different locations and how performances at the same location signify different matters when there is a different action. For example, if the forefinger is extended with the rest of the fingers and thumb in a fist, and that position is held without movement in the shoulder area a foot or two in front of the person, it signifies 'just a minute.' If the location is shifted to finger perpendicular to and touching the lips, it signifies 'be quiet.' If the location is held constant about shoulder height in front of a person, but an action is added in which the finger beats down in space repeatedly, the message now becomes 'making a point emphatically' or 'bawling out'. While this approach has not been applied to the emblem repertoires of other groups, it seems that it will be fruitful in understanding the nature of emblems.

Emblems are the most language-like of the facial and body movement activities. They are symbolic. What a movement symbolizes is learned in a culturally variable fashion. The movements are employed specifically for the purpose of communicating. In all these respects emblems are like language and different from the other nonverbal activities we will describe next. Emblems also differ quite considerably from language.

Words are typically employed during conversation in strings or sequences, governed by a syntax. American Sign Language and Indian Sign Language also usually involve the emission of a string of signs. Emblems are usually not employed during conversation in strings, but singly. We have only seen emblems employed in strings of three or four in a sequence when verbal conversation is in some way constrained. For example, if while talking on the phone you notice a person come to the door of your office due for his appointment, you may emblematically signal that he will have to wait just a minute, and may follow this with an emblem that requests he come in, and another emblem that directs him to be seated. When two people are not constrained about the use of words, however, we have rarely observed such a sequence of emblems. Eco (1975) said that Italians differ, typically employing emblems in strings during speech. Certainly when speech is constrained as among the sawmill workers studied by Meisner and Philpott, strings are employed. When emblems occur in strings the syntax may be similar to a path expression (Fillmore 1971), with the order of performance paralleling the sequence of events referred to.

Many of the most important questions relevant to an under-

Fig. 3. *Finger-wag emblem.*

standing of emblems require developmental study. There has been only limited, usually incidental study of the ontogeny of emblems. Are the emblems first shown likely to be similar for all infants, due to common limitations in capacity, needs, and problems in communicating with a caretaker? What are the ages when various emblems become established in the repertoire? How are they acquired? Do they develop out of intention movements, or instrumental acts, or are they specifically taught? Are there emblems used by children but not by adults? Certainly, there are emblems used only by adult to child (See Fig. 3). Do children recognize emblems before they can utilize them to communicate to another?[6] When bilinguals switch language, do they switch into a different emblem vocabulary? What about male/female differences in usage of emblems? In New Guinea we found that there were emblems referring to sexual activity, menstruation and pregnancy utilized and known only by women.

This is only a partial list of the variety of questions that need to be investigated. Elsewhere, (Ekman 1976) I have discussed these issues in more detail, raised other questions, and reported findings on how emblems vary with psychopathology and with interpersonal deceit. Let us turn now to a second type of facial or body movement, a type of movement which in almost all respects is maximally different from emblems.

BODY MANIPULATORS

Our earlier descriptions of this behaviour (Ekman & Friesen 1969a, 1972, 1974) used the term 'self-adaptors', which we are abandoning because that phrase is too theoretically laden. (Although we do not abandon our theory about the adaptive origin of these actions; see Ekman & Friesen 1974a:212-213). Body manipulators are movements in which one part of the body does something to another body part. Scratching the head, picking the nose, wringing the hands, licking the lips are examples. Also included would be the use of a prop in other than an instrumental act; for example, playing with a pencil, twisting a book of matches, using a paper clip to scratch an ear.

Typically, it is the hands which manipulate the other part of the body, or some part of the hand itself. The tongue can manipulate lips, teeth can bite lips, and feet can rub or scratch the legs or each other. When we measure body manipulators we distinguish body geography in terms of what part of the body is in contact with what other body part, and the nature of the action - rubbing, picking, scratching, or a miscellaneous action we call a 'play'.

It is important that actions are not classified as body manipulators just because there is contact among body parts. In some cultures putting the fore finger to the temple is an emblem for thought, and so it would be classified. If while speaking about how hard it was to look at something a person illustrated his speech by putting his hand in front of his eyes it would not be classified as a body manipulator, but as a kinetic illustrator (discussed later). The very same action, if not illustrating what is being said, would be called a body manipulator.

Another problem in defining the boundaries of body manipulators' activities is what to do with instrumental actions. Taking off a shoe is not a body manipulator, but playing with a shoelace is. Striking a match to light a cigarette is not, but fingering the match is a body manipulator. Our treatment of grooming activities as body manipulators may seem to contradict such a distinction. It might be argued that rearranging the hair which changes overall appearance should not be called a body manipulator, and only when the hand to hair movement results in no enhancement of appearance should it be classified as a body manipulator.

Our studies suggest that people are usually not aware of engaging in these body manipulator actions. If you ask the person to repeat what he has just done, he often won't remember. There is little reason to expect that these movements are ever used deliberately to transmit a message to another, other than by an actor or perhaps a psychopath. This does not mean that people do not gain information when observing these behaviours. We have found that observers characterize people who show many body manipulator actions as awkward, tense and untrustworthy, (correlations between .50 to .75; Ekman & Friesen 1974; Ekman, Friesen & Scherer 1976). As mentioned earlier the observer of the body manipulator action does not believe the action was performed to provide a message. In contrast, just that presumption is made about the emblem.

Parents, of course, hold their child accountable for body manipulators, but it is for being uncouth, impolite, etc., not for having said something with this type of activity. Despite such attempts to suppress these activities, they are more common in adults than one might expect. What we seem to learn is not to look when the other person cleans his ear, picks his nose, scratches his crotch, etc. The disattending seems unwitting, but quite consistent. Rudeness seems to reside as much in watching someone engage in these bathroom behaviours, as in performing them.

Another difference between body manipulators and emblems is in regard to the specificity of what is signified. Emblems are movements which have quite precise meanings, well known to all members of a culture or sub-culture. What observers might infer from a body manipulator is more vague, imprecise, and not necessarily agreed upon. As mentioned earlier, the occurrence of body manipulators does influence how a person is perceived, but it is in global terms. Generally, they are interpreted as signs of discomfort, nervousness, etc. Indeed, our studies have found that while the absolute frequency varies from individual to individual body manipulators do increase with discomfort.[7] Paradoxically, body manipulators also increase when the person is quite relaxed and not engaging in any censorship over these actions. Certain types of body manipulators have specific meaning but not as precise as an emblem. We found that scratching occurred more often with patients judged by their psychiatrists to have problems with hostility. Eye covering occurred more often among patients judged to have guilt (Ekman & Friesen 1968, 1974a).

Still another way these movements differ from emblems, and from the next type of action we will describe (illustrators) is in their relationship to the flow of speech in conversation. Earlier we described some regularities in the location of emblems within the flow of speech. Few regularities are found for the location of body manipulators. They have little respect for so fundamental a conversational feature as speaker turn. A body manipulator begun when the person is speaking is just as likely to continue beyond as to end within his speaking turn. The occurrence of body manipulators also has little to do with the beginning and end of phonemic clauses, nor with whether the juncture is rising, falling or sustained. Brief body manipulators (and, they seem to divide naturally between ones that last less than two seconds and ones lasting much longer) often will occur in phonemic clauses where there are many speech disruptions. These findings suggest to us that unlike emblems, or illustrators, body manipulators are not directed by the mechanisms which are responsible for the flow of speech or by listener rules. This is not to deny that what is said has an influence. If the content of conversation arouses discomfort, or the process of articulating becomes difficult, body manipulators are likely to increase.[8]

We know of no cross-cultural research on body manipulators. When working in New Guinea we noted some similarities in some of these activities occurring under situations of some discomfort, but our evidence was anecdotal. Which body manipulators are prohibited in which social contexts might well vary from culture to culture. We would expect that an increase of these actions with discomfort might, however, be universal.

Let us turn now to a third type of facial or body movement, movements which illustrate speech.

ILLUSTRATORS

These are movements which are intimately tied to the content and/or flow of speech. We distinguish a number of different types of illustrators in terms of how they relate to the simultaneous speech:

Batons:	Movements which accent a particular word;
Underliners:	Movements which emphasize a phrase, clause, sentence or groups of sentences;
Ideographs:	Movements which sketch the path or direction of thought;
Kinetographs:	Movements which depict a bodily action or a non-human action;
Pictographs:	Movements which draw the shape of the referent in the air;
Rhythmics:	Movements which depict the rhythm or pacing of an event;
Spatials:	Movements which depict a spatial relationship;
Deictics:	Movements which point to the referent.

Four of these categories were proposed by Efron. He found Sicilian immigrants to the U.S. used quite different types of illustrators than Lithuanian Jewish immigrants, although their assimilated offspring did not. Efron's study of literature suggested that over the past few hundred years the use of what we term illustrators was sometimes fashionable and sometimes déclassé. To my knowledge no one has pursued the study of cultural, ethnic, or social class differences in illustrators. Efron's work supports the interpretation that illustrators are learned, that the type shown is determined by environment, and not just family environment. It would be interesting to study the ontogeny of illustrators, but no one has in any systematic fashion.

Typically illustrators are performed with the hands and arms. Any part of the body could be employed, although not as readily for each of the eight illustrator sub-types in the list. For example, facial movement can easily perform a baton or underliner illustrator. With a limited set of referents facial movement can perform kinetographs or rhythmics. Ideographs and deictics are performable with the face but it is not common in most cultures. Pictographs are next to impossible to execute with the facial muscles.

We distinguish five functions served by illustrators. A particular type of illustrator is used when a person cannot find a word. The movement looks as if the person is going to pluck the word out of the air; sometimes a finger-snapping movement is used instead. Such movements inform the other person that the speaker has not abandoned his turn, that he is intending to continue. This is a specific instance of the illustrator function of maintaining or increasing attention from the listener and holding the floor. Illustrators can help prevent or squelch interruptions, or capture lagging attention.

Word search illustrators draw our attention to a second function, self-priming. Waving an arm in the air or snapping the fingers might conceivably help the person find the word, although the mechanism which would explain such help is obscure. Cohen and Harrison (1973) suggested self-priming to explain why people sometimes illustrate when the listener isn't present (e.g. over the telephone). Habit is an alternative explanation; but, let us grant self-priming of the speech-making process as a second possible function of illustrators.

A third function of illustrators is to explain through movement matters difficult to put in words. Pictographs or kinetographs can be helpful in giving directions about how to find a place, how to take something apart, in defining concepts such as zig-zag, helicopter, etc.

A fourth function of illustrators is the punctuating of speech, adding emphasis, underlining, tracing the flow of thought, to mark off clauses, etc. Presumably this helps the listener understand what is said.

Although not a function in the same sense as the other four, it is worth noting that illustrators increase with enthusiasm and involvement in what is being said. They decrease when there is either disinterest in what is being said or with communicating at all, or when there is conflict about what to say and words are chosen with unusual care. Consistent with this interpretation we found neurotic depressives illustrated more than psychotic depressives, and both types of depressives illustrated less in the acute phase of their disorder than when in a state of remission (Ekman & Friesen 1974a). Among non-patients we have found that when persons were required to engage in deception about their feelings, when they were in conflict about what to say, illustrators decreased (Ekman, Friesen & Scherer 1976).

Both of these findings would not have emerged from our data if we had failed to distinguish between illustrators and emblems. Most other investigators who have done quantitative studies of body movement have failed to so distinguish emblems from illustrators. Instead they simply counted hand in space movements in contrast to hand to body movements.[9] Perhaps they overlooked the differences between illustrators and emblems because both show the attribute of occurring when with another person, not alone. One could use an emblem or illustrate when alone, but apart from rehearsals or hallucinations, these behaviours occur in the presence of others with whom some attempt is being made to communicate.[10] In this way emblems and illustrators differ from body manipulators and emotional expressions, which occur often when alone as well as when with others.

Emblems differ from illustrators, however, in that they are often used in place of conversation or when conversation is not possible or not chosen. Illustrators by definition are movements which must be placed within speech. A second difference between illustrators and emblems is in regard to the specificity of what is signified by the movement. Emblems have a precise meaning or a limited set of alternative meanings, each of which is precise. Certain types of illustrators (batons, underliners, ideographs) have no semantic content. Even those illustrators which have semantic content (such is the kinetograph or pictograph) are more imprecise than the emblem, and their meaning is usually vague when seen without hearing the words they accompany.

A third difference is in the precision of the movement itself. There are variations in the performance of any emblem, but there are also limits for each. In using an emblem the person is not inventing on the spot, but drawing upon a known action to convey a message. Just the opposite happens with illustrators. If, for example, an American was to use a kinetograph to illustrate falling down stairs, a large variety of actions could be used. An American would have to invent his performance, having no standard movement pattern for that message to draw upon. In ideographic illustrators, almost any movement can be used to trace the flow of thought.

A fourth difference between emblems and illustrators is in regard to the performer's awareness of what he is doing. Illustrators are intermediate, somewhere between emblems and body manipulators. The emblem is deliberate, as deliberate as choice of words. The use of batons, underliners, and ideographs is more habitual, while the other types of illustrators are a little closer to the emblem.

Despite these differences, certain emblems may be difficult to distinguish from certain illustrators. When an emblem is used with speech it may not illustrate what is being said but instead comment on what is being said, or conduct separate discourse about matters unrelated to what is being said. In such instances there should be little confusion with illustrator activity. It is when an emblem replaces, repeats, or expands upon what is being said that the distinction between emblem and illustrator becomes more tenuous. I believe it is useful on such occasions to distinguish between emblems-used-to-illustrate and an illustrator which is not an emblem. An example will show the difference between the two. Suppose you observe a man speaking about an attractive woman while he traces an hourglass figure in the air. It could be either a pictographic illustrator or an emblem-used-to-illustrate. To be certain it was the latter and not the former you would need to know that this hourglass movement stands for 'shapely or attractive woman' independent of any particular speech context, and is so known by most members of a society (e.g. Americans).

Earlier I described the placement of a particular emblem, the shrug, within the speaker's words. Recall it usually occurs at the beginning of his turn, before he speaks, or with a verbal statement of the message, although other locations were mentioned. Also when an emblem repeats a verbalized message there was no fixed rule for where the emblem would be placed, before, simultaneous with, or after the word. By contrast the body manipulators occur with little regard for the flow of conversation, unrelated to speaking turn, phonemic clauses, junctures etc. It is not possible to generalize about the location of illustrators within the speaker's conversation, for the location depends upon the type of illustrator.

The baton coincides with primary voice stress within a phonemic clause. It seems extraordinarily difficult to place intentionally the primary voice stress on one word and a baton on another word. The neural mechanisms responsible for emphasis apparently send impulses to both voice and skeletal muscles simultaneously when both modalities are employed. We would expect this relationship between baton and primary voice stress to be maintained across languages and cultures, but have not yet looked.

The illustrators used during a word search often occur within a pause, often a filled pause (*Aaahh*). Ideographs tend to begin and end with a phonemic clause, or a group of phonemic clauses. Kinetographs and pictographs tend to coincide with what they are representing unless they replace the words. These observations about the location of illustrators within speech are preliminary, based on study of only a half-dozen conversations among Americans. They certainly suggest the usefulness of distinguishing among various types of illustrators.

EMOTIONAL EXPRESSIONS

These have received the most attention of all the nonverbal behaviours, and are the subject of continuing controversy. Most of the discussion has been of facial movements (facial expressions), with less attention to body movement.[11] At issue is whether there are universals in facial expression or whether that which is signified by each facial movement varies from culture to culture. There have been persuasive but contradictory reports from the two sides: those claiming universals (e.g. Darwin 1872; Eibl-Eibesfeldt 1972; Izard 1971; Tomkins 1962) and those claiming cultural differences (Birdwhistell 1970; Klineberg 1940; LaBarre 1947; Leach 1972; Mead 1975).

The difference in what has been found may have been due in part to differences in what has been examined. Facial movements can be recruited into a variety of quite different activities. Earlier I described how facial movements can function as emblems (e.g. a wink), and as illustrators (e.g. movements serving as batons or underliners). Although not described here, we have distinguished (Ekman & Friesen 1969a) still another type of facial or bodily activity, called regulators, which serve to manage the back and forth flow of conversation between speaker and listener. Confusion might not occur if different facial movements were recruited into these quite different facial activities - emblems, illustrators, regulators and emotional expressions. Instead, some of the same facial movements appear in each. For example, the raising of the brows (*frontalis* muscle) can be employed in an emblem for greeting or an emblem for 'no'; a baton illustrator; a questionmark or exclamation-point regulator by the listener; a surprise emotional expression, (see Ekman & Friesen in prep. (a) for a detailed description of this example).

The contradictory observations of the relativists and universalists can be resolved, in part, if we presume that the relativists were describing the cultural variations in emblems, illustrators and regulators, and not distinguishing these from the universals in emotional expression. Conversely, the universalists may have focused on the emotional expressions to the exclusion of recognizing cultural variations in the facial actions which function as emblems, illustrators or regulators.

Avoiding such confusions not only requires a conceptualization that allows for a number of differently organized facial activities, but also much more careful description of facial movement itself. The facial musculature allows for an extraordinary number of visibly and anatomically different movements which may on first impression seem similar to the untutored eye. Terms like frown, smile, play-face and even brow-raise are much too gross. Each could cover dozens of different behaviours which may or may not have the same function. Such imprecise descriptions increase the likelihood that two observers would needlessly argue because their descriptive language does not reveal that they are each talking about a different facial action.

Friesen and I (Ekman & Friesen 1976; in press) have been working for the past four years on an anatomically based system for describing facial movement. Our Facial Action Coding System purposefully offers more precision and distinctions than may actually be needed, so that empirical results can determine which visible distinctions should be disregarded. The Facial Action Coding System allows description of both the configuration (the 'expression') and the timing of facial movement. Even if exactly the same muscles were to be involved in an emblem, illustrator, regulator and emotional expression, it is likely that the onset of movement, its duration at apex and its offset - the time it takes to disappear - would not be the same.

The contradictory reports on facial expression may not have been due solely to a failure to distinguish adequately facial emblems, illustrators and regulators from facial expression, abetted by reliance on imprecise descriptions of the face. It reflects also a fundamental difference in interests. One side has been primarily interested in evolutionary explanations, in the biological contributions to behaviour. While cultural influence is acknowledged, it is seen as an overlay, and only lip-service is paid. The other side has been interested in the power of culture in determining social behaviour. It borrows from a linguistic model and emphasizes what is different in each culture. While biological contributions were acknowledged, biology was seen as only providing the equipment which was shaped by culture.

The core of the confusion about facial expression is probably the failure by both sides to specify what they have meant by emotion.[12] Most shied away from describing internal processes or even distinguishing emotional from nonemotional behaviour or situations. Each side took for granted that others knew what they meant by emotion. Such vagueness concealed oversimplified models about the biological and social influences on emotional behaviour. It allowed each side to gather evidence which failed to change the other side's view or force revision or expansion of their theoretical model. Instead, contradictory evidence has fueled the argument, which has spilled over into false disputes about what counts as evidence and what credentials are required to investigate.[13]

To resolve the dispute, or at least clarify its basis, I will go out on a limb, describing different aspects of the phenomenon we term emotion. What follows is not a theory of emotion, but only a discussion of those matters which need to be considered to deal with the contradictory evidence and to show where both relativists and universalists have been right, and wrong, in their claims. Many issues central to a theory of emotion are not discussed here but left for presentation elsewhere (Ekman & Friesen, in prep.). What I present here is what a theory of emotion needs to consider, given the evidence on the face, rather than a specific theory of emotion.

A NEUROCULTURAL ACCOUNT OF EMOTION[14]

The term 'neurocultural' emphasizes our interest in both biological and social determinants of emotion. I will start with a characterization of emotional responses, both observable and nonobservable responses. This characterization will require that I hypothesize internal mechanisms - an *affect programme* which directs emotional responses, and an *appraisal system* which determines when the affect programme becomes operative. *Elicitors* will be described as those events which are appraised quickly as the occasion for one or another emotion. *Display rules* for managing the appearance of the face in various social contexts, and efforts to *cope* with the source of emotion will also be described. This account will allow *one* answer to the question of what is emotion, and how we know when emotion is occurring. It will provide the basis for discussing pitfalls in cross-cultural research on emotion, for integrating seemingly contradictory findings, and for describing the many ways emotional expressions vary with culture.

This account of emotion, however, risks a good deal. It is too long, perhaps, for this paper, and certainly too short to be complete. It deals with nonobservables reaching beyond data to explanatory mechanisms which may help in understanding the data. It may seem wrong, misguided, mechanistic, full of jargon, etc. Nevertheless, it should provide the terms and raise the issues for exposing the *variety* of points for possible disagreement which have too long laid hidden.

Emotional Responses are brief, often quick, complex, organized and difficult to control. Let us consider each of these characteristics. Emotions can be very brief. It is not uncommon to be angry or afraid or happy for only a few seconds.

Surprise is always brief, while the other emotions vary in duration depending upon the circumstances. If an emotion lasts for many hours, or days, the language of everyday life would utilize mood terms father than those of an emotion. The person is said to be irritable or hostile, not just angry; blue, not just sad, etc. A full discussion of the difference between mood and emotion would take us far afield, yet there is probably some merit in the notion that typically emotions have fairly brief durations, shorter than moods, attitudes, traits, or many beliefs or values.[15]

Quickness refers just to the onset of an emotion. Emotions can become aroused in a fraction of a second. Not that they always must be aroused so quickly, but as will become apparent later, the potential for speed in onset time, for a very quick response, is important in explaining and distinguishing emotional from other kinds of behaviour.

Emotion is complex, entailing a number of different response systems. Only some of them can be directly observed. There are skeletal muscle responses such as flinching, thrusting forward, turning away, overall relaxation, etc. Facial responses include the expressions I will describe in detail later, as well as vascular and muscle tonus changes. Vocal responses include sounds such as screams or groans, as well as a tensing of the vocal apparatus, with consequent changes in voice quality. These are all likely to be very quick, initial responses. Somewhat longer and more elaborated are the coping behaviours directed at whatever has set off the emotion. Included would be fighting, fleeing, denying, apologizing, etc. Specific changes in the autonomic and central nervous system are also involved, in ways I will not detail here.

The subjective experience of the emotion, usually neglected by modern psychology, is another important emotion response system. The subjective experience includes - but may not be limited to - sensations which are the result of feedback from changes occurring in the already named response systems. Also included are memories, images and expectations associated with one or another emotion, and with the very specific circumstances of the occasion for an emotion. One important characteristic of the subjective experience of emotion is the awareness that the changes occurring are not easy to control consciously. I will return to this point later.

The various response systems mentioned are organized in two ways. The activity in each response system is interrelated rather than independent. *And* the changes occurring within each (or most) response system are distinctive for one as compared to another emotion. In a preliminary study we (Ekman, Malmstrom & Friesen 1971) found different patterns of heart rate acceleration and deceleration to occur simultaneously with different patterns of facial movement. This study showed organization within each response system and in the interrelationship between response systems. Admittedly, there is little evidence one way or another for such interrelationships among all the response systems we propose. And the evidence for distinctive patterns for each emotion is presently limited to facial expression.

Affect Programme. For there to be such complexity and organization in various response systems, there must be some central direction. The term 'affect programme' refers to a mechanism which stores the patterns for these complex organized responses, and which when set off directs their occurrence. I am not concerned with where in the brain this programme is located. (Lower areas must be involved, but I do not presume either a single location or involvement of only one neural mechanism.) Instead, I will describe what is assumed about how such an affect programme must operate.

The organization of response systems dictated by the affect programme has a genetic basis but is influenced also by experience. The skeletal, facial, vocal, autonomic and central nervous system changes which occur initially and quickly for one or another emotion, we presume to be in largest part given, not acquired. For example, habits would be unlikely to determine just which pattern of impulses are transmitted to the facial nerve, although we will later describe how habits, what we term display rules, develop to interfere with the operation of these responses dictated by the affect programme. Experience, of course, plays an important role. The emotional response systems change with growth, disease, injury, etc. They are not constant through life.

Through experience, with sufficient time and learning, habits become established for how to cope with each emotion. I do not believe that such coping behaviours are part of the given affect programme. These habitual ways of coping may become so well learned that they operate automatically and quickly in conjunction with specific emotions. Given our lack of knowledge about the operation of the central nervous system, it matters little whether I say that those habitual ways of coping become governed by the affect programme, or that they operate automatically in conjunction with it. Memories, images, expectations associated with one or another emotion are, like coping, not given but acquired, and can similarly become habitual, automatically involved when the affect programme is set off.

Thus, I postulate that when the affect programme is set off, a number of other things happen in addition to the responses immediately governed by the programme. Memories, images, and expectations associated with the emotion and the circumstance come into play. Coping behaviours associated with the emotion begin, and habits directed at managing emotional behaviour may become operative. All these related changes can occur automatically with great speed, rather than deliberately.

Management of the responses governed by the affect programme is usually not easy and not always entirely successful. Some of the responses under the command of the affect programme begin to change in fractions of a second. Deliberate or habitual interference is more successful with some of the emotional responses governed by the affect programme than with others. For example, it is far easier to inhibit or squelch a facial movement than to change respiration or heart rate. The difficulty experienced when trying to interfere with the operation of the affect programme, the speed of its operation, its capability to initiate responses that are hard to halt voluntarily, is what is meant by out-of-control quality to the subjective experiences of some emotions.

I have been working backwards in time. I started with a description of the brevity, complexity and organization of emotional responses. Then I described an affect programme which directs those responses. Now I must take a further step back, to consider what happens to call the affect programme into operation. There must be an appraiser mechanism which selectively attends to those stimuli (external or internal) which are the occasion for activating the affect programme. Otherwise the complex organized emotional responses directed by the affect programme would occur randomly. Since the interval between stimulus and emotional response is sometimes extraordinarily short, the appraisal mechanism must be capable of operating with great speed. Often the appraisal is not only quick but it happens without awareness, so I must postulate that the appraisal mechanism is able to operate automatically.

Automatic appraisal mechanism.[16] It must be constructed so that it quickly attends to some stimuli, determining not only that they pertain to emotion, but to which emotion, and then activating the appropriate part of the affect programme. The automatic appraisal may not only set off the affect programme and the responses it directs, but it may initiate also the processes which evoke the memories, images, expectations, coping behaviours and display rules relevant to the emotion.

Appraisal is not always automatic. Sometimes the evaluation of what is happening is slow, deliberate and conscious. With such a more extended appraisal there may be some autonomic arousal, but perhaps not of a kind which is differentiated. The person could be said to be aroused or alerted, but no specific emotion is operative. Cognition plays the important role in determining what will transpire. During such extended appraisal the evaluation may match to the selective filters of the automatic appraiser, and the affect programme may be set off. It need not be, however; the experience may be diffuse rather than specific to one emotion. I suspect that if the emotion is not specific, if the affect programme is never involved, there are limits to how intense the more general emotion arousal state may become.

When the automatic appraiser responds to a stimulus, sets off the affect programme which directs a particular complex, organized set of emotional responses, appraisal does not necessarily stop. The event may continue to unfold, new stimuli may be emitted which again are subject to automatic appraisal → affect programme → emotional responses. The same emotion may be repeatedly triggered, or a number of different emotions may be triggered sequentially or simultaneously. If the event does not continue to unfold, as in a near miss car accident, the person will usually realize and consider what has happened. He may realize there was no 'need' to have become emotional and struggle to stop the operation of the affect programme. Or, the realization of having had an emotional response may itself become the occasion for another emotional response. For example, responding to a criticism immediately and automatically with anger once realized may be the elicitor for shame or disgust with oneself. Or, the situation may become ambiguous and the extended appraisal system may work in trying to figure out what is happening and going to happen next.

Elicitors. One further step backwards leads to the elicitors of emotion, what characterizes the events appraised as emotional. There is variation in the particulars of what elicits a given emotion, yet there are also common features in what are identified as elicitors for an emotion. We use the term *elicitor* to refer only to those stimuli which are identified by the automatic appraiser as specific for one or another emotion. Elicitors call forth emotion quickly, but what occurs is not a reflex arc. The connection between specific stimulus and response is not given, and it is not fixed. Probably there is no emotion for which there is a universal elicitor, uniform in its specific details, which always calls forth the same uninterruptable set of emotional responses.[17] Certainly there is no empirical evidence of such. One possibility would be that the sight of a missile travelling directly towards the eye at a given speed would be automatically appraised so as to set off the fear affect programme. Even here interference or interruption of the fear response is probably possible. At best there would be only few if any such innately wired elicitors.

Cultural anthropology is replete with accounts of great variety in the specifics of what elicits one or another emotion. Most of the specific elicitors for emotion are variable, learned through different experiences. You cannot know what made the man in Fig. 4 disgusted, it depends upon his particular experiences. What is disgusting varies within as well as between cultures. Yet you could know something about the other responses likely to occur since you can see from his face that he is disgusted. For example, he is more likely to say something sounding like '*uuchk*' than '*mmh*', more likely to move backwards or away than to approach, more likely to describe his experiences using gustatory images than explosive ones.

Fig. 4. *Person of the South Fore of New Guinea expressing disgust. Copyright Paul Ekman 1977.*

While the specific elicitors for emotion vary with individual experience, there is also some commonality apparent in the general outline of what elicits, through automatic appraisal → affect programme, the emotions. Disgust elicitors share the characteristic of being noxious rather than painful; and, in all cultures taste stimuli are among the elicitors of disgust. Surprise depends upon the details of what would be unexpected for a particular person in a particular situation, yet surprise elicitors share the characteristics of being unexpected, novel and usually sudden rather than gradual. People learn how to recognize impending harm, to avoid danger. While the specific stimuli so identified as potentially harmful will vary, fear elicitors share the characteristic of portending harm or pain. One of the common characteristics of some of the elicitors of happiness is release from accumulated pressure, tension, discomfort, etc. Loss of something to which one is intimately attached might be a common characteristic of sadness elicitors. Interference with ongoing activity might be characteristic of some anger elicitors.

These are but tentative examples. My point is that the elicitors for each emotion share some characteristics. I propose that the automatic appraiser is sensitive to these general characteristics. With experience such sensitivity may become sharpened and more precise.

To point to similarities in what elicits surprise, fear, etc. does not mean that those characteristics will be found for every elicitor for every emotion, only that some will be similar. Through repeated experience specific elicitors become identified by the automatic appraiser as relevant to one or another emotion. We expect that many but not all of these specific elicitors will share common characteristics.

Recall that I have been describing as elicitors only those stimuli which are identified by the automatic appraiser. There are many other more complex and variable stimulus situations which are not so identified but instead become subject to the more extended appraisal process. Perhaps it is just because they do not obviously share one of the common elicitor characteristics (or because they combine elements of too many elicitors) that such stimulus configurations are subject to extended rather than automatic appraisal.

What is emotion? It certainly is not any one of the elements I have described: response systems, appraiser, programme, or elicitors. Nor is emotion just the combination. Instead emotion refers to the process whereby an elicitor is appraised automatically or in an extended fashion, an affect programme may or may not be set off, organized responses may occur, albeit more or less managed by attempts to control emotional behaviour. What gives an emotion its particular flavour, is the particulars of the elicitor, the appraisal, the part of the affect programme set off, and those parts of the response systems which go unmanaged or managed.

Is there a *sine qua non* for emotion? I disagree with past theorists who would say it is visceral reactions, or cognitive appraisal, or facial responses. My answer that there must be an appraiser, a programme and response systems more or less capable of action for emotion to occur, is only to say that the organism is alive. If the affect programme is not set off, if the appraisal is only of the extended not of the automatic kind, it is still emotion but not the same as when those things do happen. If the anger part of the affect programme is set off, for example, but there is interference with the activity of the facial muscles, it is still anger, but not the same anger as when facial activity is not so squelched.

Looking from the outside, without access to know what is occurring in the central nervous system, how can an observer tell when emotion is present, The estimate that emotion is present is more likely to be correct when:

- The response system changes are complex, when it is not just a facial, or skeletal, or vocal, or autonomic, or coping response, but a combination;
- the changes are organized, in the sense of being interrelated and distinctive for one or a combination of emotions;
- the changes happen quickly;
- some of the response system changes are ones common to all people;
- some of the responses are not unique to *homo sapiens*.

This is *not* the only time emotion occurs, but when an observer's estimate is *most* likely to be safe.

I will now explain more about display rules and coping behaviours, to provide the background necessary for then discussing the pitfalls in cross-cultural observations of emotional expression.

Display Rules. When an elicitor is automatically appraised, the affect programme is set off, and organized complex emotional responses begin, interference is still possible. The emotional responses may be interrupted, diminished, amplified, or masked with the appearance of another emotion. Some emotional response systems may be more difficult to manage than other, (e.g. heart rate more than facial movement), but people often try to control their emotional behaviour.

We coined the phrase *display rule* to refer to the conventions, norms, and habits that develop regarding the management of emotional responses.[18] A display rule specifies who can show what emotion to whom, when. These rules are often learned so well that they typically operate automatically, noticeable only in the breach. For example, the prohibition against showing anger, or the rule to substitute sadness for anger, is learned so well by some middle-class American girls, that later if liberated it requires some struggle to 'get their anger out'. Other display rules are learned more by example, by observing what others do or following implicit instructions of those who manage events when emotion is made the occasion for public ceremony. The performance for such display rules may not be as good, but errors are usually overlooked. An example of this type of display rule is that at beauty contests a winner may cry but not the losers. At funerals, one can note almost a 'pecking order' of grief expressions based on the rights to mourn. A man's secretary cannot look sadder than his wife unless she intends to state something quite different about the true nature of their relationship.

There are also *personal display rules*, habits learned about managing emotional expression which do not reflect a cultural norm, but more individualized experience. The extent to which one follows a cultural display rule may depend upon the extent to which it conflicts with a personal display rule. For example, a woman may have the personal display rule never to show her feelings of distress, which depending upon her culture, could put her in conflict in the mourning situation if just such expressions are required by the widow.

The management of emotional responses may occur also by deliberate choice of the moment, for a particular advantage, rather than as part of a long standing personal or cultural display rule. Then the person is more aware of what he does and is likely to be rather slow and inept.

Note that although the origin differs - cultural or personal display rules or deliberate choice of the moment - management of emotional expression may be either the product of well learned habit or a more deliberate process. This has implications for what we have termed 'leakage', (Ekman & Friesen 1969b), when the emotional responses escape attempts to conceal them. When the management is deliberate, either because of choice of the moment or when following cultural rules never practised so as to become habitual, it is likely that some of the emotional response systems governed by the affect programme may be noticeable at least briefly, unless the person was alerted ahead of time and prepared himself. When the management is habitual due to a culturally shared or personal display rule, the management efforts may be set into operation just about as quickly as the emotional responses dictated by the affect programme. Better concealment should occur.

Some emotional responses may be more subject to modification and disguise than others. Furthermore, display rules, personal or cultural, may focus more upon one response system than another. We (Ekman & Friesen 1969b, 1974b) have suggested that there is usually more management of words spoken than of voice, more of voice than facial movement, more of facial movement than body movement. We attribute such differences in what is managed to differences in the amount of attention which is typically paid by others to what a person does. While we have some evidence to support our hypothesis that there is more leakage in body movement that in either words or facial movement, it is limited to middle-class Americans. There are individual variations and there may be cultural differences in which emotional responses are most subject to management attempts. Even though facial expressions are managed, mistakes are made, and leakage may be detected (Ekman & Friesen 1975:Ch. 11).

Coping. This term refers to attempts to deal with the emotion felt and its source; to increase, diminish or sustain what is occurring. Coping includes various cognitive activities as well as such organized behaviours as attacking or fleeing. As the system where learning contributes the most, coping is the most elaborated of the emotional response systems.

Compare with coping the initial skeletal muscle response directed by the affect programme when an anger elicitor has been identified by the automatic appraiser. The immediate skeletal muscle response might be a slight movement forward. Coping could vary - attack, flight, denial, appeasement, etc. We discover how to cope with our emotions, what is likely to be successful, proper or improper. When angry, our likelihood of fighting or scratching our face, depends upon what we have learned about how to deal with the particular kind of anger elicitor.

Once coping techniques have been acquired, they can become so well learned that they operate automatically and are called forth when the affect programme is set off. They are notoriously difficult to change. Coping also occurs when extended appraisal is the manager. Then, more alternatives and flexibility in coping may be possible.

Biology may provide some predispositions affecting the likelihood of one versus another type of coping behaviour being developed for an emotion. For example, the skeletal muscle response for anger suggests that attack may become more frequent for coping with anger than flight. Yet, this predisposition is relatively fragile. Experience can overcome such predispositions and institute diametrically opposite coping. Coping involves a wide range of elaborated activities, and biology at best gives only a tap in a direction. Culturally and individually variable learning is the overwhelming contributor to coping.

Pitfalls in Cross Cultural Research on Facial Expressions. Before turning to describe facial expression let me describe some pitfalls which await the observer of facial expression who does not consider variations in elicitors, display rules and coping. Suppose different facial expressions are observed at funerals in two different cultures (e.g. LaBarre). In one culture let us say, the mouth corners are down and the inner corners of the brow are drawn up and together, while in the other culture the lip corners are pulled up obliquely, causing crowsfeet to appear at the side of the eyes. To conclude that this proves facial expression is culture-specific, would be to ignore the elicitor pitfall. Funerals are not necessarily a universal elicitor of grief, sadness, distress or any such feeling. If it is not an elicitor for such an emotion in two cultures, then the observation of different facial actions does not prove cultural determination of the emotional expression. Paradoxically, it was those who were wedded to a culturally relativistic viewpoint, who sought evidence that culture was the sole determinant of facial expression, who were the most 'biological' in assuming that the elicitor for an emotion was standard across the cultures compared. Many of the observations of differences in facial expression in the same situation across cultures are ambiguous because there was no independent evidence that the situation evoked the same emotion in each culture.

Suppose there was. Suppose it was known that the death of a child was the elicitor for sadness in each culture. Would the observation of different facial expressions by the mourners in two cultures prove that facial expressions are only arbitrarily associated with emotion? Only if we ignore the display rule pitfall, by failing to recognize that the cultures could differ in their norms about masking facial expressions of emotion. The different facial behaviours observed might not mean that the down-cornered mouth is associated with sadness in one culture while up-turned mouth is associated with sadness in another, but only that one culture has the display rule to cover sadness with a happy mask at a funeral. Again, it was those who viewed facial expression as an arbitrary set of movements whose relationship to emotion was solely determined by culture, who treated the facial expressions as if they were reflexes, not capable of management by habit or choice.

In a similar fashion cultural variations in coping can lead to another pitfall. The association of the same facial response with different coping activities does not mean *ipso facto* that the facial expression is arbitrary, solely determined by culture.

The elicitor, display rule, and coping pitfalls admittedly make it more difficult to obtain conclusive evidence to prove the arbitrary nature of facial expression of emotion. It would be necessary to verify by some means independent of the face that the situation compared in two cultures elicited the same emotion, that in neither culture was there a display rule to disguise the emotional expression, and that the emotion was associated with the same coping behaviour. Only then would evidence of quite different facial expressions unambiguously suggest that the emotion signified by a facial movement is totally arbitrary, different from culture to culture. No such evidence exists, but just the opposite. In an experiment I will report shortly the universal facial expressions first occurred and then disappeared when culturally different display rules came into operation, disguising facial expression more in Japan than in the United States.

Facial Expressions. In contrast to coping, there is a major biological contribution to facial expressions of emotion. Biology has shaped the affect programme, determining which facial movements are likely to occur with one or another of the emotions, and perhaps also the timing of those movements. The particular combination of facial movements and the corresponding changes in visual appearance are not arbitrarily associated with each emotion but, for at least some emotions are the same for all people. Shortly I will review the evidence for that assertion. First, let us consider just why it might be that one set of muscle actions rather than another occur with a particular emotion. Why is it that the lip corners go up not down with happiness, that the brows lift not lower with surprise, etc.?

The biological contribution may be extensive, involving an innate association between muscle movements and emotion, or more modestly biology may only predispose the organism to acquire these associations through common learning experiences. Species constant learning as the explanation for universals in emotional expression was suggested by Allport (1924). Let us consider the example of the raised brow (inner and outer strands of the *frontalis* muscle) in surprise. All that biology may contribute is how the face is constructed. Raising the brows does increase perception of the top part of the visual field. Infants encounter unexpected events in which they would raise their brow to see what is happening above them. (One could even argue that the unexpected is more likely to be above than below the infant.) Over time, perhaps abetted by the signal value of the movement, brow raising and surprise would become associated. In the strictest version of this explanation the infant would have to learn, presumably by trial and error, that brow raising increases his visual field. Alternatively that might be given, and what he learns is to make this movement when trying to see what has unexpectedly happened. To grant even more to biology, the infant could be born equipped to raise his brows when visually scanning unexpected sudden *visual* events. What he needs to learn is to generalize this response to *any* unexpected event, regardless of whether it is visual. The more extreme innate explanation would be that as a consequence of evolutionary processes, humans are constructed so that brow raising is built into the affect programme for elicitors automatically appraised as pertaining to surprise.

These explanations differ in a number of ways: in how much emphasis they place on ontogeny as compared to phylogeny; on whether current adaptive function is relevant; and whether signal value is implicated in development. I have not described completely any one explanation, nor all of the explanations that are possible. My purpose is to show that there is more than one way in which biology could contribute to facial expression. In all likelihood, more than one such explanation is relevant to account for the different universal facial expressions. It is even possible to design studies which could help choose among these explanations for particular emotions, e.g. studies of early development in blind and sighted infants.[19]

Quite apart from which way biology contributes, the evidence I will now describe of a universal association between emotion and particular facial expressions can only be explained by acknowledging a major substantial biological basis to facial expression of emotion.

Evidence of universals in facial expression. One type of evidence is based on studies in which people in different cultures were shown faces and were asked to interpret the emotion shown in each face. If what a facial expression signifies is totally arbitrary, variable from culture to culture, then the interpretations so obtained should also be variable. This type of study has been carried out by more than a half dozen investigators, the majority of whom were trying to prove that facial expression is culture specific. In thirteen different countries, where nine languages were used, the same emotional interpretation was obtained for the emotional expressions (see reviews of these studies in Ekman, Friesen & Ellsworth 1972:Ch. 19; Ekman 1973:Ch. 4).

One problem with this evidence is that all the people who interpreted the facial expressions were members of literate cultures subject to mass media influences. It was possible to argue that these people might have totally different facial expressions but had learned how to interpret each other's unique expressions. Or, they have different expressions but had all learned to recognize the stereotyped expressions shown in the theatre. Or, they have the same facial expression because they had all learned their facial expressions from cinema, television and magazines.

To meet this criticism we carried out studies with more visually isolated people in the South Fore of New Guinea (Ekman, Sorenson & Friesen 1969; Ekman & Friesen 1971). Studying only those who had not seen any mass media and few outsiders, we found that they interpreted faces as showing the same emotions as did the people from literate cultures. The one exception was that fear and surprise faces were not distinguished from each other, although they were each distinguished from anger, sadness, happiness or disgust. Another type of study was conducted with the South Fore in which members of this society were asked to show on their own face how they would look if they felt one or another emotion. The expressions they showed were once again the common universal expressions, with the exception of surprise and fear which showed elements of both emotions. (See Fig. 5.)

Doubtful about our results, not believing that there were any universal expressions of emotion, Eleanor Rosch and Karl

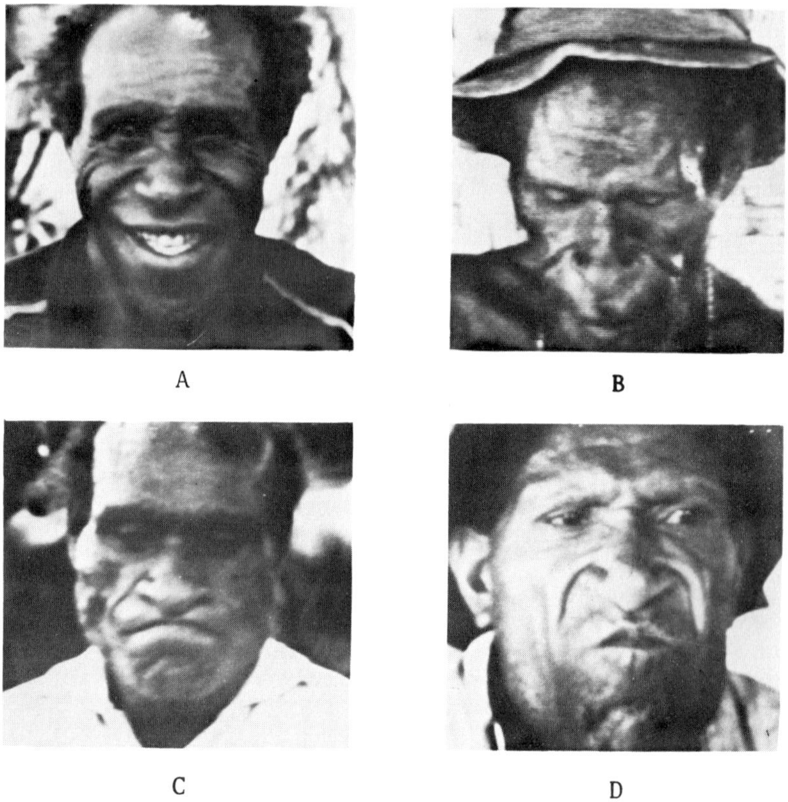

Fig. 5. *Video frames of attempts to pose emotion by persons of the South Fore of New Guinea. The instruction for* (A) *was 'your friend has come and you are happy'; for* (B) *'your child has died'; for* (C) *'you are angry and about to fight'; for* (D) *'you see a dead pig that has been lying there for a long time'. Copyright Paul Ekman 1975.*

Heider repeated these studies among the Dani of West Irian. They studied people more isolated than those we had seen. Again, the same results were obtained.

One of the more remarkable interpretations of this evidence was recently proposed by Margaret Mead.

> Ekman has demonstrated... [that] it is possible to persuade members of different cultures to produce simulations which are mutually intelligible between these cultures. If simulations are mutually recognizable there must be some universal element involved such as Ekman postulates,

> in a connection between the nervous system and specific facial muscles.... [There is a] possibility that human beings may share a core of innate behaviours which are usually highly modified, but which are available for simulation (Mead 1975:212).

By proposing that only simulations are universal while actual emotional experience is culturally variable, Mead attempted to reconcile the contradiction between our findings and Birdwhistell's assertion that he never found any evidence of universals in facial expression.

I disagree with Mead's interpretation of our work and of Birdwhistell's claims. It seems strange to suggest that what is innate would be apparent only in simulation and not in actual emotional experience. We *assume* that people readily interpreted the facial expressions and readily posed them because they have seen these expressions, and experienced them in social life. The answer of course should come from data, from studies of actual spontaneous expressions of emotion.[20] Such evidence was reported in the book reviewed by Mead but she ignored it.

We chose to study spontaneous facial expressions in Japan because of the popular notion of the inscrutability of the Oriental. We hoped to demonstrate that this was due to display rules about masking negative affect in the presence of authorities. Another reason for choosing Japan was that Lazarus, et. al. (1966) had previously determined through psychophysiological measurement that certain stress films had a similar impact on the Japanese as they did on Americans. Thus, we could avoid the elicitor pitfall by utilizing an elicitor known to produce some similar effects in two cultures. We avoided the display rule pitfall by videotaping when the subjects thought they were completely alone and unobserved.

The identical experiment was performed at Waseda University in Tokyo and the University of California in Berkeley[21]. Subjects came into a laboratory and watched both a travelogue and stress inducing films while measurements were taken of their skin resistance and heart rate. The videotapes taken with a hidden camera unknown to the subject were later measured by persons not knowing which film was seen when the facial expressions occurred. We found a better than .90 correlation in the particular facial movements shown by the Japanese and by the Americans. Virtually the same repertoire of facial movements occurred at the same points in time. (Reported in detail in Ekman 1972). We believe this is definitive evidence that facial expressions of emotion are not limited to simulations, but do occur spontaneously as well.

Later in this same experiment we brought a research

assistant into the room, garbed in white coat, to sit with the subject while he watched a stress film. Now we expected that display rules for managing facial expression in the presence of an authority figure should be operative, more so in Japan than in the U.S. Measurement showed that the facial movements were no longer the same. The Japanese looked more polite, showed more smiling than did the Americans. Examining these videotapes in slow motion it was possible to observe sometimes the actual sequencing in which one movement (a smile for example) would be superimposed over another muscle action (such as a nose wrinkle, or lower lip depressor).

If only the latter part of the experiment had been performed, if we had studied the Japanese only when they knew we were doing so (likely conditions in many anthropological studies) and we did not do micromeasurement to look for the possibility of masking, we would have been misled. Instead, guided by our theoretical foundation about emotion we were able to arrange an experiment to show the biological basis of facial expression and how cultural influences modify expression. When alone, Japanese and Americans showed virtually the same facial expressions. When in public, due presumably to display rules, Japanese more than Americans controlled and masked negative expressions.

In writing about evidence of universals in facial expression Leach has said:

> Now, since natural languages do not in general associate particular sounds or combinations of sounds with particular meanings, the linguistic analogy does not suggest there will be any consistent relationship between nonverbal signal and response when such signals are observed in differing cultural environments. And it leads to the further conclusion that any claims to have demonstrated such consistencies in experimental situations must be considered highly suspect. Man admittedly is a single species, and despite all the differences of culture the things that man is capable of doing with his limbs and his facial expressions are everywhere limited by the same set of muscles. The same units of expressive gesture will crop up again and again in all kinds of cultural situations (Leach 1972:239).

This quotation draws attention to how the evidence directly contradicts what would be expected from a linguistic analogy. Such an analogy is not relevant to understanding the facial expressions of emotion, although it has utility in understanding facial illustrators, emblems, and regulators. Taking the evidence I have summarized, together with the observations of

Eibl-Eibesfeldt (1972), studies of blind infants and children, expressions in young infants, and work tracing facial expressions phylogenetically, we can conclude there is *some* universal association between emotion and facial movements.

The evidence is consistent with our neurocultural account of emotion outlined earlier, although it does not establish all that was proposed. Some such explanation, positing both biological and social factors in emotional expression, is required to deal with the evidence. Now let me delimit the universality of facial expression by describing, on the basis of our theory, some of the ways in which culture moulds the biologically based facial expressions of emotion.

CULTURAL INFLUENCES ON FACIAL EXPRESSION

Most of the cross-cultural evidence supporting universality was about expressions which were manifest across the entire face. The faces were instances in which muscular action produced a distinctive appearance in the forehead, eyelid and mouth areas. This need not always happen. Often only part of a total expression may occur, muscle groups acting in only one part of the face with the rest of the face not active. Friesen and I (Ekman & Friesen 1975:Ch. 11) have hypothesized that such *partial* facial expressions may occur when a person attempts to control or manage facial appearance. Such attempts to control can be the product of choice of the moment, long established personal display rules or cultural display rules. As discussed earlier, display rules should vary across cultures.

Another reason for partial facial expressions of emotion would be learning that has emphasized one part of an expression more than another. Families may pay attention to and otherwise reinforce one part of a facial expression of emotion more than another. A child may learn that one part of the facial signal brings forth more response from others, and he may see that part of the expression more often on the faces of those around him. Based on such experiences, one or another part of the biologically based total facial expression could become more prominent in the person's repertoire. There is little reason to expect uniformity across cultures in these experiences.

Still a third reason for partial facial expressions is the possibility that when emotion is felt only slightly it may register in only some muscle groups rather than across the entire face. There is as yet no evidence whether this is so. If it is, we do not know whether slight versions of an emotion tend to recruit consistently the same muscles. For example, in slight fear would it be the *frontalis* and *corrugator* affecting the brow, or the *risorius* and *platysma* affecting the cheek and mouth, or the upper eyelid levator, which would most likely act? Are there consistencies for a given person across such slight fear experiences, across persons within a social grouping, across groupings within a culture, etc.?[22]

Cultural differences in facial expressions of emotion may also be apparent in the most common *blends* of emotion. A blend is a compound facial expression in which the muscular actions for two or more emotions combine in a single facial expression. (See Figure 6). Blends sometimes occur because the elicitor calls forth more than one emotion. For example, a surprising event may be also frightening, or the surprising event may be pleasing. Blends also happen because of habits that associate one emotion with another. One such habit is affect-about-affect; for example, one person may always feel disgusted with himself when he feels anger, another may habitually feel afraid of his own anger. Another type of habit relating affects is mixture of two feelings towards a class of elicitors; for example, one person may habitually feel disgusted by what makes him angry, another may habitually feel afraid of whatever makes him angry. Due to either type of habit a single elicitor of anger may result in one person showing an anger/disgust blend, while another shows an anger/fear blend. These habits associating two emotions may become established through individually idiosyncratic learning, or through learning experiences common among some social group.

Cultural variations in facial expressions manifest in blends could be the result of differences in the likelihood of encountering elicitors which call for a particular blend; or, variations in experiences leading to habits associating one emotion with another. Even if the same blend is shown in two or more cultures, it may not always have its own linguistic label. For example, facial expressions showing blends of happiness/anger or happiness/disgust, might occur, but not every culture need have a single word label for them, such as 'smug.' We are planning to test some of these hypotheses about cultural differences in partial and blend expressions. For the present, they rest only upon casual observation and logic.

The evidence of universality in emotional expression is limited to the emotions of happiness, surprise, fear, anger, disgust and sadness. We expect that there are also universal facial expressions for interest and, perhaps, shame. There may be other emotions for which there is also a universal facial expression, but probably not many. Presumably, there are emotions for which there is no specific facial expression consistent within a culture; or for which there is a consistent expression in one culture but not in another. Take the example of contempt shown in Fig. 7. We suspect that the expression for this emotion will be different in some cultures. In still other cultures there will be no consistent expression at all. There may not even be a verbal label in every culture with the denotation and connotations of contempt. It is reasonable to presume that contempt, and any other emotions for which there is no universal distinctive facial expression, is not the product of the automatic appraisal → affect programme outlined earlier. Contempt might still be related to certain types of stimuli, but appraised more slowly, with a less direct linkage with simultaneous changes in other emotion response systems. Alternatively, it is possible that an emotion without any distinctive facial emotional response may still be part of the automatic appraisal → affect programme system. There is no data to provide a basis for choosing between these possibilities.

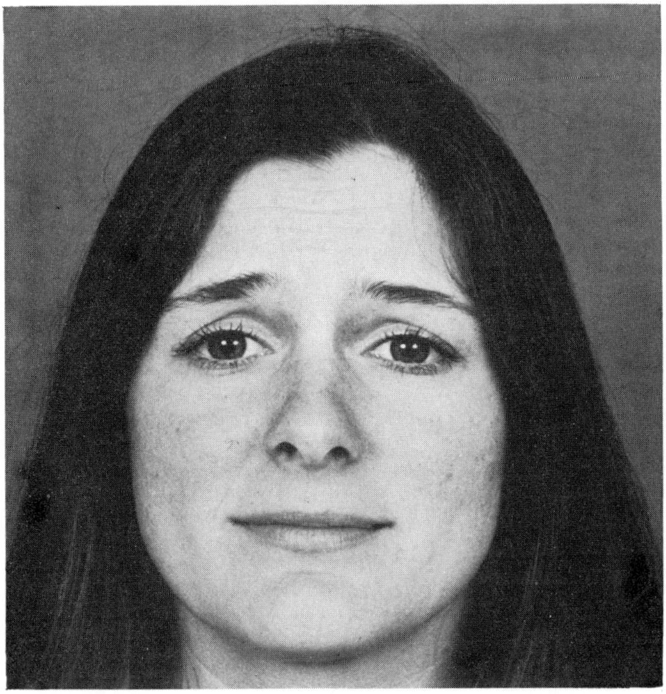

Fig. 6. *A blend of sadness (eyebrows, forehead and eyes) and happiness (lips). Copyright Paul Ekman 1977.*

Let us return to those emotions for which there is evidence of a universal expression: happiness, surprise, fear, anger, disgust and sadness. Just because there is one (or more) universal facial expression for each of those emotions does not prevent additional culture-specific facial expressions for each of these emotions. For example, in addition to a universal facial expression of fear (see Fig. 8) there might be additional facial expression for fear common in one culture but not apparent in another. Acquired culture-specific facial expressions for an emotion might receive sufficient attention in a particular culture to become more frequent than the universal expression for that emotion. Theoretically, the culture specific expression for an emotion should not be as quick to occur as the biologically based expressions.

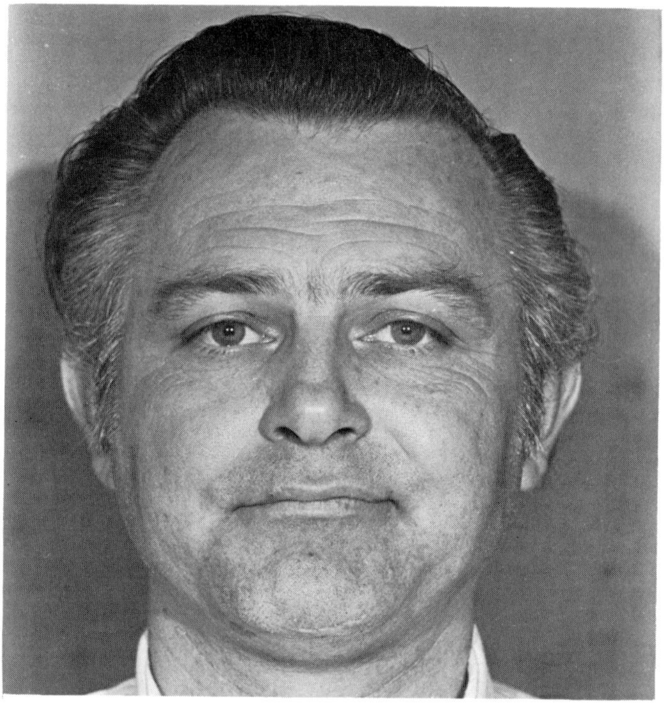

Fig. 7. *Facial Expression of Contempt which we hypothesize is not universal. Copyright Paul Ekman 1975.*

All of the sources of cultural variations in facial expression so far mentioned have been about the *morphology* of facial expression, not the *timing*. So little is known about the timing of facial expression that it is difficult to speculate about whether there would be much variation across persons within a culture, or between cultures. Is there a fixed *onset* time for the actions involved in any of the universal facial expressions to reach the point of maximum muscular action (*apex*)? Or, does onset vary with the particulars of the elicitor, cognitive appraisal, etc.? What about the duration of the apex? Is there any fixed minimal length of time the action must remain on the face before decaying (*Offset*)? Does apex duration vary from emotion to emotion, or is it solely a function of the particular elicitor, appraisal, etc.? Is offset time fixed, or does it too reflect differences in circumstance, display rules, etc.? The timing of facial expressions of emotion has not been investigated within a given culture, (with the exception of the startle,[17] let alone across cultures.

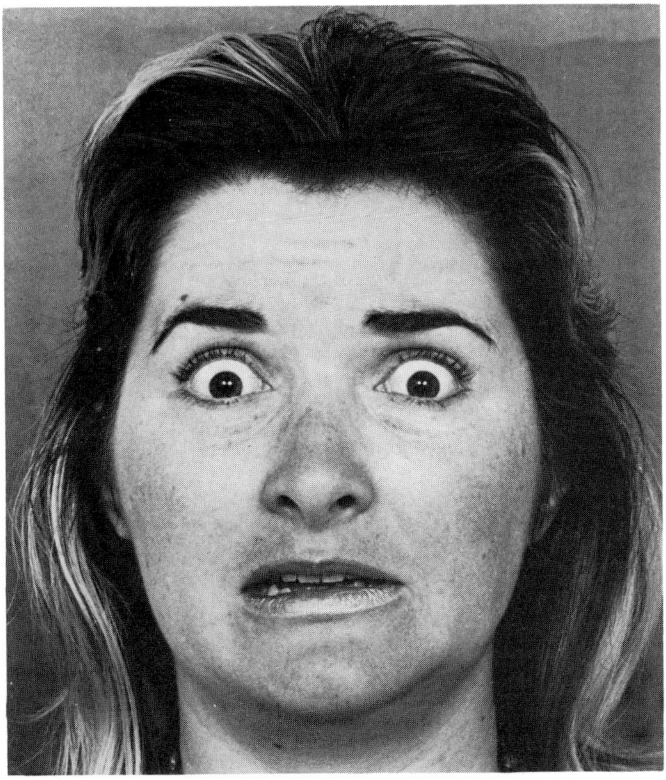

Fig. 8. *Facial Expression of Fear which evidence suggests is universal. Copyright Paul Ekman 1975.*

Another aspect of timing has to be with the sequence of muscular actions during onset or offset. Take the example of raising the brows (*frontalis*), drawing them together (*corrugator*), and raising of the upper eyelid (*levator palpebralis superioris*) in a partial fear expression. Does *frontalis* always start before *corrugator* or vice versa? Or does it depend upon the person, the specifics of the elicitor etc.? We suspect that there is probably some biological contribution to timing, both to onset, minimal apex, offset, and to sequencing, but that as with morphology the biological contribution is modifiable by experience. It is only now that precise methods for measuring facial movement have been developed that these questions about timing can begin to be examined.

Having taken some time to explain a number of different ways in which culture and various experiential factors could shape facial expressions, let me say that I do not think there is complete plasticity. For example, it would be unlikely for the *zygomatic* muscle (involved in the smile) to become a culture-specific expression for sadness. It seems unlikely that experience will be organized and maintained directly in opposition to biological predisposition. However, there is ample room for enormous cultural difference in emotional expression due to:

- variations in the specifics of the elicitors, display rules and coping;
- variations in partial or blend expressions;
- additional non-universal expressions for emotions for which there is also a universal expression;
- some emotions having no universal expression, but only culture-specific, if any, consistent facial expressions;
- variations in timing, (onset, apex, and offset, as well as sequencing), of facial actions.

There is also ample need for research to explore and document how these cultural differences might develop.

CONCLUSION

In discussing the conflicting viewpoints of Leach and Eibl-Eibesfeldt, Hinde (1972) suggested that we should expect similarities among cultures in signals that concern the personal or the emotional and we should expect differences in movements which take the place of words, which symbolize aspects of culture or which depend on languages. His interpretation agrees with our view that nonverbal communication is not a unitary phenomenon, but a term used to cover a number of different phenomena, activities which differ in origin, usage, coding, and in the relative contribution of biological and social factors.

While we can easily say that culture has more influence on emblems and illustrators, and biology on facial expression, we must be careful not to be misled by such a simple generalization. Even with illustrators and emblems there are some commonalities across cultures, albeit often hidden by the enormous differences. While the biological contribution to facial expression is dramatic, the number of ways in which experience produces differences within and across cultures is also astounding.

Discussion of the interplay of biological and social factors in facial expression, makes apparent that these are not single forces operative by but one mechanism. A variety of non-exclusive processes may be at work.

It is to all these complexities that we should attend.

FOOTNOTES

[1]The preparation of this report was supported in part by a grant from NIMH, MH 11976 and from the Harry F. Guggenheim Foundation. The research described was supported by those grants and by grants from ARPA, AF-AFOSR-1229 and by a Career Development Award and Research Scientist Award MH 6092. Wallace V. Friesen has collaborated on all of the research and in the development of the theoretical framework.

[2]There are problems with all the terms used to refer to the study of facial and body movement:

Nonverbal behaviour: that implies that what distinguishes the phenomenon is that it isn't words, and that is not necessarily what motivates much of the interest or theory; further, it is strange to use a term that defines a phenomenon by what it is not.

Motor behaviour: that implies an interest in skills.

Kinesics: that is identified with but one theoretical and methodological viewpoint (Birdwhistell 1970).

Expressive behaviour: that term implies the action is a manifestation of some internal affective state or personality characteristic, which is probably appropriate for only some facial or body movements.

Visually observed behaviour: awkward and odd to define a range of phenomena by how they are sensed by the observer.

See discussion of terminology by Sebeok (in press) and Ekman 1977.

[3]A fifth type of activity not discussed in this paper is what we (Ekman & Friesen, 1969) have termed 'regulators', movements which manage the back and forth flow of conversation. See Dittmann (1972), Duncan (1970) and Scheflen (1964) for examples of this work.

[4]None of the others who have been interested in symbolic gestures have explicitly described their methods for discovery of the emblem repertoire. Reading their listings it would appear that they include not only emblems but what I later describe

as illustrators. See accounts by foreign language teachers of French (Brault 1962), Spanish (Green 1968; Saitz & Cervenka 1971); also what is said to be a dictionary of emblem of all cultures over all of history by Bauml & Bauml (1975). This is a compilation from many sources. While suggestive as leads for establishing the emblem repertoire the use of these materials is limited without verification. If the same method for surveying the emblem repertoire is not used, contrasts between cultures in their emblem repertoire are hazardous.

[5]Once the repertoire is established then observations can be made of the natural occurrence of emblems during social interaction. Unless the repertoire has been first identified, naturalistic observation may be like hunting for a needle in a haystack, since emblems may often have a low rate of occurrence. Note that Desmond Morris (1975) is currently studying the distribution of a number of emblems across various European countries using quite different methods for identifying emblems.

[6]Kumin & Lazar (1974) began work on the decoding of emblems, but they only attempted to show that more emblems are decoded at later than earlier ages.

[7]Although using different terms to describe the activity a number of other investigators have also found that body manipulators are related to negative affect (Freedman & Hoffman 1967; Knapp, Hart & Dennis 1974; Mahl 1968; Mehrabian 1971; Rosenfeld 1966). In his most recent, but unpublished work, Freedman claims body manipulators are not a discomfort sign, but instead are self-stimulation required by difficulty in articulation and information processing.

[8]Consistent with this formulation, Schegeloff (1976) has suggested that body manipulators may occur during what he and Sachs term conversational 'repairs'. These are points in conversation where some attempt is being made to prevent disagreements from crystallizing, and discomfort would be expected.

[9]See references cited in the previous footnote. Weiner Devoe et al. (1972) are an exception also distinguishing what we term emblems from illustrators.

[10]Some people illustrate or show an emblem when talking on the phone. Under those circumstances the illustrator cannot serve any function relevant to the listener. Such illustrators may be due to habit or serve a self-priming function.

[11]We agree with Hinde's (1974) criticism of our earlier formulation which restricted emotional expressions to the face. In our current formulation we have also included vocal changes and body movement among emotional expressions.

[12]Tomkins's theory of *Affect, Imagery and Consciousness* (1962,

1963), is a notable exception. Paradoxically, the very complexity and length of his theory may have discouraged others from any such attempt.

[13]Margaret Mead (1975), in reviewing a book I edited, entitled *Darwin and Facial Expression* (Ekman 1973), raised a number of such pseudo issues. The argument about universality of facial expression of emotion is not between:
- anthropologists and psychologists: an interest in the biological basis of behaviour is found in both disciplines, and advocates of each side hold credentials in each field;
- quantitative and qualitative methods: both types of evidence have been gathered by advocates on each side;
- naturalistic and experimental methods; again, both types of studies have been cited by each side;
- those spending a few months and those spending a few years in another culture; both are found on each side.
- those... 'who are more interested in validity and reliability than in what they are actually studying' (Mead, 1975 p. 211); everyone is interested in his observations being correct and repeatable.

[14]Our account owes a great debt to Silvan Tomkins who first considered many of these issues. In general our views are the same, although we differ on some specifics, (e.g. role of facial feedback, elicitors). The description of emotional responses is based on Tomkins, and the term and treatment of the 'Affect Programme' is taken from him. Our account has similarity also to Izard (1971), as his description repeats Tomkins. Our account also borrows a few ideas from Lazarus and Arnold; quite different theorists.

[15]To distinguish emotions from moods is not to imply that moods are bereft of emotion. Quite the contrary; moods are highly saturated with one or another emotion. Someone is said to be in a hostile mood if he shows anger very frequently, or very easily, or seems ready to become angry. In a similar fashion, while I do not consider jealousy as an emotion, jealousy is characterized by strong emotions (anger, fear) within a particular interpersonal context.

[16]Mandler (1975), Lazarus (1968) and Tomkins (1962) have all talked about appraisal in different ways. I attempt here not to choose among nor integrate their views, but instead indicate what might be assumed about appraisal mechanisms to account for evidence on facial expression. Like the emotion theorists, I am quite vague about the 'mechanisms' limited by lack of knowledge.

[17]The startle response seems to be an exception, and for that reason, I question whether it should be considered an emotion. The early work of Landis and Hunt (1939) showed that a sudden loud noise (e.g. gun-shot) anticipated or not, resulted in

highly regular changes in facial and body movement despite attempts to prevent those responses. While their work was not cross-cultural, their studies strongly suggested that uniformity would be found.

[18]When we originally proposed this phrase it was meant to apply just to the management of facial expression. Now we broaden its reference to include habits about managing *any* aspect of emotional behaviour, posture, body movement, what is said, etc.

[19]If blind infants do not show the brow raise to sudden expected sounds or touches, then at least we could assume that the brow raise is not wired in for surprise, nor for scanning if the eyes are not operative. Unfortunately the data is equivocal (C.F. Charlesworth 1970; Eibl-Eibesfeldt 1972; Goodenough 1932).

[20]Let me note that in addition to the evidence reported we also gathered qualitative data as well, photographs, film and videorecords of naturally occurring facial expressions during social interaction among the South Fore of New Guinea. We studied those in detail, using methods much like those of Birdwhistell. Qualitative methods are useful, we believe, for discovery, not for proof, for they are subject to problems of sampling and observer bias. When wise observers using qualitative methods disagree, (e.g. Birdwhistell versus Eibl-Eibesfeldt) it is difficult to resolve the dispute without *ad hominem* or *ex cathedra* argument. Quantitative methods offer some hope of improving if not resolving such argument.

[21]We are grateful to Richard Lazarus, University of California, Berkeley; Masatoshi Tomita, Waseda University; Ned Opton, Wright Institute; and Jim Averill, University of Massachusetts, for their help and cooperation in making this study possible.

[22]My discussion of partial facial expressions assumes that the affect programme sets off a total expression, involving a number of different muscles, and that experience can limit how many of those muscles act. An alternative possibility is that the affect programme sets off a number of different muscles, and that experience determines whether they come to be recruited together or not. Cross-cultural developmental studies of facial expression would be relevant to determining whether one or both of these possibilities actually happens.

REFERENCES

Allport, F.H. (1924). *Social Psychology.* Boston: Houghton Mifflin.

Bauml, B.S. and Bauml, F.H. (1975). *A Dictionary of Gestures.* Metuchen, N.J.: Scarecrow Press.

Birdwhistell, R.L. (1970). *Kinesics and Context.* Philadelphia: University of Pennsylvania Press.

Brault, G.J. (1962). Kinesics and the Classroom: Some Typical French Gestures. *French Review, 36:* 374-382.

Broide, N. (1977). Israeli Emblems: A Study of One Category of Communicative Nonverbal Behavior. Unpublished Master's Thesis, University of Tel Aviv.

Charlesworth, W.R. (1970). Surprise Reactions in Congenitally Blind and Sighted Children. NIMH Progress Report.

Cohen, A. and Harrison, R. (1973). Intentionality in the Use of Hand Illustrators in Face-to-Face Communication Situations. *Journal of Personality and Social Psychology, 28* (2): 276-279.

Darwin, C. (1872). *The Expression of Emotions in Man and Animals.* Chicago: University of Chicago Press. 1965.

Dittmann, A.T. (1972). Developmental Factors in Conversational Behavior. *Journal of Communication, 22* (4): 404-423.

Duncan, S. Jr. (1970). Towards a Grammar for Floor Apportionment: A Systems Approach to Face-to-Face Interaction. *Proceedings of the Second Annual Environmental Design Research Association Conference.* Philadelphia: Environmental Design Research Association, 22-35.

Eco, U. (1975). Personal communication.

Efron, D. (1941). *Gesture and Environment.* New York: King's Crown, (current ed.) *Gesture, Race and Culture,* The Hague: Mouton.

Eibl-Eibesfeldt, I. (1972). Similarities and Differences Between Cultures in Expressive Movements. In R. Hinde (ed.), *Nonverbal Communication.*

Ekman, P. (1972). Universals and Cultural Differences in Facial Expression of Emotion. In J. Cole (ed.), *Nebraska Symposium of Motivation, 1971.* Lincoln: University of Nebraska Press.

Ekman, P. (ed.) (1973). *Darwin and Facial Expression: A Century of Research in Review.* New York: Academic Press.

Ekman, P. (1973). Cross Cultural Studies of Facial Expression. In P. Ekman (ed.) *Darwin and Facial Expression: A Century of Research in Review.* New York: Academic Press.

Ekman, P. (1975). Movements with Precise Meaning. *Journal of Communication, 26* (3): 14-26.

Ekman, P. (1977). What's in a Name? *Journal of Communication.* 27(1) 237-239.

Ekman, P. and Friesen, W.V. (1968). Nonverbal Behavior in Psychotherapy Research. In J. Shlien (ed.) *Research in Psychotherapy,* Vol III, American Psychological Association pp. 179-216.

Ekman, P. and Friesen, W.V. (1969a). The Repertoire of Nonverbal Behavior: Categories, Origins, Usage, and Coding. *Semiotica, 1* (1) 49-98.

Ekman, P. and Friesen, W.V. (1969b). Nonverbal Leakage and Clues to Deception. *Psychiatry, 32:* 88-105.

Ekman, P. and Friesen, W.V. (1971). Constants Across Cultures in the Face and Emotion. *Journal of Personality and Social Psychology, 17:* 124-129.

Ekman, P. and Friesen, W.V. (1972). Hand Movements. *Journal of Communication, 22* (4): 353-374,

Ekman, P. and Friesen, W.V. (1974a). Nonverbal Behavior and Psychopathology. In R.J. Friedman and M.M. Datz (eds.) *The Psychology of Depression: Contemporary Theory and Research.* Washington, D.C.: Winston & Sons.

Ekman, P. and Friesen, W.V. (1974b). Detecting Deception from the Body or Face. *Journal of Personality and Social Psychology, 29* (3): 288-298.

Ekman, P. and Friesen, W.V. (1975). *Unmasking the Face: A Guide to Recognizing Emotions from Facial Clues.* New Jersey: Prentice-Hall.

Ekman, P. and Friesen, W.V. (1976). Measuring Facial Movement, *Journal of Environmental Psychology and Nonverbal Behavior, 1* (1), 56-75.

Ekman, P. and Friesen, W.V. In preparation. Complexities and Confusions in Facial Expressions of Emotion.

Ekman, P. and Friesen, W.V. In press. *Facial Action Coding Manual: A Technique for the Measurement of Facial Movement.* Palo Alto, CA.: Consulting Psychologists Press.

Ekman, P., Friesen, W.V. and Ellsworth, P. (1972). *Emotion in the Human Face.* Elmsford, N.Y.: Pergamon Publishing Co.

Ekman, P., Friesen, W.V. and Scherer, K. (1976). Body Movement and Voice Pitch in Deceptive Interaction. *Semiotica, 16* (1): 23-27.

Ekman, P., Malmstrom, E.J. and Friesen, W.V. (1971). Heart Rate Changes With Facial Displays of Surprise and Disgust. Unpublished manuscript, University of California, San Francisco.

Ekman, P., Sorenson, E.R. and Friesen, W.V. (1969). Pan-cultural Elements in Facial Displays of Emotion, *Science, 164* (3875): 86-88.

Fillmore, C.J. (1971). Some Problems for Case Grammar: In Georgetown University Nongraph Series on Languages and Linguistics: 22nd Annual Roundtable, 50-51.

Freedman, N. and Hoffman, S.P. (1967). Kinetic Behavior in Altered Clinical States: Approach to Objective Analysis of Motor Behavior During Clinical Interviews. *Perceptual and Motor Skills*, 24: 527-539.

Goodenough, E. (1932). Expressions of the Emotions in a Blind-deaf Child. *Journal of Abnormal and Social Psychology,* 27: 328-333.
Green, J.R. (1968). *A Gesture Inventory for the Teaching of Spanish.* New York: Childton Books.
Hinde, R.A. (ed.) (1972).*Nonverbal Communication.* Cambridge University Press.
Hinde, R.A. (1974). *Biological Bases of Human Social Behavior.* McGraw Hill.
Izard, C. (1971). *The Face of Emotion.* New York: Appleton-Century-Crofts.
Johnson, H.G., Ekman, P. and Friesen, W.V. (1976). Communicative Body Movements: American Emblems. *Semiotica, 15* (4): 335-353.
Klineberg, O. (1940). *Social Psychology.* New York: Holt.
Knapp, M.L., Hart, R.P. and Dennis, H.S. (1974). An Exploration of Deception as a Communication Construct. *Human Communication Research, 1* (1): 15-29.
Kumin, L. and Lazar, M. (1974). Gestural Communication in Preschool Children. *Perceptual and Motor Skills,* 38: 708-710.
Labarre, W. (1947). The Cultural Basis of Emotions and Gestures. *Journal of Personality,* 16: 49-68.
Landis, C. and Hunt, W.A. (1939). *The Startle Pattern.* New York: Farrar.
Lazarus, R.S., Opton, E. Jr., Tomita, M. and Kodama, M. (1966). A Cross-cultural Study of Stress-reaction Patterns in Japan. *Journal of Personality and Social Psychology,* 4:622-633.
Leach, E. (1972). The Influence of Cultural Context on Nonverbal Communication in Man, In R. Hinde (ed.) *Nonverbal Communication.*
Mahl, G.F. (1968). Gestures and Body Movements in Interviews. In *Research in Psychotherapy,* Vol. III. Washington, D.C.: American Psychological Association.
Mead, M. (1975). Review of Darwin and Facial Expression, P. Ekman (ed.). *Journal of Communication,* 25 (1): 209-213.
Mehrabian, A. (1971). Nonverbal Betrayal of Feeling. *Journal of Experimental Research in Personality,*
Meisner and Philpott. (1975). The Sign Language of Saw Mill Workers in British Columbia. *Sign Language Studies.* 9, 291-347.
Morris, D. (1975). Personal communication.
Rosenfeld, H.M. (1966). Instrumental Affiliative Functions of Facial and Gestural Expressions. *Journal of Personality and Social Psychology,* 4: 65-72.
Saltz, R. and Cervanka, E. (1972). *Handbook of Gestures: Colombia and the United States.* Approaches to Simiotics 31. The Hague: Mouton.
Scheflen, A.E. (1964). Communication and Regulation in

Psychotherapy. *Psychiatry,* 28, (3): 245-257.
Schegeloff, I. (1976). Personal communication.
Sebeok, T.A. In press. The Semiotic Web: A Chronicle of Prejudices. *Semiotica.*
Stokoe, W.C., Jr. (1974). *Semiotics and Human Sign Languages.* Approaches to Semiotics 21. The Hague: Mouton.
Tomkins, S.S. (1962). *Affect, Imagery, Consciousness,* Vol. 1. New York: Springer.
Tomkins, S.S. (1963). *Affect, Imagery, Consciousness,* Vol. 2. New York: Springer.
Trupin, C.M. (1976). *Linguistics and Gesture: An Application of Linguistic Theory to the Study of Emblems.* Unpublished dissertation. University of Michigan.
Wiener, M., Devoe, S., Rubinow, S. and Geller, J. (1972). Nonverbal Behaviour and Nonverbal Communication. *Psychological Review, 79* (3): 185-213.

THE BODY AS AN INDICATOR OF HEALTH AND DISEASE

Ivan Polunin

This paper is about the flow of information on the state of health and disease from one person to others, by nonverbal means. It can be treated from the point of view of a particular medical system, such as a 'traditional' system, or 'modern' medicine.

A system of medicine is a set of ideas, values and practices concerned with health and disease, involving adepts whose special competence is recognised by at least a section of the population at large.

I am using the term modern medicine to describe that system which is taught in most of the world's medical schools and which is in most countries the officially recognised system. It is in large part scientific, Western in origin, international and modern, and it could be called SWIM medicine as an acronym. Of course, modern medicine strongly maintains some ancient beliefs and values, while traditional systems are adapting to present conditions.

HEALTH AND DISEASE

It is obvious that the body provides information on its state of health or disease, which is perceptible to others. With the development of modern technology, our ability to 'read the body' is generally improving, even though there is an erosion of skills in applying simple methods where these have largely been supplemented by technical methods.

However, there still are inescapable difficulties in determining the state of health of the body. Though people have traditionally been divided into two mutually exclusive operational categories, the healthy (who need no remedy) and the sick (who do), we cannot confidently ascribe everybody to one of these categories.

For those bodily characteristics which can be expressed as continuously variable measurements, we can probably all agree that there is a range of measurements which is compatible with health. Furthermore, outside some extreme limit, we would all agree that there is disease. But what of values which fall between the limits of what everybody agrees is normal or diseased?

Any limiting value used to divide healthy from sick must be arbitrary. This is suggested by the wide range of arterial blood pressure values which have been used by different people as the borderline between normal and raised arterial blood pressure, or for the blood haemoglobin value dividing the 'normal' from the anaemic. The limits chosen are commonly expressed in numbers ending in zero or five. This is evidence of our pentadactyly rather than of real boundaries between health and disease! Seen this way, disease represents the arbitrarily chosen end of a frequency distribution curve which includes the normal. We should be more interested in the values themselves than in using them to distinguish 'healthy' from 'diseased'. This is not to deny the usefulness of such arbitrary divisions. However, their danger is that we tend to think that they correspond to real divisions in the outside world.

Single measurements can only give information on one aspect of health. Where many measurements or other observations are made, the probability of finding evidence of disease is increased, but the integration of the information into a single statement about the person's health becomes even more difficult.

There may be a marked disproportion between the evidence for disease and the underlying disease activity. The subject may appear healthy even though there may be underlying disease activity, and this is especially serious where diagnostic facilities are lacking.

Supposing a person starts to lose weight due to disease. The weight may still be above average because the person was of above average weight for height. This simple situation shows us that comparing a person with a set of other people may not suggest disease, while a comparison with that person's previous condition may do so. Thus, for characteristics which can change with time, comparing them with those of other people may not be a stringent enough measure. It may be more useful to compare the present and previous state of a person.

There is doubt whether some characteristics should be considered as manifestations of disease, as behavioural or social phenomena or as complexes of these. Elsewhere in this volume, Skultans has traced historical changes in ideas on what constituted insanity. There have also been changes in what are considered to be physical diseases and signs of disease.

Health and disease can be defined broadly or narrowly. Thus the World Health Organization all embracingly defines health as 'a state of complete physical, mental and social well-being, not merely the absence of disease or infirmity'. Such a definition is useful in validating an interest in all aspects of human well-being, but one doubts whether any of us could be considered healthy by these criteria for more than very short periods.

How should we view responses to threatening circumstances, such as enlargement of spleen, lymph nodes or increased in antibody concentrations? As morbid manifestations, or as adaptive responses? Audy and Dunn (1974) regard health as a continuing property potentially measurable by the individual's ability to rally after insults, while Dubos (1965) observes that 'states of health and disease are the expressions of success or failure... to respond adaptively to environmental challenge'. On these views, 'abnormalities' like lymph node enlargement or increased antibody concentration could be taken as indicating a heightened level of health.

Morbid signs may be so common that they may be thought of as normal. For this reason, it has been said of trachoma in some West Asian countries that the mild chronic conjunctivitis it produces was not considered a sign of disease. Until recently smoker's cough was thought of as a direct effect of exposure to smoke rather than a sign of disease, while chronic insidious industrial diseases tended to be considered as the expected results of work rather than as disease. The pustule which follows vaccination is thought of as a preventive measure rather than an infective lesion. The context in which a sign is shown is important, especially in mental disease, where the signs are largely behavioural. What is considered 'normal behaviour' shows wide variations in different cultures; so mental illness is revealed by what is considered deviation from normal human behaviour in a particular cultural context. The question of context also arises with physical signs. To take a very obvious example, severe breathlessness after mild exertion is taken as evidence of disease, while severe breathlessness after severe exertion is considered normal.

Morbid characteristics may sometimes form part of an aesthetic ideal, and may lead to the production of deformities and mutilations. Characteristics judged to be beautiful are probably less likely to be thought of as morbid. Degrees of fatness, thinness, and thyroid enlargement, which are usually considered morbid, have been held up as aesthetic ideals at various times and places.

MANIFESTATIONS OF DISEASE

In modern medicine it has long been the custom to divide disease manifestations into symptoms and signs. Thus we have standard books with titles like *Symptoms and Signs in Clinical Medicine* (Chamberlain and Ogilvie 1974) and *Signs and Symptoms*

(MacBryde and Blacklow 1970). Symptoms are morbid manifestations of which the sufferer is aware, while signs are morbid manifestations which can be perceived by an observer. Usually symptoms and signs occur together, but either may occur in the absence of the other. Feinstein (1967) has pointed out that some abnormalities are both sign and symptom. Some signs can be recognised by observation alone, while others (iatric signs) need to be elicited by a special means of examination. We could speak of the symptoms and signs of health as well as of those of disease.

Symptoms are known from the patient's own description, so they are largely outside the scope of this paper. Signs are elicited by observing and examining the patient. I am most concerned here with signs that are apparent without special measures being taken to elicit them.

Besides the signs elicited by examining the patient, further information can be obtained by using more complex techniques, often biochemical, microscopical, radiological or immunological.

SYSTEMS OF MEDICINE AND BODILY SIGNS

Other things being equal, physical signs are more likely to be noticed if they are obvious. If they are not noticed, they will not influence the assessments of observers. However, what is observed and noted depends also on what the observer believes to be significant, and this depends on the observer's ideas about health and disease.

Each individual develops a system of ideas, values, experiences and behaviour involving the body in health and disease. The most obvious and highly organised of these are the minority subcultures of medical practitioners, which for the present purpose should include all who are recognised by those that seek their help as having special expertise, and not only graduates in modern medicine.

Besides the medicine of the practitioners, there are systems of folk medicine, the medical systems of laymen. These are less organised but bear a relationship to the professional system that has evolved in the same culture. Thus modern medicine has Western folk medicine as one of its sources of origin, and Chinese medicine has a similar relationship to Chinese folk medicine. Lack of awareness of the importance of different systems of medicine, or a hostile attitude towards them, can give rise to many problems. These are particularly obvious when modern medicine is newly introduced to non-Western cultures.

Systems of medicine differ greatly in the extent to which signs are considered useful. Very roughly it might be said that in the more traditional systems, diagnosis tends to be made more by intuitive integrative judgment on the patient's illness as a whole, usually by observation of the whole patient. This, however, does not preclude the use of detailed examinations, such as the Chinese examination of the pulse. In modern medicine physical examination (at least ideally) is detailed, involves examination of many parts of the body, and is highly specific, in that each part of the examination tends to yield a particular type of evidence. At the limit of specificity there are signs that are indicative of a single disease.

There is a tendency to pay less attention to vaguer and more general physical signs, about which agreement between observers is of a lower order, and to concentrate on observations where there is less room for differences of opinion. Less importance is also being placed on physical findings as a whole, while greater value is placed on observations made by special technical means. This is perhaps the inevitable result of change from a situation where there is little but the unaided senses to help in diagnosis, to a situation where a vast amount of information is potentially available from special techniques. 'Old-timers' sometimes complain that young doctors cannot diagnose simple conditions without being given the answers by the laboratory. To some extent this is the lament of the old-time craftsman whose skills are losing their importance. More serious criticisms are that they are losing sight of the patient as a whole person and his problems, and that many people and communities cannot afford such expensive investigations.

Ideas about diseases influence how the body messages are interpreted. From the cultural point of view we might consider a disease as a recognisable pattern of characteristics considered to involve deviations from health which are sufficiently distinctive to be thought of as an entity. Certain circumstances of life come to be associated with certain diseases, and are more likely to be looked for and found. When one characteristic of a disease is recognised, the others tend to be looked for.

The process of learning involves for the professional, the development of familiarity with such patterns, and of the ability to recognise or suspect them. This may be so for the layman, too, particularly with respect to common diseases.

For less common diseases, the question the layman must answer is not 'what disease am I suffering from?' but 'does my illness need outside help?'

The importance of experience is shown by observations that laymen often become more expert in recognising diseases common among them, than medical practitioners without such experience. I have observed this in Peninsular Malaysia for yaws and Malayan filariasis, which used to be common in remote rural communities where few doctors saw them. Browne (1976) has noted great skill in diagnosing the earliest clinical signs of leprosy in villages in the Congo Basin where leprosy was prevalent. Gajdusek (1957) mentions that the diagnosis of Kuru was as reliable when made by the inhabitants of highly endemic areas of New Guinea, as when it was made following modern medical investigations. Just previous to the time of writing, Kuru had not even been seen by a modern medical practitioner!

Opportunities for the perception of signs of disease are often limited under ordinary social circumstances, particularly if clothes hide much of the body. Conditions for observation within family or peer group may be better, particularly if there is responsibility for determining whether a person should receive special care. In the medical examination there is usually a relaxation of rules of ordinary behaviour to promote diagnosis. Thus the modern physician examines for signs of disease not only by inspection (in a good light), but by palpation, percussion and auscultation, with the help of a few instruments.

The information content of most speech is contained largely in its verbal sequences, and this is the only part which is easily transcribed and interpreted. The patient communicates his symptoms by means of words. However, speech can communicate much information besides what is encoded in its word sequences.

It is a useful source of physical signs, as it demonstrates the functional competence of all bodily systems used in speech. The quality of the voice depends on the condition of nose, throat and larynx and their neuromuscular control system, while the words used may indicate abnormal intellectual processes. Speech is a good indication of the speaker's state of vigour or fatigue. The timbre of the voice is probably the most sensitive indicator we have of a person's emotional state, and it can have a powerful effect on a listener.

THE EVOLUTIONARY SIGNIFICANCE OF BODILY SIGNS

The state of health of an individual at any one time can be thought of as the effects of past and present interactions between the individual, his behaviour, and the total environment. Adaptation may be fast, occurring within the individual's lifetime, or it may be a slow evolutionary process involving mutation, hybridisation, genetic selection, and genetic drift in breeding populations over vast periods.

When a wound is suffered, it is fairly easy to distinguish the injury, which occurs instantaneously, from the responses of the tissues, which last the duration of the healing process. With most disease, this is much more difficult, as the direct effects of an agent of disease become inextricably confused with the responses of the host. The situation becomes even more complex where no single factor appears to play a central and necessary role in causing a disease.

Can we think of disease manifestations as adaptations to pathogenic circumstances with a communicative function? One of the definitions of function given by the shorter Oxford Dictionary (1933) is 'the special kind of activity proper to anything, the mode of action by which it performs its purpose'. Function in biology, as opposed to mere activity, might be considered to exist where a significant advantage accrues from a structure or an activity over a significant period of evolutionary history, or where it can be shown that genes responsible for the structure or activity have been subjected to significant positive selection.

The sensory system generates information about the surroundings and about the individual himself, which is important in determining adaptive behaviours. The pain system is undoubtedly important in protecting the body from damage. As an indicator of certain sorts of insult and damage to tissues, it is potent in forcing the individual to behave in such a way as to avoid or minimise pain, and hence damage. Serious damage can occur to parts of the body which have become insensitive to pain from disease, injury or local anaesthesia, and severe and fatal effects can be suffered by people with congenital insensitivity to pain (Sternbach 1968). Pain also has a training function as a kind of negative reward whereby the individual learns to avoid painful stimuli.

Though a symptom, pain can influence behaviour in recognisable ways which become signs of pain. Thus, crying out or screaming are types of response to sudden or severe pain which may warn others of damage and danger. Darwin (1872) pointed out that certain mute species vocalised only in the extremity of suffering, and suggested that vocal communication evolved from such vocalisations.

Pain is minimised by immobility, which can give rise to the appearance of paralysis (pseudoparalysis) in the absence of any interference with the sensorimotor neural pathways. A posture in which pain is minimum tends to be adopted. Thus the injured or infected hand tends to be immobilised by the sufferer in a neutral position, in which all tendons, ligaments and muscles are moderately relaxed. Pain is least in this position, and adoption of this 'position of function' has the advantage that in the event of a permanent restriction of mobility, the loss of function is least.

Painful injuries and inflammations of tissues, such as pleura and peritoneum, also lead to comparative immobilisation. This has the effect of minimising pain and also minimises the risk of spread of infection in the respective body cavities. We can say that the function of pain is to act as a personal warning system. However, the immobilisation of painful regions of the body, the posture and movements of the face in pain and the response to pain when another person moves the damaged region of the body constitute situations in which evidence of pain is obviously communicated.

The compelling influence on an observer of a face racked with pain, or of screams of pain, is an important determinant of social behaviour. However, it would seem far-fetched to hold that the benefits from such communication were a reason for the evolutionary development of the pain system.

The face is the most important site from which visual information on the bodily state can be conveyed. The facial muscles play an important role in the control of orifices, regulating light entering the eye and the intake and retention of materials in the mouth. Incidental information on activities like looking, eating, and speech, can be obtained by watching the face.

Secondly the smile face and the laugh face, and particularly the facial configurations characteristic of terror and rage, exert a potent influence upon those who are exposed to them. For these reasons it seems proper to speak of a facial communication system with communicative functions. Not surprisingly in man the communicative content and its context are considerably influenced by learning and culture.

The face is also of special importance because it is usually uncovered and it is close to the eyes of the observer, as it tends to be on the same level. But probably most important is our tendency to observe the face, because it is such a useful indication of the subject's mental state. The closest visual contact is achieved when there is eye to eye 'contact' between two persons. We come to know people's faces so well that we become sensitive to minor changes in the appearance of faces due to disease.

The face is a focal point in nonverbal communication, with the rest of the body playing a peripheral role. The hands and legs are centres which communicate information in manipulation and locomotion when the characteristic activities of the limbs are evoked, while the trunk contributes importantly to posture.

There remains the more specific clinical information which can be determined from the face. Some diseases such as thyrotoxicosis, Down's syndrome (Mongolism), and tetanus, can readily be recognised because of the characteristic facial appearances associated with them. Some skin or eye diseases are recognisable by observing the face, as are general bodily states like jaundice, cyanosis, or anaemia. The comparatively thin and translucent skin of the face, especially around the eyes, facilitates the recognition of signs. It does not seem reasonable to me to speak of a communicative function for such physical signs even though they constitute early warning systems that can lead to therapy that saves life or preserves fertility, because generally speaking effective therapy is so recent in history that genes underlying such signs have not been subjected to significant genetic selection.

THE CONTROL OF COMMUNICATION OF PHYSICAL SIGNS

So far we have considered the communication of information about morbid physical states as if these states were the only factors concerned. We should also consider the mental state of the affected individual and behaviour which can have the effects of amplifying, attenuating, or changing the message. We should also consider the effects that acculturation and other learning have on this.

Where the subject is unaware of disease, the signs are largely the direct effects of the abnormalities present, but where there is awareness, the situation becomes complicated by personal, cultural and situational factors.

Physical signs may be associated with values. Thus tolerating pain and suppressing the signs associated with it, may be differently valued characteristics in different communities.

Sternbach (1968) believes that ethnic variations in pain threshold are mythical, but that the response to pain in different ethnic groups varies because of different values associated with the expression of responses to painful stimuli.

Hardy, Wolff and Goodell (1952) have clearly described circumstances under which the expression of pain or incapacity may be encouraged. They point out that the experience of pain in an individual may bring with it a highly developed response in certain cultures, by which the sufferer is 'protected' or 'cared for'. The individual who cannot express his need to be cared for when free of pain, because of cultural demands for self-reliance, initiative, and independence, may find himself in a socially or culturally acceptable position on experiencing pain. Thus, being 'disabled' allows or even puts a value upon being dependent and willing to be cared for. Indeed, such a person may even be disapproved of if he is seen to be impassive, 'uncooperative', or a 'poor patient'. Under such circumstances the ideal behaviour may become one of almost complete passivity in which the individual remains immobile, allowing himself to be fed, turned, and bathed. Such a pattern has been rationalised as an attempt on the part of the culture to remove all burdens from the 'sick' individual ostensibly fighting for his life.

Some individuals with more need of support than the culture will allow them when healthy, may thus find in 'pain' a suitable means for fulfillment of deeply felt needs, which bring greater rewards than freedom from pain. Pain from wounds of battle may then actually be far more acceptable than the sustained anxiety and fear induced by battle conditions, i.e. 'combat fatigue', and this can lead to self-inflicted wounds.

Awareness of morbidity may lead a sufferer to consult a lay person or a professional, and here the threshold of reporting and the action taken are influenced by culture. In general the threshold of morbidity which must be exceeded before the person seeks a consultation is decreasing for various reasons. People are beginning to feel that health and freedom from pain is their right. Secondly there is a feeling that people walking around with overt disease ought to seek treatment.

In Western culture generally, there is a growing tendency to hide signs of morbidity. For instance, invisible plasters are being marketed, and one of the functions of make-up is to hide skin blemishes.

Measures taken with therapeutic intent may call attention to the existence of a morbid state, particularly where external measures are applied. Thus, among South East Asian populations, the application of pieces of adhesive plaster to the temples of people with headaches or the raising of bruises by pinching the skin over painful areas is a common indication of pain.

The individual can control the likelihood of other people recognising him as sick. On the ond hand these signs and complaints may be played down. Where the abnormality is localised, it may be sufficient to cover the affected area with clothes or hair. Or the visible signs may be obscured by plastic surgery, cosmetic skin preparations, or by realistic prostheses. Situations in which the abnormality is easily recognised may be avoided. For dysfunctions or disabilities, situations which bring out the abnormality are avoided, though avoidance of danger may be another obvious motive. The signs of disease may be attenuated, as when a person tries to achieve a normal looking gait, in spite of pain. Alternatively, the signs of disease may be exposed to view, exaggerated, or even simulated. Hence there is a wide range through which communication of the signs of disease can be influenced. They vary from suppression of evidence of disease at the one extreme, to a situation where disease is simulated, as shown below.

The way in which the apparent physical state of the body can can be influenced is illustrated in the simple example of 'taking the temperature'. The mouth temperature reading can be falsified by using warm or cool water directly or held in the mouth just before the determination, and the reading can be influenced by the position and duration of the thermometer in the mouth, and by whether the mouth was open or closed before the determination.

Suppression of evidence of disease	Playing down evidence of disease	Neutral presentation of evidence for disease	Exaggerating evidence of disease	*Hysteria* (Subconsciously motivated evidence of disease) *Malingering* (Consciously fabricated evidence of disease)

Figure 1

STATUS AND SIGNS OF DISEASE

The Maintenance and maximisation of a person's status in the group may influence the communication of signs of disease. Generally speaking with increasing awareness in the modern world of disease as preventable and curable, together with the feeling that the individual is to some extent responsible through his actions for his state of health, the tendency is for health to be associated with higher status and disease with lower status. Thus in the general social arena, there is a tendency to play down manifestations of disease. The extent of this tendency depends, however, on prevalent attitudes to the disease. Thus venereal diseases tend to be hidden, as they are transmitted by disapproved behaviour and even where there is permissiveness of extramarital sexual intercourse venereal disease carries a stigma, perhaps because it represents a failure to prevent the preventable. At the other extreme, we have war wounds, evidence of the individual's sacrifice for the community. Where passing a medical examination is a necessary condition for obtaining a benefit, such as a job, signs of disease will amost inevitably be minimised. The opposite situation occurs where a benefit may accrue from the disease, as in compensation cases.

BENEFITS FROM SICKNESS

People adopting the sick role are generally acceded certain rights. These confer advantages that can motivate people to manifest evidence of sickness. Duties such as military service, or the obligation of people to work for their employers or to support their families, may be waived for persons because of ill health. In the case of beggars who exploit disabilities, the demonstration of the disability may be a necessary condition for success. Sickness may provide a socially acceptable explanation for poor achievement, and this may promote the manifestation of signs of disease.

Much awareness of illness and much sick behaviour among people exists in the absence of objective signs sufficient to explain them. This may be due to a lack of such signs or of skill in recognising them. However, most of such troubles are physical expressions of psychological maladaptations. In so far as they have a social aim, they carry the message that a social inadequacy has a medical cause, which is therefore not the fault of the sufferer, and that there are social obligations towards the sufferer.

Hysteria is a condition in which the subject manifests morbid phenomena which are subconsciously motivated simulations of organic disease. The subject obtains some advantage from the illness without being aware of this motivation. It differs from malingering, which is the conscious simulation of disease to obtain a benefit, but there are gradations between the two. Hysterical manifestations are affected by the person's own concept of disease manifestations. These are likely to be bizarre unless the subject is well informed about the disease which is mimicked. Hysterical manifestations occur in contexts and have characteristics which maximise their effectiveness as communications in furtherance of their purpose.

Hysteria and trance states have much in common. In both there is an alteration of the functional state of the body, induced by suggestion. The range of manifestations is much wider in hysteria, which of course may involve trance-like states. The uses of hysteria are largely restricted to achieving the sick role under conditions where it can be beneficial, while trance states are often entered so that the entrance can communicate with Gods or Spirits, often for socially desirable aims such as therapy.

COMMUNICATION OF THE SIGNS OF HEALTH

Although the observation of many people merely fails to reveal any signs of disease, and some sceptics believe that health *is* no more than the absence of disease, some people show a collection of signs which together communicate the 'picture of health'. Such a person would show signs (when awake) of mental alertness and equanimity; also physical grace, speed, strength, stamina and skill. It must be admitted that a person with these signs might also show some signs of disease, and there has been less success in measuring the degree of health than in the measurement of disease. One might speculate that such people are not only physically but also mentally healthy. However, it should be remembered that signs of health, as well as disease, can be counterfeit.

There is some relation between signs of status and of health. High status and health are generally related to high stature, normal muscle tone and extended posture. Low status and illness is generally related to low stature, a generally flexed posture ('lets adopt a low posture on this matter'), low muscle tone ('feeling your knees giving way under you'), and disease. This pattern may be modified in cultures where old age confers high status. But it is the positive qualities of experience and wisdom which are valued, and not the loss of physical and mental competence that are often found in old age.

It seems to me that one useful way of viewing much behaviour of a more formalised kind is as a sort of health display in which signs associated with health are communicated, thereby raising the performer's status with people, who may be cast in the role of enemies, potential mates, peers, patrons, subordinates or others. Communication of evidence of health may not be a principal purpose of such behaviour, but it is certainly an important effect of it.

Such displays include the marching of soldiers, with their puffed out chests and swinging arms, which is a sort of caricature of the gait of healthy young men. Gymnastics, field games, and much dancing, have a similar effect, as effective performance depends on characteristics that depend on health. Agonistic behaviour, shown in the ritualised behaviour of the gun fight in the Wild West film or Oriental martial arts displays, also imply attributes of mental health in the conquest of fear.

REFERENCES

Audy, J.R. and Dunn, F.L. (1974). Health and Disease and Community Health. In Sargent, F. (Ed.) *Human Ecology*. Amsterdam: North Holland Publishing.

Browne, S. (1976). Personal Communication.

Chamberlain, E.N. and Ogilvie, C. (1974). *Symptoms and Signs in Clinical Medicine*, 9th Ed. Bristol: Wright.

Darwin, C. (1872). *The Expression of the Emotions in Man and Animals*. London: Murray.

Dubos, R. (1965). *Man Adapting*. New Haven: Yale University Press.

Feinstein, A.R. (1967). *Clinical Judgement*. Baltimore: Williams and Wilkins.

Gajdusek, D.C. (1957). Enclosure 14a in Correspondence on the Discovery and Original Investigations on Kuru 1955-8. Bethesda, Maryland, Study of Child Growth and Development and Disease Patterns in Primitive Cultures, National Institutes of Health.

Hardy, J.D., Wolff, H.G. and Goodell, H. (1952). *Pain*

Sensations and Reactions. Baltimore: Williams and Wilkins.
MacBryde, Cyril M. and Blacklow, Robert S. (Eds) (1970). *Signs and Symptoms: applied pathologic physiology and clinical interpretation*. London: Pitman.
Sternbach, R.A. (1968). *Pain, a Psychophysiological Analysis*. New York and London: Academic Press.

WHY IS SHAME ON THE SKIN?

Andrew Strathern

Kona kwi waka ile
pipil køng ile
øit kong ile monom
Wua Kamb Ruri
dola kømb tilpa
Kømb ronom.

Out on this open grassland space
there is shame on our skins
and we shiver with cold.
Ruri, the man of Kamb, is counting
dollars, is counting out silver.

This was one of the songs sung by women performing a special dance on October 6th 1971 to anticipate and greet the arrival of a hoped-for treasure of cash in the Mount Hagen area of the Western Highlands District, Papua New Guinea. Would the cargo come or not? A large crowd of people who had helped to finance activities in the cult and expected returns for their support soon to materialise, had come to watch the men and women of this particular group dance and to see the money revealed. The women sang of their clan leader and of his place, where dozens of red boxes full of money were said to be stored. Because of his claims to have the cargo and to be counting it out, they were dancing now out in the open on the ceremonial ground of the clan, with all eyes upon them, and 'shame was on their skins'. The dance performed for the cargo, and the cargo cult itself, were an innovation in the area. The dancers' performance and appearance would be taken as a sign of the cult's success or failure, and if it failed, all of the people of the group would be deeply embarrased and would suffer a loss of standing with the groups around. One of the rival big-men in a neighbouring clan was already speaking scornfully of their action in decorating even the old women of the group with bright face-paint and plumes and pushing, he said, tree-fruits into their mouths to fill out their gaunt cheeks and make them look young and healthy - a hint on his part that he thought the whole cult a piece of elaborate deception that was destined to failure.

It is not surprising, then, that the women should feel shy and embarrassed about the public stage and their performance on it for this occasion. The word 'shame' in their song certainly refers to all of this, and in addition it may also refer to the idea that ancestral ghosts were expected to be present, though in the background, supporting human dancers at festivals, ensuring that they look attractive and that the gifts they give or receive will be sufficient to maintain prestige. They were crucially involved in the present dance, for it was to them that the people were looking to receive the money-cargo. In the presence of ghosts and other supernatural powers, people are expected to feel another kind of 'shame', which we can perhaps translate best as 'awe'. It is not regarded as being the same as ordinary fear, although it may be accompanied by a feeling of fear or apprehension. Strauss, a Lutheran Missionary scholar who has worked in the Mount Hagen area, translates the Hagen word in this context as '*scham-furcht*' (shame-fear). He writes: 'People feel "shame" in front of other people but also, for example, before the spirits. *Kor pipil iti*, "to feel *pipil* before a spirit", is an often heard expression and here the meaning of *pipil* is not quite that implied in "shame" (*Scham* in German). There is clearly a sense of "fear" involved in it here and so I use the term "shame-fear" (*Scham-Furcht*)' (Strauss 1962:30, translated from the German original.). Strauss also refers in this passage to the phrase *minngøn mundi* which he says relates to alarm at a manifestation of numinous power, 'dread' or 'awe' ('Angst vor dem Numinosen, die Scheu oder "awe"'). *Pipil*, too, as I have mentioned, carries something of this sense of 'awe'. As Strauss says, people may speak of feeling *pipil* when faced with something unusual which they perhaps have not seen before ('etwas Ausserordentliches'), and which for them is a manifestation of power to create or do things ultimately deriving from spirit forces. He cites the coming of the first aeroplanes to the Hagen area in the 1930s as having set off a feeling of *pipil* among people. Nowadays older Hageners who visit Port Moresby for the first time, use the same expression when they see large buildings of permanent materials such as the University of Papua New Guinea's main lecture theatre. 'How did they build that?' they ask, and one replies rather inadequately that it was done by 'carpenters' (*kamda*, a general term for 'builders' in Melpa, derived from the English word). In these contexts, part of the meaning is that the speaker feels inadequate, almost caught off guard by the sight. In such a case he has not, of course, done anything wrong, but his feeling of inadequacy may be thought of as akin to the feeling of someone caught out in an act which is socially disapproved of.

Further connected to this usage is one to which Strauss also (*op. cit.*: 129) refers: if someone dies, people may say both that they are grieved and that they feel *pipil*.

Why should they feel the latter? Because the death may have been caused by hostile magic of one's enemies. 'One must be ashamed, then, in front of one's enemies, for their powers are obviously stronger than one's own. One has to be afraid in case the clearly irritated and angry powers [of the enemies] may cause deaths to take place in one's own group. Or even worse, that perhaps one's own group's powers have themselves set the death-dealing forces of sorcery and sickness against the forces of life. A sufficient reason for feeling *pipil*.'

All these observations may be helpful preliminaries towards determining the semantic range of the word that I have glossed in the song as 'shame', *pipil* in the Hagen or Melpa language. They leave out, however, the interesting question of why 'shame' should especially be spoken of as 'on the skin'. The general context of the song supplies a clue to the answer as well as an entry into speculation on the old matter of the distinction between 'shame' and 'guilt' in various cultures.

The women who sang the song were exposed to the collective scrutiny of groups in the surrounding community. The skin, their outer self, is the immediate point of contact with the physical world outside them and can also conveniently symbolise the point of contact between themselves and the social forces that surround them. The feeling of 'shame', to which they refer, has both a psychological and a social reference. When persons feel 'shame' in Hagen, they say, their skin breaks out in a sweat (*kur ropa etim*), which can produce either a hot or a cold sensation, or a combination of these. Similarly English people often say that when they are embarrassed or ashamed, they blush. In both cases the internal feeling triggers off an external sign which is 'on the skin' in a direct physiological sense. However, explanations given to me in Hagen about 'shame on the skin' tended to stress the notion that 'shame' has to do with the individual's reaction to his community relationships. This, of course, is like the classic characterization of 'shame-cultures' which has been delineated for Melanesian cases by Hogbin and others. (See, for example, Hogbin 1963:76, where he speaks of '*maya*, described as the discomfort a person feels when he has been found out in an unworthy act', and adds 'I shall translate the word as "shame"'.) This composite, folk-cum-analytical explanation does not cover the whole range of usage for *pipil* (see below), but it does appear to apply well in those cases where people stress that the 'shame' which they felt is 'on the skin'.

I put the question which is posed in the title of this paper to Ongka, a leader of a particular group (Kawelka Ngglammbo sub-clan) in Mount Hagen. First, I asked him to describe various different kinds of *pipil*, and he told me as follows (from notes made at the time, not from a tape-recording):

> If we stand up in public and say something, then someone challenges us and says we are wrong and we realize that they are right and that other people are watching us and can see this too, then we are a bit ashamed about this, it is a *pipil kel*, 'small shame'.
>
> Another *pipil*, and this is a bigger one, is if a man who is really a rubbish man with no resources, dresses up and goes to a distant place. There he says he is a big-man, and people believe him and give him many things. He brings these home and he hands them out again to people. Later, his creditors come to visit him and they see that in fact he has nothing. They tell him that they are going to sleep in his house until he finds something for them. He doesn't know what to do, he pretends he is sick, he goes off and hides, and he suffers a *pipil mam*, a 'big shame'.
>
> Another kind of 'big shame' happens when we think there is no-one around and we defecate or urinate by the pathway or road, then someone comes and sees us there, we look up and see them watching us, that makes us very ashamed and we say 'Oh, have they come to kill and eat me? Shall I go and hang myself?' We hide our head in our hands. [By custom, if a man sees a woman urinating or defecating like this, he may ask to have intercourse with her, to 'finish the shame'.]
>
> Another situation is if we find another man's wife and she agrees and we have intercourse with her, then we hide and do it, but if someome sees or hears us doing it and comes up and says 'aha! I've found you out' then we die of shame (*pipil korimon*), and we hide away. If they are sorry for us, they say 'don't let's beat them, let the man pay a pig in compensation', so we are grateful for that and give a pig, and afterwards we ask people not to mention the matter because we feel *pipil* over it. Or if we have intercourse with our own wife too and someone sees us, then we and the woman go off in different directions suffering from *pipil*. [Intercourse often takes place outside, at the edges of gardens or elsewhere.]
>
> These are kinds of *pipil* which are on people's skins (*wamb king ile pipil petem pipil køkl*).

For some sexual offences, however, shame goes deeper than this, for I mentioned the question of incestuous relationships and Ongka said, 'Yes, that is when a really big shame comes out completely. This is a kind of shame which is inside and it comes out. The shame cannot be got rid of, so they might hang themselves. They say that a person is not a dog or a pig and they cannot hide themselves indefinitely, so they may hang themselves.' (In fact the disapproval of ancestral ghosts is involved here. An alternative to suicide is to make an expiatory sacrifice of a pig and to pray to the ghosts.) Ongka continued:

> Another kind of *pipil* is the *pipil* that has to do with *tipu* or spirits. Our skin breaks out in a sweat (*kur ropa enem*), the hairs on our back stand up, our teeth are set on edge, we say that a spirit is going to kill us and eat us. This happens when we go to a cemetery place or a house where a man has died and we hear a bat or an owl calling, or we see a marsupial and we think it is a spirit doing this.

I asked Ongka about the difference between this and ordinary fear (*muntmong*), and he said:

> This (*muntmong*) is when we see a tree is falling on us, or a river will sweep us away, or a car will hit us, or we will fall down a slippery place, or that some people will kill us with 'poison' (sorcery-stuff). We don't say that we feel *muntmong* of spirits, only of people and things.

I asked if other words could be used as well as *pipil*. He said that *nggørli* was an equivalent term:

> We say we feel *nggørli* (*nggørli enem*, literally 'it makes *nggørli*') when we see a snake or another thing which we dislike, say a place where people defecate, and we don't want to tread on it or see it or hold it. If we see a rotten or a stinking thing, too, we say we are *nggørli*. [It is my impression that *nggørli* is more often used in those contexts which imply a direct physical revulsion from an object or an action. It may also be used to indicate a reluctance to harm, or to make marks on, someone's skin. People say *wamb king - enga pipil enem*, 'we feel ashamed because it is someone's skin'.]

I then asked directly why Ongka had spoken of some kinds of *pipil* as being on the skin, and he replied:

> It is when people see us, it is not that there is anything

inside (*mel ti rukrung pei na petem*) it is outside (*ekit oronga*) only, it is when people see us doing these things that we feel *pipil*, when they see our skin, and we feel *pipil* on the skin.

All sensible people, he added, feel this *pipil*. If someone does not, his relatives will tell him 'you have no shame on your skin, you are crazy (*kupørl roron*). People have shame on their skin, but you have none. If you had a soul (*min*) as other people have, it would have given you a good social attitude (*noman kae*, the *noman* is thought to be inside a person's chest and to guide his thoughts, feelings, and actions). But you have no *min*, and so you do not feel shame.' He explained further that a man who does not feel shame lacks a true *noman* because the *min* in him has been lured out by a wild spirit (*kor rakra*) which lives beside streams. The spirit steals the person's *min* and feeds it and talks to it, so the person becomes crazy and doesn't behave normally.

As Ongka himself explains, his generalization applies only to some of the instances of *pipil* that he discusses, those in which the shame is 'on the skin'. So we cannot, on the basis of his account, dub Hagen simply as a 'shame culture' in the sense of the phrase used by Hogbin of the Busama. Clearly, matters are complicated. There seem to be delicate semantic connections or ramifications between separable clusters of usage of *pipil*, thus:

a. 'shame' at being caught out in a culpable act, e.g. theft, adultery, incest.
b. 'shame' at being seen while engaged in a private act, such as defecation or sexual intercourse.
c. 'shame' before one's enemies as a result of a death caused by their magic.
d. 'shame' before the wider community as a result of being proved wrong or inadequate in a discussion; or as a result of potential or actual failure to keep one's promises and meet exchange obligations.
e. 'shame' that indicates respect for someone else's body or skin.
f. 'shame' that is physical revulsion at something unpleasant. (In both e. and f. a notion of squeamishness may be involved.)
g. 'shame' that is awe in the presence of spirits, or admiration at something which obviously reflects spirit-derived power. In the case of some manifestations of spirits there may be a further sense of revulsion as well.

In usages a, b, c, d and g there is a notion of a loss of prestige or of inadequacy caused by a confrontation with the power of the community or the power of spirits. That one is found in wrongdoing or in a private act means that one's 'powers', for example powers of luck (*kil kø̸i*), have temporarily left one. Similarly, in the presence of spirit force one feels that one's own forces have been removed from one. So the situation is threatening. This idea of threat is partly carried over into e and f usages, which can be summed up as 'shuddering' to do something physically unpleasant, just as a moral person would 'shudder' to be seen doing something considered morally wrong by the society.

In his exegesis Ongka took the discussion one step further than this kind of listing of interconnected clusters of meaning. After stating that *pipil* is on the outside only, he contrasted it with sickness. People do not get sick because of shame on their skin, he said: sickness comes from inside, just as does *popokl* (*anger or* frustration). 'When we are *popokl* about something we hide it, we keep it to ourselves and don't tell people. We preserve a good exterior. With *pipil*, on the other hand, we admit it to the outside world, we say that we are *pipil* and ask them not to be hard on us, but inside ourselves we may be feeling quite good, and when we go off by ourselves we feel all right; it is only when other people are there that we actually feel ashamed about it.'

Ongka's formulation here is consistent with a view that 'shame' can be regarded as a kind of social 'cosmetic', a means of dressing up one's behaviour so as to appease others and forestall their anger. Possibly, indeed, an interactional paradigm can be set up, in which we see that 'shame' behaviour is a means of averting the 'anger' of others, and conversely 'anger' may be a means of creating 'shame' in others, so that they will take notice of one and be sorry for what they have done. This could be so in cases where an aggrieved person communicates the fact that he is *popokl* to the person he thinks has done him wrong. (See M. Strathern 1968; A. Strathern 1968; Valentine 1963).

In his analyses of people and events, Ongka is very much of a social realist. He is also somewhat sceptical about the alleged existence of the spirit world. At the same time, he is in many ways a firm traditionalist. It was he who made the scornful remarks, quoted earlier, about rouging up old ladies to take part in the cargo dance, and his grounds were that in the past prayers to ghosts had not produced wealth in the direct way that the cultists were claiming their rituals would do, so why did they imagine that such prayers would be successful now? His explanation of the relationship between *pipil* and *popokl* goes a good deal further than cynicism about people's actions, however. He is setting up a contrast between *popokl* as anger resulting from others' wrongdoings, which the angry person feels inside him but may keep to himself, and *pipil* as a perfectly genuine feeling that is triggered off by encountering external social disapproval of what one has done. Both feelings are, he tells us, in a sense guided by the *noman*, which is inside one, and so in that sense both have internal origins. But *pipil* is 'on the skin' because it is felt when people see one or find one out.

Popokl stays in the *noman*. It may sometimes not be revealed. It may not show in any obvious external signs. Hageners stress that one cannot see inside the *noman* of another person. 'Can I turn you inside out and see what is in your *noman*?' they say rhetorically. 'No, each one of us has his own "power" and his own *noman*. You have yours, and I have mine.' So whether someone is *popokl* or not cannot always be seen as easily as if the person is *pipil*, ashamed. *Popokl* can be inferred, however. Sometimes it is thought to follow closely on the feeling of shame itself. Thus if someone makes another ashamed by asking him a question he cannot answer, or refutes what he has hazarded to say, the ashamed person is likely also to feel angry, *popokl*, with the interrogator. 'Shame' and 'anger' may thus in this instance be felt together by the same person, and in this context both are said to be felt internally. In usages of this kind we are closer to the ordinary English sense of the word 'shame', applied to an internal psychological condition of the individual resulting from recognition of guilt in some matter. However, we may still note the public context: an individual's 'shame' is triggered off when he is tonguetied by a question and many people are watching or listening. I am grateful to Mr Hugo Akie, a Melpa student at the University of Papua New Guinea, for giving me this example about being unable to answer a question in public. He stressed the point that both 'shame' and 'anger' here are felt internally. In order to sort out different kinds of linguistic usages we need to distinguish between those which are intended as direct representations of 'how it happens', as in Mr Akie's example, and those which have become cultural metaphors, as in the phrase 'shame is on the skin'.

Although this feeling of *popokl* may not immediately show, it is likely to reveal itself over time in the social relationships which have been disturbed by the *popokl*-producing events. The usual way in which it is revealed is through sickness. This is why in his account Ongka goes out of his way to say that with shame there is nothing 'inside', it is all externalised. By contrast, sickness and *popokl* both come from inside, and sickness is in fact the means whereby *popokl* is shown to the outside world and moves can be made to setting the situation to rights. *Popokl*, in fact, points out the guilt, and hence the responsibility, of others in relation to a harmed person. Either the person harmed falls sick in protest against the wrongdoing, or the wrongdoer himself falls sick as a result of the wronged person's anger. The latter process is similar to the process of witchcraft, but both are actually thought to be governed by the actions of ghosts ancestral to one or other of the persons involved, or to both of them, and the ghosts' actions are held to be justified in this context. In some cases, the ghosts themselves are said to be *popokl* with a wrongdoer and to make him sick, whether or not anyone has specifically complained to them. They would do this, for example, in a case of incest between close kin, such as father-daughter incest. The only way to put matters right and avoid serious consequences from the sickness is to make an expiatory sacrifice as fast as one can. If a person has done something very shameful, such as incest, it is said that the ghosts may also make them mad, and again a sacrifice of a pig is the only way to deal with such an emergency. If a person becomes sick and some social complication or wrongdoing is suspected by the kinsfolk, they will kill a pig and ask both the ghosts and the sick person why it has happened. The sick person in particular is asked to reveal what is in his *noman*, to tell the kinsfolk if he or she has done wrong, or if there is some anger against others for wrongdoing. After the talk has been revealed, the patient's sickness may be healed, since confession is a step towards the removal of *popokl*, either the patient's own or that of someone else who is angry with him. (These ideas antedate the missionization of the Hagen area, although they fit in quite well with certain mission-inspired notions of confession. The general idea of linking physical misfortune or sickness to moral or social causes is, of course, world-wide.)

Popokl thus pinpoints responsibility or guilt for wrongdoing, brings about, as people see it, punishment, and stirs people into restoring the situation through a compensatory and expiatory sacrifice. This does not always happen in practice, of course, and a severe impasse may be created if the *popokl* remains unrevealed or the wrongdoing is locked in the private *noman* or conscience of the person.

Responsibility and liability in Hagen culture at the group level are symbolised also in the notion of the *mi* or totemic divination substance of the tribe (cf. Strauss *op. cit.*, esp. pp. 206-85 where case histories of the use of *mi* in conflict situations are given). An individual who is made *popokl* by some slight or offence against him may swear an oath on the *mi* of his group not to see the offender again. (A. Strathern 1972:154-6). The offender may then feel uneasy, thinking that the *mi* may punish him, especially if the offended man is from his own clan or sub-clan or is a close maternal relative. Accusations of sexual delicts or criminal acts of sorcery can be tested by making people swear on or promise by the *mi* that they are innocent.

Because they have not actually been seen by others, the accused may choose to deny the charges and thus avoid shame and the need to pay compensation. However, it is thought that the *mi* will find them out. They are required to swear by eating a piece of it (usually the leaf of a particular kind of tree or plant, or it may be a kind of stone). Sometimes a big-man will decide that the group must jointly burn their *mi* object (*mi køkli*) as a test of a member's innocence or guilt in a matter involving the safety of the whole group. For example, in one case during 1974, Ongka threatened to do this if young men of his own sub-clan did not reveal whether they had committed sorcery against a big-man of a neighbouring group who had died. Leaders of the alleged victim's group were threatening Ongka and his group with reprisals at the time.

Once persons have taken the *mi* inside themselves, then if they have lied and presented a false exterior, the *mi* will discover the truth that is inside them and will kill them. So their denial enables them to avoid ordinary social contumely and shame, but they are punished supernaturally for not revealing their guilt. Finally, the *mi* also stands for the notion of in-group solidarity and strength, derived from its members' proper moral and ritual connection with the spirits, especially ancestral ghosts. The *mi*, like the ghosts, particularly dislikes offences within the group which are not revealed and settled by compensation. If a son steals from his father, the father may make him swear on the *mi* that he did not do so. Unless the boy confesses quickly, it is thought that he will die, so it is a severe move for a father to make in order to teach his son a lesson. In one case a boy was said to have gone mad after denying such a theft, and to have confessed just in time. Similarly, the *mi* dislikes open fighting and killing between men who belong to the same clan, or closely related clans within the wider tribe. These men are 'brothers', and while they are expected to quarrel and fight, they should not kill one another by sorcery or with weapons of war such as the spear and arrow. If they do, and do not settle this, ill-luck, sickness, and death may befall them, and the strength of their group will be lost. In one case early in 1975, there were accusations of killings within a particular tribe. Following the death of a man in one clan, his clansmen, who suspected sorcery by their neighbours, rushed out to take physical revenge and were met by a big-man of their own group. He urged them not to take violent action in this way against 'brother' clans, but they would not heed him and swept past, stoning his car, which he had placed across the road. They had eaten the *mi* of their group and sworn an oath to fight. Not long afterwards an extensive landslide occurred in their area, sweeping away most of the houses they had built around a church, and the church itself. The landslide carried away the body of the man whose death had sparked off the clansmen's anger. The local Member of the House of Assembly was reported to have interpreted all this as a punishment from God (*Papua New Guinea Post-Courier*, Wednesday, 26 February 1975: 21). In more traditional terms it was seen by many people as a punishment from the *mi*. The clansmen had sworn by it in error, forgetting that it stood for solidarity within the group. They had gone out to-destroy their brothers, and in that sense, to destroy themselves. They had transgressed their own basic values and had refused to heed the secular restraining influence of their own big-man, so they brought on themselves retribution from a source of power higher than ordinary political authority.

To conclude. I have moved from an initial question, posed in the title, about a statement which Hageners make with regard to 'shame', to a further question about 'shame' and 'guilt'. We quickly saw both how complex in fact are the various referents of the Melpa terms which I have glossed as 'shame', and why a certain kind of shame is by Melpa speakers particularly said to be 'on the skin'. We have in this expression a conjunction of physiological and sociological appropriateness. 'Shame' does show on the skin, and the communication of it is heightened by explicit gestures such as hiding the face in one's hands; but more crucially, it is caused by others seeing one, by being found out. It is thus, in some cases, 'only skin deep', and that is why, for the Hageners or Melpa, it cannot produce sickness in people. So far, it appeared that we were dealing with a 'shame-culture' as has been described for some other Melanesian cases. But again, that would be a superficial view, itself 'only skin-deep', for it leaves out the whole dimension covered by the ideas of *popokl* and the *mi*, to which Ongka explicitly drew attention in counterpoint to his explanation of *pipil*. It is in the realm of *popokl* that we find the counterparts of ideas of responsibility and guilt which are supposed to characterise 'guilt-cultures'. To paraphrase John Barnes's review of Meggitt's book on the Mae-Enga (Barnes 1967: 43), neighbours of the Hageners to their west: if we were to set up a continuum between shame-cultures and guilt-cultures, the Hageners would be found at both ends.

ACKNOWLEDGEMENT

I am grateful to all the institutions which have enabled me to work in the Hagen area at various times between 1964 and 1975: that is, to Trinity College, Cambridge and the University of Cambridge; the Royal Anthropological Institute; the Australian National University; and the University of Papua New Guinea. Most of all, I am grateful to Ongka, for explaining so many things to me.

I am indebted to Marilyn Strathern for the ideas about shame as appeasement for anger.

REFERENCES

Barnes, J.A. (1967). Agnation among the Enga, a review article. *Oceania, 38,* 33-43.

Hogbin, B.I. (1963). *Kinship and marriage in a New Guinea village.* London.

Strathern, A. (1968). Sickness and frustration: variations in two New Guinea Highlands societies. *Mankind, 6,* No. 2, 545-52.

Strathern, A. (1972). *One Father, one blood.* Canberra: ANU Press.

Strathern, M. (1968). *Popokl:* the question of morality. *Mankind, 6,* No. 2, 553-62.

Strauss, H. (1962). *Die Mi Kultur der Hagenberg Stämme.* Hamburg.

Valentine, C.A. (1963). Men of anger and men of shame: Lakalai ethnopsychology and its implications for socio-psychological theory. *Ethnology, 2,* 441-77.

FEAR OF SORCERY AND THE PROBLEM OF DEATH BY SUGGESTION

Gilbert Lewis

It must be uncommon to go through the ethnography on a people and not come across something on their beliefs about witchcraft or sorcery. The more one reads the harder it becomes to single out one example from another. I often heard about sorcery where I did fieldwork - The West Sepik District, New Guinea, on the inland site of the Torricelli range overlooking to the south the plains of the Upper Sepik River. I heard a lot of talk about sorcery, but although I stayed among the Gnau people for almost two years, I saw few people struck ill by sorcery.

Apart from the ingenious and scholarly books or articles which analyse the ideas involved or the web of relations between accusers, victims and accused, there are some other features of sorcery that are curiously obscure. Is it really the case that healthy people have died in a day, or in three days, because they knew they were victims of sorcery? Who has seen this happen with his own eyes? Is there no explanation for it but sorcery? Marcel Mauss, in his essay on the 'Effet physique chez l'individu de l'idée de mort suggérée par la collectivité', urged anthropologists to reflect on the death by suggestion (Mauss 1926). Cannon (1942) and others after him, (Richter 1957, Griffiths 1960, Engel 1970 and Wolf 1971, Lex 1974) have sought to explain in terms of physiology how fear of sorcery might have caused death. That area of the subject is uncertain, holding a morbid fascination to find if it is true or not.

There is the sorcerer - what does he do and think? Some anthropologists have been intrigued by that. In his essay on the sorcerer and his magic, Lévi-Strauss retells the story of Quesalid, A Kwakiutl shaman and a sceptic, who gave Boas an account of his life, and his first disbelief that turned eventually to doubt: 'Only one shaman was seen by me, who sucked at a sick man, and I never found out whether he was a real shaman or only made up. Only for this reason I believe that he is a shaman; he does not allow those who are made well to pay him. I truly never once saw him laugh'. (Lévi-Strauss 1958:196, quoting the text recorded by Boas.) What are we to make of some other statements by the sorcerer himself? For example, this excerpt of a Murngin sorcerer's account of his work in Warner's *Black Civilization:*

> Her large intestine protruded as though it were red calico. I covered my arm with orchid juice. I covered the killing stick with it too. I put the stick in the palm of my hand so that I could push the point upward with my thumb. When she inhaled I pushed my arm in a little. When she exhaled I stopped. Little by little I got my hand inside her. Finally I touched her heart. I pushed the killing stick with my thumb up over the palm, which pressed the stick against my fingers, into her heart. She had a very large heart and I had to push harder than usual.... I turned her over. Her large intestine stuck out several feet. I shook some green ants on it. It went in some little way. I shook some more on, and a little receded. I shook some more, and all of it went in. Everything was all right now. There was no trace of the wound. I took the tomahawk handle which had her heart's blood on it. I whirled it around her head. Her head moved slowly. I whirled it again. She moved some more. The spirit that belonged to that dead woman went into my heart then. I felt it go in. I whirled the stick again and she gasped for breath. I jumped over her and straightened her toes and fingers. She blew some breath out of her mouth, and it was all right. It was noontime. I said to her, "You go eat some lilies". The woman got up and walked away... I said to that woman, "You will live two days. One day you will be happy, the next day you will be sick.".... The next day she walked around and played, laughed, talked, and made fun and gathered a lot of oysters and lilies. She came back to camp that night. She brought the things she had gathered into the camp. She lay down and died that night (Warner 1937:199-200).

Could I have been the man who said that? What conceivable experiences could lead me to tell such things and believe them? Were they fabrications or fantasies?

Among the Gnau people, I was not told by a sorcerer or would be sorcerer anything that can match what I have quoted from Warner. I did not witness illnesses or deaths more devastating and intrinsically mysterious than illnesses and deaths I had seen in English hospitals. Given, then, my curiosity about these aspects of the subject of sorcery and my wish to find something I observed myself that would bear upon them, I shall consider people who know of the risk of sorcery and think they have reason to fear it.

My purposes in this paper are to provide an account of the main features of sorcery among the Gnau, and to set this account against my observations of behaviour showing fear. This will involve some discussion of fear and how it may be studied in an anthropologist's field setting. I shall discuss a possible mechanism of psychogenic death. Finally, I shall comment on the apparent discrepancy between public or commonly stated beliefs and evidence of individual belief. I shall organise my observations in terms of behaviour with respect to past, present, and future situations, and discriminate between public and individual responses.

THE TYPES OF SORCERY RECOGNISED BY THE GNAU

For the Gnau, the category of those who know of the risk of sorcery and think they have reason to fear it would include everyone grown up enough to take in common knowledge. They distinguish destructive magic from evil sorcery. By 'destructive magic', I refer to techniques of power to do harm to people by using materials and spells which are blown or whispered into them. Whether the use of these techniques against someone else is regarded as right or wrong, justified or not, depends on the particular context and personal viewpoint. Destructive magiv may be used to protect one's property or to revenge a wrong done against one, or one may see it as a secret murderous act of jealousy and black-hearted spite. They call such destructive magic in Pidgin English *poisin*. The Gnau know various techniques. When they got to know me better, some Gnau men were prepared to tell me frankly that they knew how to do them.

But as for evil sorcery, no man ever admitted to me that he could do it, and no man, except one on one occasion, ever said to me that he would in any circumstances be prepared even to try to learn it. The evil sorcery is called *sangguma* in Pidgin. *Sangguma* sorcery is sorcery in the strict sense: no one openly admits to knowing it or practising it; it inspires horror; it provides the bogeyman for small children; it is feared, reviled and despised. Its use is always antisocial and essentially illegitimate. In contrast to destructive magic, people cannot conceive of its use being justified. The distinction between destructive magic and evil sorcery is conveyed by the way the Gnau use Pidgin: it is the distinction between *poisin* and *sangguma*. In the Gnau language, *sangguma* sorceries have substantive names: *langgasutap*, *minmin*. The verbs relate the sorcery to its victim and not the sorcerer to his victim as, for example, in these expressions: *Langgasutap manem* - "*sangguma* struck him"; *Minmin watewunpen* - "*minmin* hangs up on him".

Their theory attributes to those who possess *langgasutap* abilities which ordinary men do not have: those with *sangguma* can move without trace, conceal themselves in tree cracks, become invisible; they may cut someone open and eviscerate him without leaving a mark; they may eat human flesh; they dumbfound their victims.

When destructive magic is used, they speak of the man doing or working it, not the magic doing it: most often they would say *nebariyen* - he (agent) works or does him (victim). The root - *bari* is the common verb for 'to work, do, or make.' Almost as common is the verb *nasupepen*, 'he blows or bespells him.' It is based on the root - *su*, 'to blow.' There is no noun for magic or magician/sorcerer: the ideas are expressed in verbs and they distinguish the techniques by more complex phrases (see Lewis 1975: Ch. 5). When destructive magic is used, men work it; but when *langgasutap* is used, although men bend it to their evil whims and 'hold it' (*natao*), it is the *langgasutap* which strikes, not the man. At least that is implied in verbal usage. By convention, a man who holds *langgasutap* intending to use it has, they say, become a shade of the dead (*gelputi*) rather than a human being (*metilden*). If Gnau walk by night, or even by day, and they hear suspicious rustles or crackles in the bush, they call out 'Who are you? A shade of the dead? A human being?' If, after a second chance, no voice answers 'I am a human being here', then they should and may be ready to shoot. When a girl thought she saw *langgasutap* (a man she did not know standing a little way off in a garden, sudden, silent, and immobile) she fled back into the village to be asked over and over again 'Was it a man or a shade of the dead you saw,'

But *langgasutap* is a technique requiring to be learned and not an inborn quality or substance. When there is suspicion of *langgasutap*, men may go out with bows and arrows to hunt down its bearers. The men holding *langgasutap*, though they are addressed as shades of the dead, are yet shot at as men may be. When one was shot (according to a story I heard) the body was not found, but the arrow was retrieved with blood on it and a bit of half-digested rice. When, during my stay, a *sangguma* man was shot through the thigh and caught by men in a village to the north of Rauit, where I lived, he was tied to a pole like a pig. He was naked and wounded; he spoke no language they could understand, nor Pidgin. Many wished to kill him. They proposed a test. They untied him. It was just at nightfall. They left him, but went to hide at a little distance and watch what he did. He could not walk. He lay on the ground and began to weep. So weak and human did he then seem, that doubt prevailed over the proposal to kill him on the spot and he was taken tied up to the mission centre next day. In fact he was a lost man from Waina

Sowanda, trying to escape home from the coast across a hundred miles of coastal range and land which he had seen once from the window of an aeroplane on his journey out from the Patrol Office nearest his home, to Wewak on the coast.

The work of destructive magic, it is said, will kill the victim. But it can be stopped by counteraction. There is no such remedy thought possible for *sangguma*. If *langgasutap* has struck someone, he or she will surely die. The remedy for destructive magic depends in effect on the agent negating his magic. He must be identified and induced to countermand his act. The nefarious evil of *langgasutap* might be done perhaps at whim, but it is final and absolute. The people do not speak of it as something that could be open to negotiation, to a change of plan, to an altered outcome.

I shall not describe the content of their ideas about such magic and sorcery in greater detail at this point. The Gnau use destructive magic overtly for such practical purposes as protecting property and crops from theft, and discouraging trespass, and when they do so they put up signs of it which are visible. They also sometimes use it and hide the signs or material used. Because of this, they recognise that people whom they would not wish to harm may come into contact with it inadvertently. *Poisin* is relatively open and above board. It is an admitted activity of the Gnau. It has legitimate and valued uses. When a man of consequence dies, they announce his death on the signalling *garamut* (slit gong), they beat out the list of his achievements in life; they commemorate first the number of men he shot in warfare, second the men he killed by magic, third the cassowaries he shot, then the pigs, and so on. But *sangguma* has no other face but that of death-dealing, by silent, secret, evil and unjustifiable means. The Gnau do not know how to do it. Nor do the people of other villages south east of them, with whom they tend to identify themselves in some other respects. But west of the Gnau, their neighbours know the sorcery and practise it. These are things they say.

THE QUESTION OF FEAR PRODUCED BY GENERALLY ACCEPTED BELIEFS

But are they frightened by what they say? We would expect the answer yes, and might hope to hear about the ways in which they show their fear, how often and how much it affects their daily lives. Clearly it is a field of interest to psychology and anthropology. In the absence of observers trained in psychology and choosing or prepared to work in such societies, the social anthropologist may decide to keep silent about the psychological aspects of the subject because he feels professionally incompetent. That would seem a pity, because some anthropologists have been surprised that the people who said they believed such things about witchcraft or sorcery, did not seem pervasively oppressed by them in ordinary life, or filled enough with dread, as the anthropologist translating the ideas into terms of his own life and culture would feel they ought to be. Mary Douglas has touched on this strangeness with an apt metaphor:

> Perhaps no-one has gone further than Philip Mayer in implying that witchcraft beliefs (at least in Africa in its original untouched state before missionaries, cash and colonists had disturbed the balance) were a fully domesticated species not to be compared with the wild European sort (Douglas 1972:xix).

Both by description of observed cases and by analysis of the coherent systems of underlying ideas, various anthropologists (for instance Evans-Pritchard 1937, Marwick 1965, and Turner 1957) have contributed to an explanation, implicit or explicit, for the undistressed acceptance and familiarity, the domesticity, of these ideas. Their readers can see better why the beliefs may not always be accompanied by horror, that there are recognised procedures to counteract the threats, that the suspicions and accusations are brought forward in specific restricted and intelligible situations, and that they may constitute, as it were, 'natural' rather than dreadful supernatural explanations for misfortune and illness.

THE CHARACTERISTICS OF FEAR

But what can we know of fear? English is rich in words clustered round the emotion: horror, dread, terror, fright, fear, trepidation, alarm, consternation, anxiety, apprehension, concern, panic, frantic, scared, startled, agitated. The differences among these reactions to perceived present or future danger, have to do with:

1. the severity of the emotion (shattering terror, mild doubt, diffuse perturbation);
2. its relation to the future (distress of the moment, immediately impending threat, distant possibility);
3. the definition of its object (fear referred to some external definite distinct object or situation, anxiety to a trouble or agitation not so linked to something precise and known and definite, its object vague and uncertain);
4. the timing and duration of the emotion - enduring or transitory, chronic or acute, it may be an abiding or recurrent state; its onset sudden or insidious.

These are all aspects of fear at the level of conscious

awareness. Though chiefly turned to the future, as fear must always be, it rests on past experience, and it reverts to the past to account for the troubles in store. Herein as with rationalization, and some other psychological devices, there is evident a strong desire to make things understandable in a causal nexus. The individual's expression varies with the strength of his fear and his estimation of his helplessness or his ability to counteract the threat. The observer is concerned with whether the threat is either not recognisable or is, by reasonable standards, quite out of proportion to the emotion it seemingly evokes, needless anxiety.

There are, of course, also the manifest bodily disturbances of fear.

> Some of these are of functions normally under voluntary control (e.g. running in panic, agitation, screaming, sudden defecation); others are of functions not wholly or at all under voluntary control. Of these phenomena there are many varieties, e.g. dryness of the mouth, sweating, horripilation, tremour, vomiting, palpitation, giddiness, abdominal pain; and others physiological and biochemical, that can be detected with appropriate methods of investigation.... There are subjective bodily discomforts during the period of fear. The sense of constriction in the chest, tightness in the throat, difficulty in breathing, and weakness in the legs are conspicuous; there are others mostly representing the subjective side of what is also objectively manifest. (Lewis 1967: 120).

These bodily manifestations are signs or symptoms of the neuroendocrine and motor-visceral aspects of the emotion. The emotional state has the subjectively experienced quality of fear or a closely related emotion. It is unpleasant. Its characteristics may be inferred from a person's demeanour and conduct even though he says nothing about what he is feeling, but for the most part the observer will need to be told about the feelings of the person if he is to assess the other's state satisfactorily.

I have quoted these characteristics of fear to indicate that an anthropologist could quite well report very fully on states of fear in other people without requiring special skills and training. Darwin's description of the expressions of fear (Darwin 1872:289-309) is one of the finest and most detailed I have read. It stands because he had those gifts of acute observation and accurate description to which every field anthropologist would aspire (for other descriptions of the objective and subjective varieties of fear see Janet 1903 and Jaspers 1963:365,

478, who cite examples of behaviour in earthquakes, war, fear of apparently certain death, at the moment of dying, and of terror in the mad).

The ability to recognise mood and emotion in other people underlies the anthropologist's grasp of many situations. Sometimes it is important. It is most often unstated and unanalysed. It is hard to account for the precision and immediacy of one's response. A man died after a long and agonising illness in the village in which I lived. The whole village gathered. They poured out grief, crying and wailing for a whole night. At dawn the next day, they waited for some clan relatives to come. They waited until noon. From noon, some were pressing to get on and bury him before the corpse began to rot and stink. His two elder brother's sons went into the hut with the body to make it ready for burial. I knew them both well. After a while, one went past the place where I was sitting with other people. I saw him for 4 or 5 seconds as he passed going straight to the men's house. His face was set, he was a little pale, his gaze was fixed on the ground, there was something of hurry and tension in his walk. I may have noticed he had something in his hand, but I am not sure. The quickness, the set face, the fixed gaze, gave me the sense that he was getting away from something unpleasant and shocking. Immediately I guessed specifically that he had just had to do something that I had heard of as a possibility, but which had not happened so far in my stay. I guessed he must have just before cut off the eyebrows, eyelashes, axillary hair and head hair of the corpse, and therefore must intend to divine the cause of the death by a particular method aimed to find out about sorcery. An hour or two later, I asked him if he had done that and he said he had not. But he lied. I think I can identify the reasons why. Some time after the divination he told me that I had been right.

OBSERVATIONS OF THE FEAR OF SORCERY AMONG THE GNAU

I will return now to fear of sorcery and destructive magic among the Gnau. I intend to reorganise my observations around issues of timing and relationship to particular events, including prospective, present, and retrospective views of danger or threat. I will look first at public and then later at individual doubt and uncertainty. Further on I will describe two cases that have a slender bearing on the phenomenon of sudden death. I will bring forward the distinction of despair from fear.

Acute public anxiety: Minmin

During my fieldwork a wave of fear spread from the west about an outburst of sorcery, new in some respects, called *minmin*.

Because of it, I attempted then to gather together all my notes on the times and ways in which I learnt about the new sorcery and the old *langgasutap* as a sort of subject for attention. Though unsystematic and defective in many respects, it was a sketch of the epidemiology of concern with sorcery over a stretch of time. I was interested in the progress and spread of knowledge about it, the increasing precision and elaboration of information, the differentiation of *minmin* from *langgasutap*.

When fear was greatest, garden work patterns were shortened to the safe hours of the day, distant gardens were not visited. A young husband beat his wife because she refused to fetch taro from a garden on grounds of risk. The people ran out of salt and reverted to eating ash instead, because they would not go down to the mission trade store to get it. They would not take their vegetables to sell to the mission. I could not persuade people to take my letters for posting there. At this time men were expected back from two years' absence working at plantations and people worried that *minmin* would strike them on their way back. There were many false alerts and rumours of their return which led to groups of parents and sisters or wives going down, despite their fears, to the mission airstrip. On one night there was a burst of crying and wailing from a hamlet. At first it sounded as though someone had died; then, listening more carefully, the people with whom I was sitting distinguished that it was Sakai crying for his eldest son Kamei, whose return he awaited. Sakai feared he was to be struck by *minmin* on the road from Lumi, the subdistrict centre, home; he had had a dream. Need to renew a gun licence overcame the ex-*luluai*'s fear to walk abroad. Desire to go off to the plantations led to a band of young men going off.

There is not space to attempt a full account of the detail I assembled, but the point is that a picture does emerge from such heterogeneous detail of the level of general anxiety and its change over time. In terms of spread, high public attention, movement across linguistic and cultural boundaries, growth and decline, novelty and change, it was a phenomenon with some formal points of similarity to a cargo-cult. Allusions to *langgasutap* (real *sangguma* sorcery as I first learnt about it) had been common, but because the ideas about it were so established and embedded, I rarely heard spontaneous extended discussion of it, as I came to hear them about *minmin*. *Langgasutap* was one of the prominent features in distinctions made by the Gnau to define differences of culture between the peoples of their area. The detailed data obtained on the growth and spread of knowledge about *minmin* show a progressive differentiation of *minmin* from *langgasutap*, growing knowledge about its nature, methods and distribution. *Minmin* gradually displaced concern with *langgasutap* as a danger and a cause of death. The knowledge and concern had a wave-like spread, which followed and was seen to be confirmed by the increased numbers of deaths caused, in my view, by an epidemic of influenza. The gatherings for mourning were an important means by which these ideas were disseminated and elaborated. A myth of the origin of *minmin* appeared at this time and was soon widely known.

Within two months of the influenza epidemic, fears of *minmin* were such that people were reluctant to travel beyond the close gardens surrounding their villages. The missions whose personnel depended on village garden produce, were no longer able to get satisfactory supplies. The problem was brought to the attention of the patrol officers. A treatment for *minmin* had appeared, but in keeping with their theory of destructive magic, the man who could treat could also strike down. The knowledge of *minmin* did spread, and thus with it the knowledge to kill. At Rauit, people from villages which the Gnau considered knew *langgasutap* were sought out to treat the cases presumed to have been struck by *minmin*. At the time of greatest fear, still no one in Rauit would admit to me that he might be ready to try to learn the treatment. They depended on outsiders from other villages for treatment and they paid them money for it. However, seven months after the fears had subsided, a Rauit man told me that he, with two others, had arranged to learn about *minmin* from one of his non-village clan kinsmen, so that he would be able to treat people in the village. He arranged a secret meeting in the forest to which he went with gifts, but the kinsmen of the other village did not turn up.

At the time of greatest fears, many of the deaths that occurred in the area were put down to *minmin*. Eventually, a general desire for its suppression was crystallised by a man of a nearby village who called a meeting to decide a common request that might be put to the patrol officer who was expected. There were rampant demands for public torture and mutilation of those (unspecified) who knew *minmin* and had used it. Although they looked to the Administration to suppress it, a common theme of their general view was that the rise and spread of sorcery was an ill effect of the changes that white men had brought about. For before, in the time of warfare and arrows, no one passed freely over the land of other villages. As a patrol officer came soon after this, many villages were able to bring forward their complaints, although they muted the demands for death and mutilation and all their accusations were pointed at other villages: none were made between co-members of the same village, as they had suggested earlier that they might be. The village and the man who had often been indicated to me as the first to acquire the sorcery, a man who was a leader in a cargo cult in the early

1950s, was so unwise as to boast to the Patrol Officer that he had sold a bogus magic to other men, the gullible men of villages elsewhere, and he, with certain others, was later imprisoned on this count.

The knowledge of his imprisonment, though at first garbled and the penalty exaggerated, was diffused through the villages, and after this the fears and references to *minmin* gradually declined. At Rauit, the man whose lingering and distressing illness had coincided with the rise and spread of *minmin*, died at last. His illness was never ascribed to *minmin*, but at the continued gathering for the performance of the ritual of Malyi by which they sought to make him live, the topic of *minmin* had offered an engrossing subject which in part seemed to me to serve as a distraction from the grievous distress they felt for him, his wretchedness and his expected death. After it, there were fewer occasions for public reflection on the iniquity of *minmin*. Some months later it came to be widely known that just before the Patrol Officer came, there had been an assembly of the men with *minmin* who had together by a ritual thrown off their powers, and many of these men had fled to the plantations, fearing physical retribution or gaol. The Patrol Officer's visit had in fact come soon after the departure of men for plantation labour in January 1969.

In this summary account there are things to note: the initial vagueness and uncertainty about the nature of the danger and how to avoid and counter it; the rise, spread, and sharing of the fears; the growing and more precise definition of what there was to fear; the coincidence of what was seen as proof in the deaths from influenza; the situations in which fear was overridden; in Rauit as I witnessed it, its particular relation to distress over the ill and dying man. But the account is a summary.

Persisting or chronic public fears: the question of the control and prevention of future dangers

Minmin began as a variety of *sangguma*, something close to *langgasutap*, and then gradually began to develop some of the attributes of destructive magic. A difference to note between destructive magic and *langgasutap* is the estimate of ability to avoid and counteract them. Destructive magic allows a possibility or remedy. It can be countermanded. The presumption of intent on the part of the agent doing it requires a reason for his hatred, and a reason which will account for murderous hatred, since in theory destructive magic kills and does not merely harm or cause misfortune. As few people feel that there are others with specific grounds to wish to kill them, so those who are themselves ill, or who are close relatives of the sick person, rarely suggest that destructive magic has been done on purpose. The use of destructive magic must be intelligible in terms of motive. On the other hand, those who are not so close to the patient may in a casual way, privately or speculatively, wonder about the possibility and with distancing in time and social relationship, old illnesses are more easily, less upsettingly, without particular conviction or a sense of outrage, put down to magic or sorcery.

In contrast *langgasutap* carries, so they assume, an impulse into those who learn it, to try it out to satisfy themselves with proof of lethal power of the true acquisition of the knowledge. To hold the power of *langgasutap* is considered sufficient motive for its use. It does not have to be intelligible in terms of revenge, hate, or defence. Destructive magic is not like this. It is taught to individual men at the discretion of other men who already know it. In deciding whether to teach it they are influenced by various considerations such as friendship, desire to help someone to protect or revenge himself, or recognition of the ability and good sense of a younger man who will later have to take responsibility. In each clan there is a scatter of men who know the different kinds of destructive magic, but it is difficult to obtain reliable information on the full distribution of such knowledge. Most of the men who either admitted knowledge, or were pointed out to me as knowing important techniques, were senior men. The youngest of the men who knew any of the important techniques were in their middle thirties and they were both eldest sons, notably effective hunters, serious and determined in general manner, and rather less light-hearted and talkative than some of their age equals.

With *langgasutap*, the stage would have seemed set for threat and danger from persons unknown, always possible, aimless, stealthy, invisble, lethal. But it was not really so. For in the first place there were simple rules for self protection:

1. Not to go alone into the bush. *Langgasutap* strikes the solitary person. Two in company are not vulnerable, even if the companion is only a child. *Langgasutap* cannot strike down two people at once. As a rule, women and children never go on their own to the bush. Men sometimes do, but they carry something to use as a weapon. Many of the older men carried their bow and two or three arrows, even to go to work at a nearby garden. And one of the arrows was a barbed man-killing arrow When I asked why they carried that kind of arrow, the conventional answer was, 'Because of *langgasutap*'.
2. To be alert. The Gnau are alert to sounds, the patter, swish and rustle of animal and bird movements, and they are quick to identify them. *Langgasutap* and its bearers may be close to invisible, but they make noises as they move.

3. A bird may warn of *langgasutap*. The sulphur-crested cockatoo (*igambati*) is common. It may scream sudden warning of the presence of *langgasutap*. The screech has a harsh tearing quality which fits it for a warning.
4. Dreams may provide warnings. People have dream premonitions sometimes about each other, and they go to warn the other to stay in the village (which is safe from *langgasutap*) or to take special care.
5. Alertness may also protect because of an idea that if one first looks on the man with *sangguma*, the power is diminished (I am less certain that this idea is generally held). But cockatoos often screech suddenly in gardens and in the bush, and I have never seen anyone take fright because a cockatoo screeched.

What signs are there of a prospective state of persistent anxiety? They do not show persistent anxiety, but rather a general level of awareness and the need for reasonable precaution such as might be paralleled in our society, for example, by the feeling of parents about the risk to children of road accidents. An unexpected delay, a possibility at the back of the mind - the delay drawn out, the anxiety heightening. I saw people seriously worried and agitated, discussing the possibility of *langgasutap* because members of their family or hamlet were inexplicably delayed towards nightfall or after dark on five or six occasions during my whole 22 month stay, that is, in a population of about 350 people and to my knowledge. The *minmin* scare was quite different, with a high level of fear for a few weeks, deaths heard of in villages round about, the possibility frequently springing to mind, and many signs of anxiety. Imagine the state of a small English community after two children are found in the vicinity killed by an unknown sex maniac and I think the reader will have a fairly parallel idea of the prospective state of fear for a short time.

FEARS FOCUSSED ON PARTICULAR EVENTS: PRESENT FEAR

A) *Passing thought*

The danger springs to mind as a possibility. I was sitting talking to Waleka, when Nepalkin came up walking quickly to ask me to go and see his mother who was sick. Waleka straightway called out to Nepalkin, 'Is it *langgasutap*?' Nepalkin answered 'No, she's just ill'.

B) *Acute Public Concern*

Sudden onset is important. On 19 September 1968, during the

afternoon, Maka, a woman of about 40, was spirit-possessed for an hour. During it she spoke of danger from *minmin* to her HeB and his son. She asked where they were; the danger was at a place in the bush called Wolwakut. Danger from Nembugil village. She stayed possessed and was assisted into her house. Singing for the Malyi ritual was going on in the men's house at the time, and people went on with it. Suddenly it was stopped because from Maka's house there came a sudden noise of crying and calls, 'Maka is dead'. Noises of women crying out. People were running to Maka's house, women wailing and in tears, crying out 'She's dead'. Her HyBW was throwing her own bamboos out of her house, smashing her own cooking utensils, destroying, desolate, grief-stricken. I called to Maka's HeBS who was rushing there, asking what it was. He said, his face very tense, '*Langgasutap* has struck her.' Her aged father was crouching, weeping, 'Maka! Maka!' Her husband was in the hut. He led her out to me. She was dazed, pale, haggard-looking, blank-faced. I examined her. People prepared to treat her and she became quiet for a little and then started in possession, and she began crying and announced she was going to die. She fell back. Two men raised her up and supported her walk to the porch of her house. There was wailing and weeping. She was put semi-recumbent; her husband supported her back, weeping; her head was lolling backwards, sometimes forwards, her eyelids fluttering. Wara, who was about to treat her, kept hold of her right wrist, her mother and another senior woman flung themselves wailing on her chest, other women grabbed her loose arm or sobbed down over her legs which were tense and trembling. The frantic confusion of bodies brought Géricault's painting 'Le Radeau de la Méduse' to my mind. The diagnosis was soon made of something other than sorcery. This example has described an extreme public reaction to a first suspicion of sorcery.

C) *The victim*

First case: In this example, I am concerned with the victim herself, rather than the people around her. Again Maka, on 4 January 1969, went to work a garden, leaving with Dameku (HeBW) and Rumak (HeBSW). She worked alone in her husband's garden. She returned in the afternoon and ate the evening meal cooked by her eldest daughter. At about 8.40 p.m. she became possessed for about an hour. That night she slept with Dameku and Rumak in Dameku's house. During her possession she revealed she had been struck by *minmin*. People went off at dawn next morning to another village to fetch someone to treat her. In the morning she was sitting in the front part of Dameku's house, crouched with her bent knees up to her chest and her arms enclosing them.

Duapo, a senior woman, was sitting with her daughter beside her. Maka's face looked blank and worn-out. She stared downwards, not looking at people. When I spoke to her, she looked up and stared rather unwinkingly at me. I asked whether, and where, she had a pain, and she said in her *wuna'at* (vital centre) exactly and precisely localised to the xiphisternum and epigastrium, and in her neck. I asked her about it. She said, '*Minmin watewunpeg*' (*minmin* hung up on me). How did she know, had she seen them? 'No, I did not see them. You cannot see them and they stand a long way off. I was up in the tree. My ears went deaf. I felt dizzy and sweat came out over my body. I went to wash in cold water. I felt the pain in my chest and the back of my neck'. She felt all right when she came back to the village, she had eaten and felt all right, the spirit came and said things - the same spirit as before. It was said she had been struck by *minmin* from Nembugil. What else had the spirit said? That Dauwaras would die. (She spoke and answered my questions as though she remembered directly what the spirit had said, although she may have been repeating what others told her she had said.) She did not die or go unconscious. She had slept all right that night. She did not dream. She had no headache or disturbance of vision. This morning she felt the pain in her chest and in the back of her neck, sharp and big. I asked whether it was like a fist pushing or a needle. She said a needle. She said it was severe, but she showed no tenderness or shrinking when I palpated her chest and examined the movements of her neck. The only finding was doubtful tension in the paravertebral muscles of the neck on the right side close to her occiput. She said that rotating her head to the left made the pain worse, but there was no resistance or limitation to movement and her other neck movements were full. Her pulse was normal, 68/minute, her breathing normal and regular. Her chest was normal to auscultation. Her mood was listless. Her voice low, flat, her articulation and speed of utterance normal. She only answered direct questions. She did not talk with the women sitting with her. There was no evidence of muscular tension, agitation, heightened distractability, but rather the reverse. When not talked to directly, she lapsed into looking at the ground. She took no particular notice of Waukei, (her twenty-month old daughter) when she came into the house, but held her appropriately for her to suckle.

Later that day she was treated for *minmin* and four arrow heads were withdrawn, three from her chest and one from her neck. The largest was of white wood, which looked like bone, thin, and about 1½ inches long. Maka had said there were two in her chest one at either side, and that they had gone in from her back (*minmin* shoots from behind) near each shoulder blade and high up. The arrow heads were withdrawn from in front. The stress that they had gone in there and not at a more lethal point, which would have killed her, was related to the idea, then becoming current, that adults could feel themselves being shot and could twist aside from the arrows, but that children could not, and when struck, blood would flow from their nose, mouth and anus.

Maka remained depressed for some time after her treatment. She did not complain of pain, but felt odd, as though things were running around under her skin, in her blood. She slept all right, but with some aches at night in her neck. It was in the daytime she felt odd (it seemed to me that she was more sensitive and self-conscious under the gaze of others). She did not smile or show interest in things, but lay about close to the house. After this, she went to her parents' hamlet about twenty minutes' walk away. She stayed there for another three weeks before she felt back to normal. One month in all.

I have described the usual and conventional behaviour of Gnau people when they feel they are ill in Lewis (1976). Maka's behaviour after her treatment for *minmin* followed the conventional pattern of someone ill.

Second Case: Not long after Maka had gone to her parents' hamlet on 18 January, a girl there had her puberty ceremony. In the afternoon just as this was ended, news came that some men from Winalum had turned up at the main part of the village. Later, one of them went to Maka to treat her further for *minmin*. Sugam, a woman married to a classificatory yB of Maka (she was staying near him), went to the main part of the village with a gift of meat to one of her husband's relatives. A tree had fallen across the path. It was so big that it was a nuisance to climb over it, but nearly everyone disliked the alternative of bending low to go under it, on the grounds that other men and women climbed over it and therefore to go under it entailed the risk of going underneath something which one's juniors had gone over - an elaboration of a taboo idea. On her way there, Sugam had noticed a hank of some weeds and grass just on the other side of the log, and had avoided stepping over the grass. Then in the late afternoon close to nightfall, she came back quickly and, forgetting the grass, stepped over it. Just as she crossed up over the fallen tree and was about to step down, she felt she was struck from behind. Her right ear went deaf and ringing. She felt dizzy. She got back to the hamlet still feeling odd, and there in the evening became possessed and told details of how she had been made ill. I was called to her at 9 o,clock that night.

Sugam was crouching in a corner of her hut porch with her back to everyone. Her head was drawn down to her knees, with her elbows resting near her knees and her hands up at the sides of her head, covering her ears. Three other women were sitting in the porch with her. Many other people outside the porch were watching. I went in, Sugam remained motionless.

I asked Sugam where the pain was - no answer. She did not move or turn to look. In the head? Monosyllabic grunt of assent. Nape of neck? Yes. Ear? Yes. Chest? No. I asked if her eyes felt bad. No. I put a thermometer under her arm, for which she made slight accommodation. I took her head, and after touching and palpating her neck, to which she made no response, I moved her neck slightly. There was no evidence of illness except for her complaint of pain and odd behaviour; no evidence of a stiff neck or fever. Her pulse was regular and normal. She was not sweating and did not have nystagmus. I produced an aspirin. I offered it to her and put it in her mouth which she opened. I asked for water. Wairi, her sister-in-law, and others near us objected immediately that she must not drink water, because she would be made cold by it and might die. Her husband came inside and told her emphatically to swallow the pill as it was. She tried to but could not. She was still in her crouching position, hands etc. as before. I went outside. The others were waiting rather quietly and expectantly. Sugam came outside unaided, still holding her hands to the side of her head. I then saw that she was holding the pinnae of her ears bent forward to cover her earholes. Her head was downcast. She went a little way and crouched down. She was to go to Watalu hamlet for the night. She was waiting for the others to lead her. Someone called out that the mountain was dangerous (there is a point along the path where it lies about two feet away from a sheer drop): someone must hold her. When she first came out she walked slowly, slightly meanderingly, and Gaita went to her side, others saying she might set off alone and fall over the cliff edge.

Then we set off with my light, Gaita and another immediately behind her. Sugam continued to walk with covered ears and downcast head, but she was clearly watching the path and she did not stumble or falter on uneven ground. Sugam said nothing. As we came to the fallen tree, I wondered how Sugam would approach it. She walked bending to go underneath it, with no sign of any special reaction to it. When Sugam got to the hamlet, she crouched down silent, hands still over her ears. In the dark of the house there was a tall man from Winalum whom I had not seen before. At about 11 o,clock, he performed a treatment on her to remove arrowheads. It was for *minmin*. In the process of treatment he took her hands from her ears, and afterwards she sat with her hands by her sides.

In the descriptions I have given, there is little of the signs or symptoms of acute fear. Instead of tension and agitation, of preparedness to fight or flee, the behaviour appears rather that of someone doomstruck, shattered, apathetic, hope given up, passive. It is more like the behaviour of despair, depression or shock.

Further cases: The aspect of listless despair in the two cases I have described is of interest in the light of other studies. Mauss (1926) quoted a number of reported cases of psychogenic death among Polynesian people and Australian Aborigines. His distinctions between the types of belief and setting in which such deaths occurred have been refined and extended by Ellenberger (1951, and 1965) with further cases from anthropological, historical and other sources. Ellenberger distinguishes acute from slow psychogenic death. Clearly, explanation of the mechanisms involved in these allegedly psychogenic deaths may require different answers when we seek to account for sudden, nearly instantaneous death rather than death within a few days, or death after a prolonged period of waiting. The medical possibilities are very different, given each span of time. The kinds of case quoted by Mauss and Ellenberger understandably enough contain no medical evidence of a sort to exclude the great variety of possible causes of death, such as might be revealed to post mortem examination. I have found one instance of death supposedly caused by a magical rite, for which there was also an autopsy report. The man on Goodenough Island, New Guinea knew that he

> was regarded as dead by his fellow tribesmen. On being ignored, rejected and excommunicated and after a period of panic, he had become listless, apathetic and inert. He expressed at no time a desire to live, and acted as though convinced that his end was near. He had taken to his pallet and refused food and water before being brought to the hospital .... Although he did not appear severely ill, his state varied between one of frank depression and apathy, without terror, and remaining silent and more or less immobile. His pulse rate was 65, his heart was slightly enlarged, and X-rays of his chest were not contributory. His blood pressure was within normal range or slightly elevated.... He became increasingly apathetic, seemed detached and resigned, barely moved, and his bed covers remained undisturbed for hours. His skin and mouth were dry, his urine contained a slight amount of albumin and had a high specific gravity. He was seen to pass no excreta after the first few days. He received penicillin, arsenicals, and digitalis. No one came to see him and he interested himself in no other patients. On the ninth day after admission he was found dead in bed. Autopsy revealed cirrhosis of the liver, splenomegaly, and widespread arteriosclerosis. Also, amyloidosis of the spleen, kidneys, pancreas and liver was revealed on histological examination. No immediate cause of death was discovered. The likelihood is that the death was due to rejection of fluids, brought

about by psychological reactions to tribal rejection. (Simmons and Wolff 1954, reporting a personal communication: 94-5).

In many of the cases cited by Mauss, the victim isolated himself, withdrew from social contact, seemed hopeless and helpless and soon died. How soon the victim died is explicitly stated in only a few of the instances, and most of the reports are anecdotal: the author who reported the events was not himself their witness. On the basis of psychiatric work among Australian Aborigines Cawte (1974) notes, in contrast to the anecdotes, that the sequences involving complaints of sorcery that he actually observed himself in the field 'usually involve an overt somatic or psychiatric disorder, or else a social situation seriously disturbed by tension and interpersonal manipulation' (Cawte 1974: 90). He describes the variety of patients complaining of sorcery that he examined, but no case like any of the anecdotal accounts.

It may also be noted that the folk illness *susto* (in some places termed *espanto*) which is found throughout Central America and parts of South America (Rubel 1964) is an illness whose name means 'fright', and the name reflects the folk view of the cause, which is believed to dislodge the individual's soul. The individual suffering from *susto* is reported to withdraw socially, but the illness is not associated with dramatic or sudden death. Tenzel (1970) examined several Mayan Indian patients who were held to be suffering from *susto* and they all showed evidence of physical disease. Sal y Rosas (1957) likewise found somatic disorders in many of the Peruvian patients with *susto* whom he examined. These citations are meant to instil some caution over accepting anecdotal reports of mysterious and dramatic deaths. F.E. Williams (1928:214) some time ago questioned whether deaths from sorcery really occurred. He thought it was likely that those who were said to have died of sorcery had in fact been ill: when the victim had first believed himself struck by magic, he was already in reality dying of the unsuspected disease that had him in its grip.

There is some evidence that severe hopelessness, helplessness and despair may affect morbidity and mortality. Such evidence includes the statistical studies of death following bereavement (Hinton 1967:174; Granville-Grossman 1971:188); studies of American prisoners of war during the Korean War (Chodoff 1966:387: Wolff 1960:28); and studies of the development and exacerbation of somatic illness preceded by psychological 'giving up' (Schmale 1972). Sudden death is known to occur in animals in settings of overwhelming stress from which no escape is possible, for example, in wild animals trapped. Richter's experimental study (Richter 1957) is perhaps the most relevant to the subject of sudden death. The tough, wild, urban rat is formidable in strength, attack, cunning and endurance, compared to the laboratory rat. But a greater proportion of wild than laboratory rats died suddenly when handled and then subjected to a stress for which no solution was possible (swimming in a jar against a jet of water). They died, not during the fight-flight pattern of sympathetic overstimulation, but when they 'gave-up' the struggle, at a point which long preceded actual physical exhaustion. As they succumbed, their heart rate and respiration slowed, in a way that suggested vagal parasympathetic overstimulation. Clipping the rat's whiskers, presumably thereby removing a source of sensory information, hastened 'giving up' and shortened survival. Cannon (1942) had suggested that in 'voodoo death' the heart ceased in systole in a state of sympathetic overstimulation; Richter, on the basis of his experiments, suggested that such deaths might rather occur with the heart ceasing in diastole in a state of vagal parasympathetic overstimulation. Extreme vagal parasympathetic activity would correspond psychologically to the person overwhelmed by passivity and helplessness rather than to the fight-flight tension of the overacting sympathetic adrenal side of the autonomic balance. Sudden death in acute terror is reported anecdotally less often than death in those who appeared listless and apathetic, or who appeared to feel trapped or doomed and to have given up. Just as there are occasional reports of death in a state of intense terror; so there are also accounts of sudden death in extreme rage; I have not found reports of extreme terror in psychotics which led to death.

RETROSPECTIVE VIEWS OF SORCERY

With F.E. Williams's point of doubt in mind, I will illustrate some examples of the retrospective diagnosis of sorcery. These show the diagnosis in hindsight.

The public view: The case of Sunikel was the only one in which a diagnosis of *langgasutap* was made during my stay at the village. The point of chief interest is to note how the villagers made an initial factual explanation which was later altered. Sunikel was a twelve-year old girl who for years had had occasional epileptic fits in which her face went blank, she jerked and fell unconscious, sometimes being incontinent at the same time. On 27 August 1968, early in the morning, she was sent to fetch water, but she did not return. Her father and brothers and sisters went to find her. Her father went to the water pool at Lugeban and he found her head downwards in the pool. He pulled her out by the legs. She was dead. He carried her back to the village and they began to mourn. Her father said what he had seen and done. He said that Sunikel must have had a fit and fallen into the pool. Others said that it was wrong to have sent her alone, he knew she had fits, he had plenty of other children who could have been sent. The body was taken into her mother's house to prepare it for burial. Sunikel's uncle went to the waterhole and measured its depth with a bamboo pole which he notched. It was about three and a half feet deep. On his return he stood the pole against a much younger child than Sunikel to compare the depth and the child's height. Some men sitting near me at the mourning, began to discuss *langgasutap* and to agree with each other that it must have been *langgasutap* for her to fall and drown in such shallow water. They considered by what likely paths the men with *langgasutap* might have come. This was now about two hours after the body had first been laid out in the village. The theme of sorcery was taken up. By the afternoon it was decided to fetch a clan relative from another village to perform a certain method of divination. This was done two nights later. The findings of the divination were that she had died by *langgasutap*, struck on the back of the neck and both sides of the jaw, and that this was accomplished by three named men from two other villages. Some of those who had watched the divination told me privately afterwards that they thought that the truth had not been learned about Sunikel's death. They said they doubted it, because the method of divination had not been one of the traditional Gnau methods. Although a few people were sceptical about the details revealed by the divination, I think that everyone thought that her death was due to sorcery and this probably reflected both that she was a bright vigorous girl despite her occasional fits, too young and strong to die, and that her death came without warning, without preliminary illness; it was unacceptable to see it purely as a tragic coincidence.

The victim's view: I was told that a sick man I knew had, a few years before, been struck by *langgasutap*. To be told that, contradicted theory from the start for, in theory, *langgasutap* is always fatal. I asked him about it. The victim's account, which I wrote down, but not verbatim, was this:

> He said he did not know if it had been *langgasutap* or spirit attack. He had been cutting sago for his wife to work. Late in the afternoon she returned home ahead of him. He finished cutting the sago palm, took the sago grubs he had collected earlier and went to cut some bamboo for cooking tubes. While he was cutting these, he suddenly felt dizzy. He gathered his bamboo tubes, his axe, a packet of leaves and the grubs. It was evening. He began to go home - the bush was called Pilikawun - at a stream, he fell unconscious. He must have remained unconscious for some time because when he came to his senses it was raining heavily, thunder and lightning. The stream was not one with a spirit staying there (a common

reason given for storms is that men go by certain water places associated with a spirit). He lay there. The bamboo tubes, the grubs, the packet of leaves and the axe had fallen by the water. His wife was alarmed - men had gone out to find him. His brothers-in-law found him, lying in the stream - he was not crying out or in pain. One shouted at him: *'Dji Saitan ao Sulaiman ao dji masi'?* (literally, 'You Satan or Solomon or who are you?' - a nice touch - Satan is quite commonly used of evil spirits and for *sangguma* men in Pidgin, and he was speaking Gnau to me. I never heard anyone else bring in Solomon). He replied *'Saitan gnau - deg matilden'* (Not Satan - I am a human being). They came up and carried him on their backs and put him down at the porch of his wife's house. They lighted fires around him, made sago jelly for him and scraped him coconuts and these he ate. In the morning, his wife went to fetch the things that had fallen by the stream and they were there. The others said to him 'What struck you, was it *langgasutap* or a spirit?'. He said 'I don't know. I went unconscious - I think probably a spirit. (To me) Could it have been *langgasutap?'* It only happened to him once like that.

Retrospective doubts at a longer interval: In talk, and in finding out the cause of death when taking down genealogies, I was quite commonly told that this or that person had died because of sorcery or destructive magic. Distance in time (and also distance of social relationship in talk of present illness), I think, allowed the speaker to give this view straight out as though he were sure of it, without the view causing him distress or really mattering to him. But in some of these cases, which I followed up, when I questioned those who had been (or were) close to the person concerned or had witnessed the case, I found that there was no such certainty of the diagnosis, or that in fact the death was not attributed to sorcery.

One case which illustrates such doubt well, was that of a death that had occurred seven years before I made my field study. The case was cited to me by many different people as an example of death from *langgasutap*. Duwatei was a little girl who, those seven years ago, had been playing with other children by a tree at a distant hunting camp. A branch fell and struck her on the back of her neck. She was knocked unconscious and picked up. Her mother and her brothers began to carry her back to the village, which was a full day's walk away. She recovered consciousness, but at the crossing of a river in the forest, she tensed, went into a spasm, her eyes rolled up, and she died. Her burial in the village was accompanied by dramatic expressions of grief and anger on the part of her mother, her father, and her father's brother. Suspicion quickly grew that the death was in reality one by *langgasutap*. Suspicion focussed on the village of Nembugil (known officially as Anguganak). On hearing the rumours, a deputation of men came from that village demanding to try an ordeal which might prove them innocent. The ordeal was one that would have required them to dig up her body and eat a piece of her remains. There was a fight over the spades to stop them digging up her body. Duwatei's uncle was restrained when he drew his bow in the fight, the ordeal was not performed, the accusations were retracted, and the matter rested at that. Nonetheless, the general retrospective view at Duwatei's village was that her death was due to *langgasutap*. The account that I have just given is taken from the statement of her elder brother who had been present at the time. I recorded it on a tape-recorder. Her mother came to me the next day and asked to tell me about Duwatei's death on the tape-recorder. She, in common with others, had heard how the tape-recorder played back the voices of people. In the translated excerpt of the transcription below, it will be seen how, after first having related to me what had happened, she asks the tape-recorder to answer her as to what really was the truth: she asks for her daughter's ghost, through the machine, to tell her what it really was that killed her, the branch or *langgasutap*? She turns from speaking to me to question the machine, and in so doing she expresses her own unstilled doubts after so long a span of time. I then tried to explain to her in my halting Gnau that she could have no answer from it. (On one other occasion at a later stage in my fieldwork, I was also sought out for my machine, so that someone could speak to his dead daughter.) The relevant part of the mother's account is this (she speaks first of what happened at the graveside):

> (My husband) came to strike me with his axe, crying out 'Oh my daughter! Where was I when I should have carried her back? We should have been together - we should have carried her back to the village together.' (I answered him) 'You were a long way off in the bush. I carried her back and you were over at Marep and alone I carried her back. You were there and I carried her back. I thought then that you and I should carry her back together, one behind the other. But was she well enough to wait for you? No! I had to carry her back.' I told him to kill me so that I could be with Duwatei. I told him to kill me to be with Duwatei. I grieved for my daughter telling him to strike me so that I could be with her.... She must speak to me and say if it was *langgasutap* that struck her or the tree which struck her - speak to me to say if it was the tree. Let her speak to me and say if the tree struck her then, or *langgasutap* struck her. She must say if it was *langgasutap*

which struck her then. If it was *langgasutap* which struck, let her speak. If it was *langgasutap*...

SORCERY AS A CAUSE OF ILLNESS: THE GENERAL PERSPECTIVE OF THE GNAU

By concentration on the subject of magic and sorcery, it is all too easy to give a distorted picture of its importance in the daily life of a people. If the reader were asked to weigh in his mind whether the Gnau are more or less hag-ridden by their ideas of sorcery than X or Y other people, I think it would be difficult for him to answer. To put the matter into perspective, I will mention briefly the results I obtained from my study of all illness in a community of 355 people on average during a period of twenty-two months (see Lewis 1976). During this time, I saw 274 cases of sickness. Of these 274, only seventeen were at some point diagnosed by some individual as due to destructive magic or sorcery, a small proportion. The diagnoses were applied to serious illnesses (for instance, pneumonia, incapacitating acute arthritis), usually after death; or given at the end of an illness in which treatments for other causes had been tried without success; or else the diagnosis was brought forward by the victim herself through possession in illnesses of an ill-defined nature. In some cases, the sudden onset and disastrous nature of the illness was associated with the diagnosis of sorcery, as distinct from destructive magic.

The Gnau do not lack beliefs about sorcery and destructive magic, but the way they apply them to explain illness shows some of the force in favour of village harmony. They approve of the use of destructive magic to protect property. If it is supposed that someone has come in contact with it and so fallen sick, the assumption is that this has been inadvertent and there is no malice between the co-villagers concerned. There is open discussion of the possibility and remedy by counter action, as in the case of Dauwaras, where on one day all the men in the village who knew a certain form of destructive magic, were called on and treated him lest he might have come in contact with it unwittingly. But the fact of death proclaims intended malice and hatred. The crucial issues in a diagnosis of destructive magic and sorcery after death are the intention to kill and who was responsible. The entailed view of lethal outcome in their understanding of magic and sorcery is the feature that is critical for the significance of this knowledge. An accusation of such sorcery or magic within the village, would openly disrupt relations within it. So, unless there are very pressing grounds to bring them forward, suspicions must remain private speculation or be referred to the malevolence of spirits. The diagnosis of attacking lineage spirits is one that is made during illness, and it offers the possibility of treatment by spitting and invocation to the ancestors presumed responsible. When, within the village, such diagnoses are given of illness caused by spirits related to members of the village, the treatment is readily and willingly performed, for refusal might suggest intended harm or appeals to ancestral spirits to cause the illness. When the spirits of other villages are involved, the diagnosis is likely to be replaced, if the illness leads to death or looks as though it will, by the diagnosis of intended destructive magic. Ancestral spirits may work harm because of insult or injury to their descendants either (1) without their descendants' wish or knowledge, or, (2) by their descendants' expressed wishes, furthered by special appeals to them involving certain rites; and (3) a man may practise destructive magic or sorcery. The diagnostic lines between these three diagnoses are subtle and the direction easy to shift. Accusations of sorcery are turned outwards towards other villages. Formerly accusations of sorcery and intended black magic between armed and hostile villages rarely provoked denials. There were no attempts to show innocence: the modes of divination acted to confirm suspicion and strengthen a determination to seek vengeance. In present conditions, rumour of such accusations led to the accused men hurrying to face their accusers, demanding of them to try some ordeal. Presumably the results of divination are now called into question more often.

COMMENT

Mauss (1926) collected the reports of cases in which someone had died because, placed in a particular situation, he shared and believed ideas that suggested he must die. To Mauss, these cases were of great interest, for they were cases '*où la nature sociale rejoint très directement la nature biologique de l'homme*'. They showed up in a dreadful way the power of collectively held ideas over the individual, the ability of beliefs to produce physical effects on the individual. He indicated a difference between deaths reported from Polynesian societies and those from Australian Aboriginal societies. In Polynesia, death was linked with the idea of guilt and the predominant symptom was not fear, but rather shame and humiliation. The Australian cases were marked by a belief in a clearly defined cause, conviction of death consequent on the sorcery of 'pointing the bone', and accompanied by acute, immediate paralysing fear.

The notion of psychosomatic illness and of psychogenic disease is one that receives more conspicuous attention and acceptance now than it did when Mauss wrote his paper. It is not a novel idea. Indeed almost from the earliest recorded times, men have generalised about the interaction of mind and body, and have described the familiar bodily manifestations of fear, sadness, anger, and other transient and 'normal' emotional states, but these were unsupported by evidence of the ways in which the mind could be held accountable for the occurrence or the course and form of disease. More precise studies of the physiological processes which might or might not be preludes to lasting morbid changes on the structure and chemistry of the body, were pursued in the nineteenth century. There was general acceptance of the role that worry, frustration and strain might have in hindering recovery or fostering illness of many sorts; but there were no detailed analyses of the social and psychological influences which apparently had this effect, or of the specified illnesses that could be thus affected and the mechanisms involved. In this century, the subject has been investigated experimentally in man and animals, notably by men such as Pavlov and W.B. Cannon. The terms 'psychosomatic medicine' and 'psychosomatic disorders' have come back into general use. They were initially introduced to stress that the conceptual separation of mind and body in medicine is not only unreal but also unfruitful in certain conditions (for instance peptic ulcer, asthma, ulcerative colitis). By stressing that man in health and disease functions as a psychosomatic unit, the Hippocratic and holistic approach in medicine was, so to speak, reintroduced. When we speak of psychological processes and physiological processes, we are speaking of different ways of approaching one phenomenon. The phenomenon itself is not so divided. It is evident that the term 'psychosomatic', as commonly used, has two meanings: one when applied to the general field of medicine (the holistic approach); and another when applied to disorders such as essential hypertension, asthma, or peptic ulcer, in which psychological factors are supposed to play a major role. 'Psychosomatic' is thus a vague term reflecting re-awakened concern for the social, psychological, as well as somatic, factors involved in diseases, and especially in a certain range of diseases. (See World Health Organization Technical Report Series No. 275, 1964).

The rapid or sudden death of a healthy adult caused by the certain belief that he must die, or by the wish to die, would be a melodramatic instance of psychogenic forces. I would be cautious of accepting the kinds of reports Mauss used as evidence that such deaths do take place. Most were hearsay stories, few of the cases were seen by those who reported them, all were unsupported by adequate medical investigation. Of course, it would require very special circumstances to be able to provide such information. But I stress this caution because there seems a readiness to accept such accounts as evidence when they come from exotic places, although they might not be taken with the same respect at home. It is as if because the stories are about distant and 'primitive' Aborigines, Polynesians, or Africans, we credit them as evidence of the mysterious powers of psychosomatic forces. We might be more critical and less credulous if they were about people of our own culture. Of course, it may be argued that our culture and the beliefs generally accepted in it are so different from those of the exotic societies, that there is no point in looking for analogies. So we might turn instead to psychiatric sources, expecting to find there examples of the fixed, incorrigible idea of approaching death as a delusion of the mad which led to death, or of the wish to die being realised. Few such cases have been documented and the literature is scant. Engel (1970) has reviewed it in a paper which begins with the analysis of American newspaper accounts of 160 instances of sudden death occurring under noteworthy circumstances; he inquired further into many of the cases. He divided the life-settings in which death was reported to have occurred into:

1) settings of loss (such as the sudden death of a close person, in acute grief, or in mourning, or following loss of status or self-esteem) - such settings accounted for 58 per cent of the 160 instances;
2) situations of personal danger or threat - 36 per cent of the instances;
3) situations where the people died at times of presumed pleasure or triumph - these 'happy endings' accounted for 6 per cent.

From his review of these instances and the available medical literature that touches on them, certain characteristics of the lethal life-situations seem to emerge.

> In brief, the responses range all the way from struggling to cope with what is experienced as an overwhelming situation to giving up completely, even to the point of accepting death with equanimity. This formulation receives support from those few cases of sudden death in the literature in which fuller information is provided concerning the psychologic state of the person just before dying. Further, such reports indicate that while many different affects may be manifest at different times in the period preceding death, those most consistent and pervasive are helplessness or hopelessness - the giving-up affects. (Engel 1970:532)

He points out that we can only speculate about the mechanisms of death in these cases.

> The literature indicates that up to 65 per cent of sudden deaths occur in individuals with diseases of the heart and aorta.... Be that as it may, when looked at from the psycho-

> logical perspective, it seems obvious that conditions existed which were conducive to profound neurogenic cardiovascular influences. Most writers have tended to stress excitatory phenomena, which are well known to be associated with pronounced sinus tachycardia and ventricular irritability. But such are not present in all cases and certainly not in those who accept death calmly. More consistent are inhibitory phenomena, as exemplified in the various aspects of giving-up. This leads us to suggest that some of the lethal influences may involve rapid shifts between sympathetic and parasympathetic cardiovascular effects.... Gellhorn has pointed out that under the influence of increasingly strong stimuli and states of heightened central excitability, parasympathetic responses may ultimately become dominant, with the result that stimuli which previously elicited sympathetic responses may now evoke parasympathetic ones. This would indicate a shift from a fight-flight (ergotropic) to a conservation-withdrawal (trophotropic) pattern. Such autonomic overactivity and instability provide ideal conditions for the development of lethal arrhythmia and cardiac arrest. (Engel 1970:533-4)

There seems to be some agreement among the physiologists and doctors who speculate on the cause of psychogenic death, that the mechanism in cases in which people behave as though they give up hope would be like the one Richter postulated on the basis of his studies of wild rats. The descriptions I have given of the behaviour of Maka and Sugam, when they were believed to be struck by sorcery, fit with what Engel calls 'giving-up' or the 'conservation-withdrawal pattern'. When death has not been sudden but has followed after a longer period of withdrawal and despondency, accompanied by marked reduction in eating, drinking and activity, it is reasonable to suppose that debility and supervening infection were the most immediate causes of death.

Apart from the preceding guesses about the mechanism of death when death is caused by a belief in sorcery (guesses at what might be the case if in fact such things do happen), there is another more general theme worth comment. Mauss took for his topic one which showed the striking effect that commonly held beliefs may have on individuals. The more general theme on which I wish to comment is that of the evidence for belief. People in a given society may generally assert that certain statements are true, yet the observer has doubts about how firmly they are convinced of what they say. He notes the set of stated ideas, yet finds that the behaviour which he would expect to correspond or follow from them does not show a matching depth of feeling or affect: it does not seem reasonable to him that people should be apparently so unconcerned about a threat which they say is so dreadful. The observer looks to behaviour for signs of actions that can speak to him of belief more loudly and certainly than words. Part of the fascination in the idea of a death by belief in sorcery is that it would seem to show such final and tragic evidence of true belief in what is said.

If we seek to account for the discrepancy to which I have drawn attention, there are a number of points to stress in the observations that I made among the Gnau. Firstly, it must be noted that a direct form of questioning common in the conduct of an anthropological field enquiry, may tend to isolate and magnify the set of relevant ideas. They may gain a certain dogmatic precision, fixity, and coherence by being required in answer to direct questions *in abstracto*, and they do not convey the range of qualifications and conditions on which people in fact draw when they seek to apply the outlined ideas to situations of real life. The differences, for example, between the main features of *langgasutap* as I outlined it in the beginning of this paper, using their stated views, and the features of the cases that I describe in the section on retrospective views of sorcery, are striking. The effort alone of isolating the set of beliefs in order to discuss them, may make it hard to see them in perspective - witness the effect of reading a paper such as this, which isolates its topic and then after a long account reports how small a proportion of all illness was ever attributed to sorcery. That effect closely matches the way in which the alerting bias towards what is an anthropological topic *par excellence* draws the field inquirer on, so that he gives more close attention to the subject and weights the balance by his questions and pursuit of detail.

Secondly, ideas, such as the ones of sorcery I deal with here, are not necessarily fixed to only one kind of situation or one field of relevance. In this instance, it is simplistic to take the view that ideas of sorcery are organised or have their origin entirely as a means to explain how some kinds of illness and death come about, which is what may be their most striking and surface import. For, as other anthropologists have seen, such ideas may also serve to discriminate among different types of complex social relationship. By transposing less definite and articulate views and feelings about relationships into symbolic terms, Gnau people have encapsulated a sense of the differences among types of relationship into specifically formed notions such as one about the practice of sorcery: *Langgasutap* sorcery is found in one set of villages which are in complex ways felt to be different from, and inimical to, another set of neighbouring villages with which the Gnau feel culturally most identified. Because of this, the ideas are maintained as generally accepted truths which affect people's behaviour in various ways (for instance, choices of what path to take, where to hunt, what woman to marry), though the ideas are not on the face of it directly related to these kinds of choice, explanation, or control. There may be values or reasons which indirectly support the general currency of such beliefs and may give them shape. The beliefs may also affect what people do in various situations, and they are maintained and their effects persist in a way that is relatively independent of any confirmation or direct application of the surface-import of the views. The observer then may see discrepancy between behaviour and what strikes him first about the ideas, which is their surface-import.

The weight and significance of the views that the Gnau generally affirm as common knowledge about destructive magic and sorcery, when they speak about them *in abstracto* or prospectively, is different from the weight they have when called up to explain a particular person's state of severe illness or his recent death. Such an event is revelatory, and the ideas reappear cast in its harsh light and provoke strongly emotional responses. The apparent significance of the beliefs varies with circumstance and the timing of observations and questions. I have tried to illustrate some aspects of these differences by organising my observations into situations of future, present and past fears, and by distinguishing public from individual attitudes.

We would expect to find some evidence of worry or fear when people say they think that it is possible for some to harm others by secret, invisible or mystical means. Such ideas are widespread in the world. The actual fears provoked in a given society will reflect the precise detail of assumptions about the circumstances in which sorcery or destructive magic may be used. The details differ. In the Gnau instance I have described, some points stand out as important ones to bear in mind.

A. Destructive magic and sorcery are used by human agents who aim to cause bodily harm to other people, and they aim to cause death and not some lesser harm. Because death is the aim, they are not light matters but grave ones.

B. There is a difference between destructive magic and sorcery in that it is generally said that destructive magic may be countermanded, but that nothing can be done to save the victim of *langgasutap* once he has been struck. As there may be a chance of stopping destructive magic, so there is a reason to try to recognise it and do something to avert a death. I think the Gnau rarely diagnose *langgasutap* unless the victim is dead. An individual suddenly scared may fear it possible; but only after death has given it conviction is a firm diagnosis generally agreed.

C. Destructive magic is something that people supposedly would use only with reasonable justification, or for reasons that others can understand in terms of motives such as hate or revenge. But *langgasutap* is something darker, less understandable, and less predictable. It is unjustifiable and unambiguously evil. The difference between the two in this respect colours people's attitudes to them, for the difference involves an estimation of their own ability to control, foresee, escape or avoid them. The difference becomes an issue when people judge the possibilities of a particular situation.

D. The wave of fears about *minmin* occurred recently and represented the elaboration of a less important fear which, in the present time of rapid social changes, of increased contact with outsiders, and catalysed by the coincidence of an epidemic of influenza, arose and spread to take the place of traditional fears and grew to a preoccupying pitch of general fear. But then as the new ideas were shared, the fears, at first uncertain and diffuse, were modified and given more precise form. Those features in which *minmin* had originally resembled *langgasutap* were redefined and changed so that *minmin* acquired a more distinctive, known and specific character. These changes involved a shift towards the quality of destructive magic, which made it potentially controllable and allowed treatment. The fear of *minmin* was resolved in part by local action and in part by the intervention of the patrol officers.

ACKNOWLEDGEMENT

I carried out fieldwork based on Rauit village in 1968-69, and I am grateful to the Social Science Research Council for a generous grant.

REFERENCES

Cannon, W.B. (1942). "Voodoo" Death, *American Anthropologist*, *44*, 169-181.

Cawte, J.E. (1974). *Medicine is the Law*: Studies in psychiatric anthropology of Australian aboriginal societies. Honolulu: The University Press of Hawaii.

Chodoff, P.C. (1966). Effects of extreme coercive and oppressive forces: brainwashing and concentration camps. *American Handbook of Psychiatry* III, 384-405.

Darwin, C. (1972). *The Expression of Emotions in Man and Animals*. London: J. Murray.

Douglas, M. ed. (1972). *Witchcraft, Confessions and Accusations*. London: Tavistock.

Ellenberger, H.K. (1951). Der Tod aus psychischen Ursachen bei Naturvölkern. *Psyche* (Stuttgart), *5*, 333-44.

Ellenberger, H.K. (1965). Ethno-psychiatrie; partie descriptive et clinique. *Encyclopédie Medico-chirugicale: Psychiatrie* (ed. H.Ey) tome III: 37725 B10 1-22.

Engel, G.L. (1970). Sudden death and the 'medical model' in psychiatry. *Canadian Psychiatric Association Journal 15*, 527-38.

Evans-Pritchard, E.E. (1937). *Witchcraft, Oracles and Magic among the Azande.* Oxford.

Granville-Grossman, K. (1971). *Recent Advances in Clinical Psychiatry.* London: Churchill.

Griffiths, W.J. (1960). Responses of wild and domestic rats to forced swimming. *Psychol. Reports 6:* 39-49.

Janet, P. (1903). *Les obsessions et la psychasthénie.* Paris: Alcan.

Jaspers, K. (1963). *General Psychopathology,* trans. by J. Hoenig and M.W. Hamilton, Manchester (1st ed. in German 1913).

Hinton, J. (1967). *Dying.* Penguin Books: Harmondsworth.

Lévi-Strauss, C. (1958). *Anthropologie Structurale.* Paris: Plon.

Lewis, A.J. (1967). Problems presented by the ambiguous word 'Anxiety' as used in psychopathology. *Israel Annals of Psychiatry and Related Disciplines 5,* 105-121.

Lewis, G.A. (1975). *Knowledge of Illness in a Sepik Society.* London: The Athlone Press.

Lewis, G.A. (1976). A view of sickness in New Guinea. In J.B. London (ed.), *Social Anthropology and Medicine,* ASA Monographs No. 13.

Lex, B.W. (1974). Voodoo death: new thoughts on an old explanation. *Amer.Anthrop., 76,* 818-823.

Marwick, M.G. (1965). *Sorcery in its social setting,* Manchester.

Mauss, M. (1926). Effet physique chez l'individu de l'idée de mort suggérée par la collectivité (Australie, Nouvelle-Zélande). *Journal de psychologie normale et pathologique, 23,* 653-669 (also reprinted in Mauss, M. (1950). *Sociologie et Anthropologie,* Paris, P.U.F.).

Richter, C. (1957). On the phenomenon of sudden death in animals and man. *Psychosom.Med., 19,* 191-8.

Rubel, A.J. (1964). The epidemiology of a folk illness: *susto* in Hispanic America. *Ethnology 3,* 268-83.

Rosas, S.F. (1957). The myth of "Jani" or "Susto" of the indigenous medicine of Peru, Report to *Asociacion Psiquiatrica Peruana,* 26 September 1957.

Schmale, A.H. (1972). Giving up as a common pathway to changes in health. *Advances in Psychosomatic Medicine, 8,* (ed. Z.J. Lipowski) 20-41.

Simmons, L.W. and Wolff, H.G. (1954). *Social Science in Medicine,* New York: Russell Sage Foundation.

Tenzel, J.H. (1970). Shamanism and concepts of disease in a Mayan Indian community. *Psychiatry 33,* 372-380.

Turner, V.W. (1957). *Schism and Continuity in an African Society.*

Manchester.
Warner, W.L. (1937). *A Black Civilization: a Social Study of an Australian Tribe*. New York: Harper Bros.
Williams, F.E. (1928). *Orokaiva Magic*. London: O.U.P.
Wolf, S. (1971). Psychosocial forces in myocardial infarction and sudden death. In L.Levi (ed.) *Society, Stress and Disease*. London: O.U.P.
Wolff, H.G. (1960). Stresses as a cause of disease in man. In J.M. Tanner (ed.), *Stress and Psychiatric Disorder*. Oxford: Blackwell.
Wolff, H.G. (1964). World Health Organization Technical Report Series No. 275: *Psychosomatic Disorders*.

BODILY MADNESS AND THE SPREAD OF THE BLUSH

Vieda Skultans

This paper explores some themes and episodes in the emergence and development of the idea of insanity during the early nineteenth century. It considers the social and intellectual climate which was necessary for the idea of insanity to gain a public foothold. It also describes the peculiar bodily preoccupations which this new idea of insanity generated.

Several writers have sought to explain the growth of asylums and the elaboration of ideas about insanity which took place in the latter part of the eighteenth century. Foucault for example, sees the goals of the asylum as primarily intellectual, and insanity as a residual category which is indispensable in an age glorifying reason. Rothman, in *The Discovery of the Asylum*, offers a more socially grounded view of the asylum:

> The well-ordered asylum was to fulfil a dual purpose for its innovators. It would rehabilitate inmates, and then by virtue of its success, set an example of right action for the larger society.... The well-ordered asylum would exemplify the proper principles of social organization and thus ensure the safety of the republic and promote its glory (Rothman 1971:4).

This view of the asylum as a model for society depends logically on an enlarged concept of insanity itself. Rothman does, in fact, describe such a process:

> Insanity was no longer the exclusive province of the raving lunatic or totally incompetent. Medical superintendents opened up the category and alerted the community to a whole new range of possibly deviant behaviour (*ibid.*: 123).

Such deviance was thought of as the product of society's flaws and the deviant as its victim:

> Just as the first penologists located the origins of crime within the community, so psychiatrists linked mental illness to social organization. The epidemic of insanity, like the prevalence of crime, pointed to the most fundamental defects of the system, from mistaken economic, political and intellectual practices to grave errors in school and family training (*ibid.*: 125).

This approach to insanity gives little emphasis to heredity and predisposition and much to its exciting causes. For example, Samuel Woodward, an American alienist wrote: '... the strongest predisposition need not result in insanity.'

Rothman very ably explores the relationship between this new, expanded concept of insanity and asylum organization in mid-nineteenth century America. I want to look more closely at the expansion of the concept and its consequences for symptomatology. Eighteenth century men and women suffered from an abundance of complaints such as hypochondriasis, melancholy, the spleen, and the vapours. Whilst not recognised as varieties of insanity, they quite clearly had psychological as well as physical components. Implicit in such conceptions of disease is a view of man as an undifferentiated whole. Such distinctions as are made are between the contained and containing parts of the body, not between mind and matter. Complaints such as those listed above are used more or less interchangeably. Hypochondriasis, for example, was a condition that involved both digestive disorders and discomforts, physical fatigue as well as a more generalised weariness and personality handicaps such as indecisiveness. The condition is movingly described by the anonymous author of *Confessions of a Hypochondriac or The Adventures of a Hypochondriac in Search of Health*. He writes: 'What is a hypochondriac, if not one who has no faith in himself, nor, indeed, in anyone else for two hours together? It is a state of being without a sense of security; a perpetual suspense, an expectation, a dread of an inevitable evil' (Anonymous 1849:109). A rhetorical question expresses the case of the problem even more succinctly: 'Was I for ever to feel that sense of a vacuum, sometimes in the relenting heart and sometimes in the sensitive stomach?' (*ibid.*: 179). Although published in the nineteenth century, these confessions are written in what is essentially an eighteenth century idiom.

These complaints gave the subject matter to many learned treatises. George Cheyne, Richard Blackmore, and Robert Whytt, were among the best known of English Physicians who write on nervous distempers. The syndrome acquired the title of *The English Malady*, and George Cheyne published a book by that name in 1734. This was the first, although not the last, time that a particular disease was claimed by the English as being peculiarly their own. Richard Blackmore begins *A Treatise of the Spleen and Vapours* with an affirmation of this:

> If the natives of this island either from the peculiar constitution of the air they breathe, or the immoderate quantity of flesh-meats they eat, or of the malt liquors they drink or any other secret causes, are more disposed to coughs, catarrhs, and consumptions, than the neighbouring nations; they are no less obnoxious to hypochondriacal and hysterical affections, vulgarly called the spleen and vapours, in a superior and distinguishing degree. And of all the chronical distempers that afflict the body, or disturb the mind, these two, consumptions and the spleen are the most rife and prevalent; and either directly or by their own power, or by introducing other diseases, make the greatest havoc and destruction among the people (Blackmore 1725: iii-iv).

George Cheyne, who counted himself among the afflicted, describes himself as suffering from '... head-ach, giddiness, watchings, lowness and melancholy' (Cheyne 1734:329). Unlike later writers, Cheyne sees these spiritual afflictions as inextricably bound up with physical distempers.

> I never saw any person labour under severe obstinate, and strong nervous complaints, but I always found at last, the stomach guts, livor, spleen, mesentery, or some of the great and necessary organs or glands of the belly were obstructed, knotted, schirrous, or spoiled, and perhaps all these together (*ibid.*:184).

This view of nervous distempers is in the earlier tradition of humoural pathology exemplified by, for example, Robert Burton. In his hands, melancholy becomes a catch-all of human ills and ailments, and provides the *Urquelle* of later disease categories. Burton describes the condition in moving words:

> As it is with a man imprisoned for debt, if once in the gaol, every creditor will bring his action against him, and there likely hold him - if any discontent seize upon a patient, in an instant all other perturbations... will seize upon him, and there likely hold him; and then like a lame dog or broken winged goose, he droops and pines away and is brought at last to that ill habit a malady of melancholy itself (Burton 1621:18).

This contagious, all-embracing quality of melancholy is

reaffirmed by the huge variety of causes and kinds of melancholy. Burton begins by considering the 'received' division of melancholy into three kinds: Head melancholy, melancholy of the whole body, and hypochondriacal or windy melancholy, each of which proceeds from distemper in the named part of the body. Having located melancholy in the body, Burton then goes on to give an inventory of public and private miseries, including poverty, ambition, avarice, love... in fact, a catalogue of human vices, failings and disappointments. Thus, although Burton pays lip-service to the distinctions between the infirmities of the spirit and those of the body, in practice he shows it an easy disregard.

In this approach Burton belongs to the Mediaeval tradition expounded by Aquinas. It is perhaps best understood by contrasting it with the Cartesian view of the emotions. For Descartes, in *The Passions of the Soul* (1649), emotion is a purely private mental event that is only contingently connected with its bodily manifestations and its object. For the Mediaevals, by contrast, bodily manifestation of emotion is neither the cause nor the effect of emotion: it is its necessary condition or the *materia*. Clearly, Burton is mediaeval rather than Cartesian in temper, if not in time. Burton is equally at home in writing about the liver as 'the shop of humours' as he is in writing about the peculiar vulnerabilities of women, despite his charming avowals of modesty. He writes:

> For seldom shall you see a hired servant a poor handmaid, though antient, that is kept hand to her work, and bodily labour, a coarse countrey wench, troubled in this kinde, but noble virgins, nice gentlewomen, such as are solitary and idle, live at ease, lead a life out of action and employment, that fare well in great houses, and jovial companies, ill disposed peradventure of themselves, and not willing to make any resistance, discontented otherwise, of weak judgements, able bodies, and subject to passions, such for the most part are misaffected and prone to this disease (Burton 1621:303).

But, then, suddenly recovering his sense of propriety Burton exclaims: 'But where am I? Into what subject have I rushed? What have I to do with nuns, maids, virgins, widows? I am a bachelor myself and lead a monastick life in a college. ... though my subject necessarily require it, I will say no more' (*ibid.*:304). Propriety forbids him to go on. But the next paragraph begins: 'And yet I must and will say something more, add a word or two ... ' (*ibid.*:304), and so on.

So much for *The Anatomey of Melancholy*. The writing of the eighteenth century remains within the same general theoretical framework. Cheyne, for example, introduces the all purpose term 'cachexia' to refer indiscriminately to either bad states of the body or the mind, or both. Again, as with Burton, the important distinction is between the contained and the containing parts of the body, or, in Cheyne's terms, between solids and fluids:

> The human body is a machine of an infinite number and variety of different channels and pipes filled with different liquors and fluids, perpetually running, gliding, or creeping forward, or returning backward in a constant circle and sending out little branches and outlets, to moisten, nourish, and repair the experiences of living (Cheyne 1734:4).

This anatomical picture suggests to Cheyne certain causes of insanity. These are (1) grossness of the fluids, (2) a corrosive quality of the fluids, (3) a too great laxity of the fibres. However, this location of nervous distempers does not put them beyond the individual's control. Cheyne is cautious about the common division of nervous disorders into original and acquired. He writes: '... a poor creature, born subject to nervous distempers, has no more reason to complain, than a child whose father has spent his worldly fortune and left him poor and destitute' (Cheyne 1734:19). Thus health is within the reach of all provided a correct regimen and, in particular, diet are followed. Horse-riding and a daily pint of claret were commonly prescribed. On the other hand, the hypochondriac apologist 'was advised to take champagne moderately, a nutritious diet - exercise and amusement' (*ibid.*:74).

Throughout this period the term 'complexion' refers alternatively to skin colour, disposition and character. The vocabulary of humoural pathology also follows this pattern. Words like sanguine and melancholic are used to describe complexion and appearance as well as styles of action and attitude. Such linguistic usage encourages an easy transition from emotional disposition to accompanying bodily characteristics. Throughout the seventeenth and eighteenth centuries, this is the background against which a continuing and familiar discussion of nervous distempers is set. For example, the essayists of the *Spectator* air common concerns and widespread fears. Boswell wrote about his hypochondria with an ease and readiness which has only recently reappeared in the women's pages of the national press. He prefaces his discussion with the phrase 'We hypochondriacks... ' David Hume, in a letter written to Dr Cheyne at the age of nineteen, admits to suffering from the vapours. It was subsequently diagnosed as 'a disease of the learned'. Certainly, no stigma attaches to these disabilities. Cheyne, in his essay on long life writes: 'fine folks use their physicians as they do their laundresses, send their linen to them to be cleaned, in order only to be dirtied again' (Cheyne 1724:182).

To summarize what I have said so far, terms like melancholy, hypochondria, spleen, vapours, are used interchangeably in the medical literature of the seventeenth and eighteenth centuries. General literature also contains many references to these afflictions. They are diseases with spiritual and material components to them. This double aspect makes them both more distressing and more manageable. Cheyne has a memorable opening to his book: 'The Spirit of man can bear his infirmities, but a wounded spirit who can bear?' This view of nervous distempers is, I suggest, in the tradition of mediaeval philosophy. It also makes it easier to acknowledge nervous disabilities. Hume, for example, in his letter to Dr Cheyne goes on to say:'Tho I was sorry to find myself engag'd with so tedious a distemper yet the knowledge of it, set me very much at ease, by satisfying me that my former coldness proceeded not from any defect of temper or genius, but from a disease, to which any one may be subject' (Hume 1734:14). However, the nervous distempers which were given so much attention in the eighteenth century were clearly distinguished from insanity. Hume was well aware of the difference: '... and there seems to be as great a difference between my distemper and common vapours as there is between vapours and madness' (*ibid.*:70).

Clearly, this reading of the literature goes against the interpretation put forward by Michel Foucault in *Madness and Civilization*. In a simplified form, Foucault's interpretation goes as follows: until the seventeenth century madness was a recurrent ingredient of all experience. 'In the Renaissance, madness was present everywhere and mingled with every experience by its images and dangers' (Foucault 1964:74). In the age of reason, however, the tone changes: 'All those forms of evil that border on unreason must be thrust into secrecy. Classicism felt a shame in the presence of the inhuman that the Renaissance had never experienced' (*ibid.*:68). And this shame brings about what Foucault calls 'the great confinement.' However, any review of the English literature of this period suggests, on the contrary, a personal and intimate treatment of that which is of universal concern.

In the early part of the nineteenth century the tone of writing changes. Moral management makes its first appearance, and insanity is seen essentially as a disease of the will to which anyone can potentially succumb. Until this time, conditions such as the vapours and melancholy were attributed to an imbalance of bodily humours. This imbalance could be righted by following various prescribed regimens. With the changing conceptions of insanity and its change of locus from the humours to the will, possibilities for its rapid increase were created. The nineteenth century view of man is dualistic and holds that the most important activities in a person's life are private and invisible. The workings of thought, emotion, and the will, take place in a private arena. Within this theoretical context it becomes more than ever important to identify the outward signs of inner states.

Once the relationship between a faulty will and insanity has been established, the numbers of those potentially susceptible to insanity are vastly increased. Many practical situations can be found which illustrate an inadequate or insufficient exercise of the will. In particular, civilization and an accompanying luxurious life-style present the will with temptations aptly described as 'the curse of polished life' (Burrows 1828:21). Thus Maddock, writing about the spread of insanity, says: '... perhaps the rural and agricultural districts may claim some degree of exemption, over the more densely-crowded population of manufacturing cities' (Maddock 1854:13). But then he goes on to write, in effect, that since England is the most educated and civilized of countries it suffers from the highest incidence of insanity: 'In no other country, compared with England, do we find such numerous and formidable examples of this extensive scourge. In some measure this may arise from habitual pride and hauteur of the English character; partly from the commercial greatness which it has long been her boast to enjoy; and partly from the unnatural style of living too generally adopted' (*ibid.*:13). Whatever the incidence of insanity, it is clear that at the level of belief, early nineteenth century society is seen as presenting people with very special problems. These relate in particular to the demands made of individual endeavour, or in the idiom of nineteenth century physicians, to the trials of the will and its possible collapse. It is interesting to note that the recognition of insanity coincides with the discovery of childhood. For example, Coveney writes: 'Until the last decades of the eighteenth century the child did not exist as an important and continuous theme in English literature. Childhood as a major theme came with the generation of Blake and Wordsworth' (Coveney 1957:29). The novelty of childhood as a separate category is echoed by Ariés: 'The idea of childhood is not to be confused with affection for children: it corresponds to an awareness of the particular nature of childhood, that particular nature which distinguishes the child from the adult, even the young adult' (Ariés 1960:125). These two newly discovered categories of insanity and childhood emphasize individual malleability and growth and it is, therefore, not surprising that both categories emerged simultaneously.

Thus in the early 1800's the theme of Moral management first appears. Moral managers urge the abandonment of physical restraint. Instead they advocate psychological techniques. Appeal to the conscience and will of the patient is suggested. In this way it is thought the power of self-control can be nurtured and the art of self-government furthered. In fact, the practice of moral management relies on a changed view of man and his emotions. The Cartesian disjunction between body and soul asserts itself. These theories emphasize the privacy and power of thought, feeling and emotion. At the same time they seek the external signs of these inner events. Within this context the study of facial expression assumes great importance. Many studies on physiognomy appear. Charles Bell (1806), Alexander Morison (1824), Thomas Burgess (1828) and Charles Darwin (1872) are all concerned with the physical expression of inner states. Even general texts on medicine contain sections on expression. Charles Bell writes: 'Expression is to passion what language is to thought.' The key to this grammar must therefore be sought. Paradoxically this interest in expression leads to the acceptance of a mechanistic view of facial expression and to stereotypes of physiognomy. Following the Cartesian approach, different blobs of emotion are thought to impress the face in different ways and thus result in different expressions. Johann Caspar Lavater's *Essays on Physiognomy* (1778) became a best seller of the period. It was a standard possession of every self-respecting family, who would not dream of hiring a cook or governess, let alone consider a prospective son-in-law, without consulting Lavater. This vogue for reading faces was beautifully satirized a few years later by George Lichtenberg (1783). The article, published in a German medical journal supposedly sets out to be a study in physiognomic perception. It aims to increase students' awareness of physiognomy by an examination of pigtails. Firstly, it describes them, then it sets questions for further exercise.

Within this context it is not surprising that stereotypes of the physiognomy of insanity emerge. One of the most interesting aspects of the study of expression is the interest in blushing. In Thomas Burgess's *The Physiology or Mechanism of Blushing* (1828), the blush is seen as the distinguishing mark of the human and the supreme sign of man's spirituality. Burgess argues that unlike the other expressions of emotion, blushing cannot be caused by any physical means. The mind must be affected. In fact, Burgess believes that blushing was created by God 'in order that the soul might have sovereign power of displaying in the cheeks the various internal emotions of the moral feelings; so as to serve as a sign to others, that we were violating rules which ought to be held sacred' (Darwin 1872:337-8). Others, like Charles Bell, saw the blush as the crowning touch to beauty. 'It adds perfection to the features of beauty'

(Bell 1806:95). Darwin saw the blush essentially as the product of self-awareness, and wrote: 'It is not the simple act of reflecting on our own appearance, but the thinking what others think of us which excites a blush' (Darwin 1872:327). He cited Coleridge, who said that it is sufficient to stare hard at some persons to make them blush.

However, the theory that is most relevant to this paper is that of Burgess. He distinguishes between the true and the false blush. The false blush is due to a morbid sensibility or oversensitivity, whereas the true blush emanates from the conscience. He writes:

> After the impression is made on the sensorium which is to excite this phenomenon, we become immediately conscious of what is about to take place, we feel that the will is overpowered - and, for the time being is rendered subordinate to the mental powers, and the emotions of sympathy. Now, from a feeling of our own helplessness, like a bad swimmer when out of his depth, we become flurried, and in our eager attempts to avert the threatened result, by endeavouring to expel from the mind or imagination that association of ideas which is about to bring it forth, we only fix it the more firmly, and ensure its full development, to the deep mortification and prostration of our will (Burgess 1828:134).

This connexion between blushing, conscience, and the prostration of will, is particularly interesting in relation to contemporary theories of insanity, which attached great importance to moral force and the power of the will to fight insanity. We would, therefore, expect the blush to figure also, and indeed the peculiarity of the idiot was said to be his inability to blush. The long-standing debate about whether negroes blush is to be seen against this background. The controversy is not so much a physiological one, as one about moral development. The inability to blush was said to provide the outward feature of inner failing; but excessive blushing, on the other hand, provided evidence of inner moral derangement and very often betrayed the chronic masturbator.

This emphasized another, related area of physical activity that acquired importance during this period: masturbation. Again, this interest relates very much to the view of man current at the time.

Nineteenth century medicine gained two diagnostic categories. They are spermatorrhoea and masturbational insanity. Both convey anxiety about loss of semen. The history of interest in this area is worth considering and requires some explanation. Why did masturbation come to be thought of as an activity of general importance, let alone of medical importance? And why did this happen in the mid-nineteenth century?

Literature contains few references to masturbation. John Aubrey is unusual in describing the young Duke of Buckingham's solitary pursuits (1950). Otherwise one has to wait until the twentieth century before it becomes an acceptable literary theme 'Scientific' interest in the subject can be given a precise date of origin. In 1710, an anonymous clergyman published a treatise on masturbation entitled *Onania, or the Heinous Sin of Self-Pollution.* Very many editions appeared in the course of the century. *Onania* is written with elegance and style. This terseness of style seems typical of eighteenth century writing and was los by later nineteenth century physicians. The flavour of the book may be caught by quoting the opening sentence:

> Self-pollution is that unnatural practice by which persons of either sex may defile their own bodies without the assistance of others, whilst yielding to filthy imaginations they endeavour to imitate and procure to themselves, that sensation which God has ordained to attend the carnal commerce of the two sexes, for the continuance of our species (Anon 1710:1).

The first edition is just under 100 pages long. Later editions are slightly longer and include a correspondence between the author and a critic of the book. The debate concerns the relative sinfulness of fornication and masturbation. The critic argues that self-pollution is the lesser of two evils since it avoids the debauchment of innocent women.

Onania was in many ways ahead of its time, in that it appears neither to have reflected current concerns and anxieties nor to have influenced them. An easier, more relaxed attitude to sexual activity was the prevailing norm. In England, such ideas acquired their greatest popularity in the mid-nineteenth century. During the 1840s and 1850s there is a spate of books dealing with loss of semen. The work of Ellis (1838), Dawson (1840), Lallemand in English translation (1847), Milton (1854), Curling (1856) and Acton (1865), all dealt with this subject and appeared within a few years of each other.

Masturbation, or 'the primal addiction' as Freud was later to call it, is one focus of anxiety. Another more general and basic area of concern is seminal loss, which may, but need not, be caused by masturbation. A generally accepted, if not clear, definition of spermatorrhoea is given by Curling in his book *A Practical Treatise on the Diseases of the Testis.* He writes: 'The emissions may, however, be more frequent than is consistent with health, and too readily excited, so much so, indeed, as to affect virility, and to give rise to constitutional symptoms of a serious character. These excessive spermatic discharges constitute the complaint termed spermatorrhoea' (Curling 1856:386). This loss may be voluntary, as in intercourse and masturbation, or involuntary, as in 'nocturnal pollutions' or other emissions, for example, whilst at stool. Typically the picture presented is that of a dribbling penis. Throughout the forties and fifties, physicians showed increasing concern that insufficient attention was paid to the dangers of seminal loss. For example, Milton writes:

> It has always appeared strange to me that this affection should remain abandoned by the profession to a few solitary specialists, and for the benefit of the vile harpies who prey on this class of victims. Surgery which has wrested so much from empiricism and ignorance, seems disposed to yield up this, as if it were, debateable land, to chance, philosophy, utter neglect or quackery (Milton 1854:243).

Lallemand, the French specialist on spermatorrhoea describes it as: 'A disease that degrades man, poisons the happiness of his best days, and ravages society' (Lallemand 1847:ii).

In part, these fears reflect uneasiness about sexuality as such, but a general coyness about sexuality does not explain the particular interest in seminal loss. Seminal loss in intercourse is set apart as being rather less harmful (cf. Spitzka 1887), but it is precisely the loss of semen involved in sexual intercourse, rather than the activity itself, which is considered pernicious. For this reason male masturbation is singled out for consideration. Spitzka writes: 'The effects of such indulgence are less serious in the female owing to the less exhausting nature of the discharges.'

In fact, beliefs about seminal loss constitute a distinct syndrome. Most writers on the subjects are agreed on the causes of spermatorrhoea. These include constipation, worms, piles, gonorrhoea, heat, heavy bedclothes, highly seasoned food, alcohol, intense application of the mind, and excessive indulgence in sexual intercourse, usually of a promiscuous kind. The effects of spermatorrhoea are similar to those of masturbation. In general these are debilitating. In particular, there is an intimate connection between seminal loss and the condition of the brain. There is an inability to sustain mental and bodily fatigue; a heaviness in the head; giddiness; sleeplessness. Interestingly, the appetite increases, often becoming voracious. Dawson writes: 'I have generally found unnatural seminal discharge accompanied with increased appetite, owing to the necessity which the system feels, of compensating the daily losses which it sustains' (Dawson 1840:6). Long term effects, like those of masturbation, are more serious. Parched skin, loss of hair, stammering, deafness, blindness, form part of the familiar list.

Masturbation forms a sub-category of the interest in loss of semen. The picture of the typical masturbator is one of extreme selfishness. Spitzka described a patient representing this category as follows:

> His demeanour was obtrusive, mean and selfish. He sat out all my other patients on the morning he called, withdrew to the waiting-room, under indignant protests, when I represented to him that I could not keep a physician accompanying patients, who had come a great distance, waiting any longer, he having already consumed two hours. He came in repeatedly and finally, after I had finished, he took possession of the field, and as I hurried off to my much delayed lunch, he exclaimed, 'Hurry up, doctor, do not be long; I have a great deal to tell you yet. My case is of more importance than any other you ever had; I am the most important man in my family' (Spitzka 1887:239).

Maudsley describes the consequences of masturbation begun early in life as follows: '... we have degenerate beings produced who as regards moral character are very much what eunuchs are represented to be - cunning, deceitful, liars, selfish, in fact, morally insane; while their physical and intellectual vigour is further damaged by the exhausting vice' (Maudsley 1868:156).

Another characteristic of the chronic masturbator is his shyness, timidity, and, in particular, his inability to look others in the eye. David Skae, the Scottish physician, who first identified a particular brand of insanity which he termed masturbatory, also singles out similar features of the masturbator:

> ... that vice produces a group of symptoms which are quite characteristic, and easily recognized, and give to the case a special natural history. The peculiar imbecility and shy habits of the very youthful victim, the suspicion and fear, and dread, and suicidal impulses, the palpitations and scar look, and feeble body of the older offenders, passing gradually into dementia or fatuity, with other characteristic features familiar to all of you, which I do not stop to enlarg on, all combine to stamp this as a natural order or family (Skae 1863:315).

The sad psychological portrait of the masturbator is reinforced by the physical appearance of sallow complexion and dark ringed eyes. However, masturbation is not only one among the ingredients of a character type: it is also the cause of further disabilities. General weakness is a most common result, followed by headache, backache, acne, indigestion, blindness, deafness, epilepsy, and finally death. In fact, the syndrome produced is very similar to that of spermatorrhoea. Finally, an inability to participate in social life develops.

Although interest in masturbation persists into the latter part of the nineteenth century, there is a reinterpretation of its nature. Maudsley (1868) regards the activity with the abhorrence due to a form of moral degeneracy. It is worth quoting one of his cases at length.

> As an example of the high-pitched and absurd sentiments professed sometimes by these degraded beings, I may mention the case of a gentleman who had a plan for curing the social evil. He set forth with great feeling and energy the miserable and wicked thing which it was that many of the most beautiful women should be degraded to gratify the worst lusts of men; and professed himself to be grievously distressed by the sin and evil which were caused thereby. How were so much vice and misery to be done away with? His plan, which he practised himself and proposed that others should follow was to masturbate every morning into a tumbler of water and to drink it. He argued that the lust was thus satisfied without injury to any other person, while the man himself was strengthened by the nourishment afforded to his brain. Here, then, as in other cases, was a mind enervated by vicious practices, dwelling continually on sexual subjects, and concocting, not designedly, but with unconscious hypocrisy, an excuse for the vice which wrecked his life. It is a curious thing that to such a state of moral degradation have patients of this class come, that they will actually defend their vice on some pretence or other (Maudsley 1868:160).

When Maudsley comes to reconsider the subject in *Pathology of Mind* (1879), he sees masturbation as a product of bad inheritance over which the individual has little control. In line with this reappraisal is the increasing insistence on preventive surgical measures in contrast to the earlier moral exhortations.

At first sight, the mid-nineteenth century interest in spermatorrhoea and masturbation seems incompatible with the concurrent commitment to moral management. The 1840s and 1850s are also the period during which the major works of the moral managers appear. The moral managers urged the abandonment of physical restraint which had been central to eighteenth century therapeutic practice.

Instead they advocated psychological techniques. Appeal to the conscience and will of the patient was suggested. In this way, it was thought, the power of self-control could be nurtured and the art of self-government furthered. However, both areas of interest presuppose self-control and thus complement each other. Both must be seen as an expression of concern with continence in its widest sense. The disapproval of venery is but another aspect of the pursuit of moderation and discipline. Masturbation, one of the most solitary of activities, is regarded as the arch-vice precisely because of the hopes vested in the private endeavours of the individual. In typifying loss of control, it is seen as the moral failure *par excellence*.

These ideas cannot be explained, as Szasz tries to do, as an attempt by the medical profession to control human sexuality (1971). Neither can 'the masturbatory hypothesis' be seen as a fallacy of reasoning, as Hare would like it, which is dispelled by the increasing clarity and coherence of medical thought (1962 Nor can it be explained as the product of the emotional immaturi and prurience of the physicians as Alex Comfort claims. One of the most illuminating works against which to set the masturbator hypothesis is that of J.A. Banks, *Prosperity and Parenthood* (1954). As the title suggests, Banks examines the relationship between the postponement of marriage and expectations as to what constitutes an appropriate middle-class life style. According to Banks such ideas appear from the 1830s onwards. 'A "proper" time to marry! - more and more as the century wore on this became the theme of the middle classes, until the words "prudence" and "postponement" became the two most hackneyed in their vocabulary' (*ibid.*:36). The situation is even more graphically described in an editorial taken from *The Times:*

> The laws which society imposes in the present day in respect of marriage upon young men belonging to the middle class are in the highest degree, unnatural, and are the real cause of our social corruptions. The father of a family has in many instances, risen from a comparatively humble origin to a position of easy competence. His wife has her carriage; he associates with men of wealth greater than his own. His sons reach the age when in the natural course of things they ough to marry and establish a home for themselves. It would seem no great hardship that a young couple should begin on the sar level as their parents began, and be content for the first few years with the bare necessaries of life, and there are thousands who, were it not for society, would gladly marry on such terms. But here the tyrant world interposes; the son must not marry until he can maintain on much the same footing as his father's (*The Times*, 7th May, 1857).

Late marriages may, in part, be responsible for the growth of anxiety about masturbation. However, beliefs about masturbation and the masturbator must also be set against the wider context of values and beliefs about the nature of man. Although, or perhaps because, beliefs about will-power deal with such an elusive and intangible area of human nature, they give rise to certain crude bodily stereotypes. The blush provides the outward evidence of a finely tuned moral sensibility. Similarly, the chronic masturbator embodies the antithesis of all the valued characteristics of the period. He is the polar opposite of nature's gentleman: the person who can get by on the strength of inner resourcefulness and outer accomplishments. He provides the prototype of uncontrolled and undisciplined behaviour. The full meaning of these ideas only emerges from a more general view of Victorian ideas about man and his place in society.

REFERENCES

Acton, William (1865). *The Functions and Disorders of the Reproductive Organs.* London: John Churchill and Son. Fourth edn.

Anonymous (1849). *Confessions of a Hypochondriac or the Adventures of a Hypochondriac in Search of Health.* London: Saunders and Otley.

Anonymous (1710). *Onania or the Heinous Sin of Self-Pollution and all its Frightful Consequences in both Sexes.* London.

Ariés, Philippe (1960). *Centuries of Childhood.* Harmondsworth: Penguin.

Aubrey, John (1950). *Brief Lives,* ed. Lawson Dick. London: Secker & Warburg.

Banks, J.A. (1954). *Prosperity and Parenthood.* London: Routledge and Kegan Paul.

Bell, Charles (1806). *The Anatomy of Expression.* London: John Murray, third edn. 1844.

Blackmore, Richard (1725). *A Treatise of the Spleen and Vapours: or Hypochondriacal and Hysterical Affections.* London: J. Pemberton.

Burgess, Thomas (1828). *The Physiology or Mechanism of Blushing.* London: John Churchill.

Burrows, George Man (1828). *Commentaries on Insanity.* London: Underwood.

Burton, Robert (1621). *The Anatomy of Melancholy.* London: J. and E. Hodson, eleventh edn. 1806.

Cheyne, George (1724). *An Essay of Health and Long Life.* London: Strahan and Leake.

Cheyne, George (1734). *The English Malady or a Treatise on Nervous Diseases of all Kinds.* London: Strahan and Leake.

Coveney, Peter (1957). *The Image of Childhood.* Harmondsworth: Penguin.
Curling, T.B. (1856). *A Practical Treatise on Diseases of the Testis.* London: John Churchill.
Darwin, Charles (1872). *The Expression of Emotion in Man and the Animals.* London: John Murray.
Dawson, Richard (1840). *An Essay on Spermatorrhoea.* London: Aylott Jones. Sixth edn. 1852.
Ellis, William Charles (1838). *A Treatise on the Nature, Symptoms, Causes and Treatment of Insanity.* London: Samuel Holdsworth.
Foucault, Michel (1971). *Madness and Civilization.* London: Tavistock Publications.
Hare, E.H. (1962). Masturbational Insanity: the History of an Idea. *Journal of Mental Science, 108,*1-25.
Hume, David (1734). *Letters of David Hume.* Editor J.Y.T. Greig. Oxford University Press 1932.
Lallemand, J. (1847). *A Treatise on Spermatorrhoea.* London: John Churchill. Translated by Henry J. McDougall.
Lavater, Johann Caspar (1778). *Essays on Physiognomy.* London: William Tegg and Co. Fifth edition 1848.
Lichtenberg, George (1783). Fragment von Schwanzen. *Baldingers Neues Magazin fur Aerzte 5.*
Maddock, Alfred Beaumont (1854). *Practical Observations on Mental and Nervous Disorders.* London: Simpkin, Marshall and Co.
Maudsley, Henry (1868). Illustrations of a Variety of Insanity *Journal of Mental Science, 14,* 149-62.
Maudsley, Henry (1879). *The Pathology of Mind.* London: Macmillan and Co.
Milton, John (1854). On the Nature and Treatment of Spermatorrhoea, *Lancet, 1,* 243-6, 269-70, 467-8, 595-6.
Morison, Alexander (1824). *Outlines of Mental Diseases.* Edinburgh: MacLachlan and Stewart.
Rothman, David (1971). *The Discovery of the Asylum.* Boston and Toronto: Little, Brown & Co.
Skae, David (1863). A Rational and Practical Classification of Insanity. *The Journal of Mental Science, 9,* 309·19.
Spitzka, E.C. (1887). Cases of Masturbation. *Journal of Nervou and Mental Diseases.*
Szasz, Thomas (1971). *The Manufacture of Madness.* London: Routledge and Kegan Paul.

ON BODY PRODUCTS

J.B. Loudon

This essay was originally provoked by one or two minor puzzles about human excretion as a social action.[1] The puzzles mostly arose from simple matters of common everyday observation. Admittedly clinical experience as a physician also played a part in arousing curiosity, since it gave direct if privileged access to material otherwise less easily obtained. But, in comparison with other biological happenings similarly subject to customary rules, the minutiae of this particular daily activity seemed to be largely neglected by most anthropologists.

To begin with, it is presumably accepted generally that one of the ways in which every normal human infant begins to become aware of itself as a separate individual is through its excreta, especially its faeces. The hypothetical emergence of man as an animal different from all others is re-enacted in infancy through the Weberian transition from behaviour to meaning and action. The child learns that excretion behaviour can develop into one of the actions through which it may exert influence over other people. Approval is the reward for producing faeces at the right time and in the right place; but it finds that dismay or disapproval may be elicited in others if it produces faeces at the wrong time or in the wrong place or of the wrong kind in terms of consistency and quantity. As it gains control of its sphincters the child learns that at least some of these variables are amenable to the will; in certain cases the next step is the discovery that it is possible to refuse, maybe for quite a considerable period, to produce any faeces at all. Here it finds at its command a further powerful weapon by which, in at least some societies, it may be able to inflict mental pain on those with whom it has the closest ties and on whom, in the nature of things, it is most dependent.

Some of these aspects of excretion have obvious and well recognised implications for the study of socialization. Others are encapsulated in a familiar medical rule of thumb; namely, the commonest cause of so-called constipation in young children is an anxious mother. But another quite different dimension regarding not only excreta but body products in general seems to have attracted less attention than it deserves. Most human beings develop relatively neutral or positive reactions and feelings about many of the smells and substances produced by their own bodies; at the same time they tend to have more or less negative attitudes towards the smells and substances, objectively indistinguishable from their own, which are produced by the bodies of other people. An unobtrusive part of the process of socialization is learning to recognize reciprocity in these matters, that what is true for oneself regarding others is likely to be true for others regarding oneself. But there are important reservations to be made. One which is pregnant with social and biological overtones concerns the difference, in terms of physical and chemical properties, between the body substances produced by the male and female genito-urinary systems. Another less obvious but no less important matter is the part played by body smells and body substances in the establishment and expression of critically significant social relationships. Here the positive or negative character of attitudes towards body products has a specifically social component, involving selective legitimacy of contact and prescribing intimacy, trust or friendship, on the one hand, and distance, suspicion or enmity, on the other. Fortes (1975) has noted of the Tallensi that what they regard as mystical 'dirt' (comprising intimate, private, bodily exudations) is innocuous where, as in the procreative relations of husband and wife, there is mutual trust, but potentially dangerous where no such trust exists.

Perhaps because of a tendency to concentrate attention on faeces as the most conspicuous example, body products and their smells are liable to be dismissed as a topic only worth a scatalogical bagatelle. This is understandable but mistaken. A strong case can be made for them as playing a much more important part than is generally granted in human social interaction. Throughout life the body sheds or emits a variety of substances - gaseous, liquid and solid - in varying quantities and at various times and frequencies. The physiological processes involved are subject to external and internal influences, including factors generally assumed by the words social and psychological. Conversely, some of the substances are commonly employed as indicators of the organism's relationship with the environment, whether internal or external, social or physical; they also play an important part in the expression and communication of ideas. Those which figure most commonly or prominently in the repertoire of customary behaviour do so by virtue of their physical and chemical properties and of the nature, periodicity and circumstances of their production. The repertoire is itself ultimately rooted in the neurophysiology of the human species; hence it is to be expected that certain body products, however variously interpreted and emphasized, are raw material for thought and action in all human societies.

In the literature of social anthropology the term 'exuviae' tends to be used indiscriminately to encompass all the substances produced by the body in the ordinary course of existence, especially those which are found to be prototypes of impurity in particular cultures. Biologically speaking this is of course an over-simplification. Body products comprise not only exuviae in the strict sense (that is, things which are cast off) but also various materials generated through secretion and excretion. What is more, the indigenous 'social paradigms' (Barnes 1969:100) concerning body products often if not always reflect crucial distinctions which have at least some biological basis; they include, for example, the significance attached to the sex and age of the body producing them, the occasions in the life cycle of the organism on which they are produced, and the consistency and smell of the substances.

Human exuviae proper typically comprise structures derived from the ectoderm or external layer of epithelium - hair, teeth, nails, skin and the like - which are shed by the body at various times or frequencies; skin desquamations are shed daily, for instance, and teeth are usually shed only twice in a lifetime. Exuviae may also be taken to include products of the reproductive cycle, including blood and the internal lining of the uterus shed during menstruation. Teeth and hair are among those accessories of the body which may constitute what might be called artificial exuviae - items deliberately modified or removed for social reasons: other artificial exuviae include digital phalanges, prepuce, clitoris and labia majora and minora - parts of the body removed by surgical operation in the course of rites of transition or purification.[2] Sometimes these objects are themselves regarded after removal as being of potential ritual importance, as may also be the case with some of the incidental by-products of the operation, including blood, serum and pus; more usually it is the intentional act and the resulting modifications of the body which are ritually significant rather than the parts removed.

By contrast with exuviae, excretions consist of material, separated out by glands from the blood and other body fluids, and elaborated into particular substances which either undergo excretion as waste products of metabolism or fulfil special functions within or on the surface of the body. But I am not here concerned with internal secretions produced by the endocrine or ductless glands: indeed it may be assumed on behalf of social anthropologists (and the same is true for most of their informants) that a crude criterion for distinguishing secretions likely to be of direct cultural significance is the existence of ducts or other orifices connecting the interior of the body with the outside - even though in the case of certain ducts which are invisible to the naked or uninformed eye their functions or very existence may be ignored or misunderstood. This is not of course to deny the social importance of some of the effects of internal secretions, such as adrenaline and insulin; anger and hunger are undoubtedly important considerations in the analysis of human social relations. But the direct mode of operation of these secretions is clearly of greater interest to physiologists and psychologists, for example, than to social anthropologists, for whom they are a 'given'. Secretions which are or should be of professional concern to social anthropologists include sweat, sebum, tears, saliva, milk, semen and certain other fluids produced by the reproductive system, such as prostatic and vaginal secretions and those of the aptly named greater vestibular glands.

The word excreta is generally taken to refer exclusively to faeces and urine, even though waste products are also excreted in other less conspicuous ways in breath and sweat and through intestinal gases emitted as flatus. Excreta and the act of excretion assume social importance for partly biological reasons which seem obvious enough at first sight. The organs involved in the discharge of faeces, flatus and urine are closely associated anatomically with the genitalia - the frontiers *par excellence* of the self. They raise basic questions about identity. Urethra and anus are clearly parts of the body: what comes out of them is, and yet is not, part of oneself. Furthermore, human excreta and at least some of the secretions mentioned have specially characteristic smells. In some cases they are more noticeable when the substances are fresh, in others when the substances are stale or undergoing bacterial decomposition. In any event, for reasons which are partly biochemical and partly no doubt the result of socialization, some of these smells seem to be peculiarly perceptible to members of all societies and peculiarly liable to endowment with special meanings.

Among lower vertebrates behavioural reactions are dominated by stimuli arising from the sense of smell and accordingly the cerebral cortex is almost entirely olfactory in its afferent connections. The process of prosencephalization which accompanied the evolutionary development of higher mammals meant that non-olfactory areas of the brain increasingly overshadowed the olfactory centres. This is particularly the case among primates; and the special senses most relevant for anthropologists, and most usually discussed by them - either explicitly or (more usually) by implication - are vision and hearing, neurologically the most highly developed senses in mammals and fundamental to any study of verbal and nonverbal communication.

As lower controlling centres, including the olfactory, were superseded by the evolution of higher centres, the former did not simply atrophy and disappear even though their activities were subordinated and even overwhelmed by the controlling and inhibitory influences of the latter. If 'the cerebral cortex of the human brain is the organ of liberty of action and its correlate, liberty of thought' (Le Gros Clark 1939), it may still be said to be dogged by such lower centres as the olfactory, which may be sluggish but is by no means dead; and if it is true that 'the history of the behavioural sciences has been to reclaim bit by bit and make significant to us our common animal nature' (Douglas 1975:3), then it is perhaps time we gave appropriate attention to the sense of smell which is among the most sensitive discriminants of *hama*, of things inspiring disgust, the importance of which 'as a dominant theme in Lele culture can hardly be exaggerated' (*op.cit.*:12).

Among mammals there is no great variation in the importance and relative size and development of the rhinencephalon, the olfactory portions of the cerebral hemisphere. Rodents are the outstanding example of macrosmatic mammals, classified as those which depend very greatly for survival on a highly developed sense of smell and on other associated feeding reflexes arising in the snout or muzzle, including touch, taste and muscular sensibility. Man puts himself among microsmatic mammals, whose olfactory centres are less important. Nevertheless those parts of the human body most plentifully supplied with sensory nerve endings consist of the nose, lips tongue and finger tips; and it is a matter of common observation, attested by a flood of writers of childhood memoirs, that the senses of smell and touch are those most powerfully evocative of profoundly important (and usually, though not always, pleasurable) memories.

Early accounts of tribal societies often contain statements about the remarkable acuity of the sense of smell to be found among 'savages'. Johnston (1908:508), for example, writes that it is so extraordinarily developed among the Congo Pygmies as to be 'nearly as keen as that of a dog'. No doubt such reports, supporting an overspill from rudimentary evolutionary theory, are responsible for the frequently encountered suggestion that 'primitive' man has a more highly developed olfactory sense than is found among members of 'advanced' societies. One authority, repeating what is to be found in other earlier treatises, writes as if it were a matter of common knowledge that 'in civilized man olfaction is a sense much in abeyance' (Reading 1950). The opposite view is, of course, as frequently encountered in other contexts - that civilized man is more fastidious about smells than his primitive brethren. Some writers have been at pains to correct both extreme views. Bryant is a good example: in a characteristically quizzical passage of carefully recorded detail he writes that 'probably not more than ½ per cent. of the Zulu possesses a body-smell beyond the normal. But the occasional exceptions are certainly impressive, and are as disliked by the Zulus as much as by ourselves. They distinguish at least four varieties of disagreeable body-smell, difficult to explain in words, but termed by them - the *uHlofu* (the smell of a bitter pumpkin)[3], the *iNgoso* (the shrew-mouse smell), the *iQuqu* (the he-goat smell), and last, but least-loved, the *umSanka* (the nauseating umSanka-bush smell). This worst type, to us, suggests Irish stew, flavoured strongly with onions and asafoetida. To the Zulus, perspiring Europeans are classed as perfumed *à l'Ququ*' (1949: 117-8).

Freud was more squeamish, no doubt because he was less secure, than Bryant. 'It is plain' writes his biographer 'that there was for Freud a security in knowledge of the anatomy and physiology of the nervous system.... It was scientific. assured, and a necessary check on "speculation"' (Jones 1964: 259). Freud felt himself particularly reluctant to dispense with a physiological anchor at the time he was grappling with the ideas and the overwhelming personality of Wilhelm Fliess, the originator of the so-called 'nasal reflect neurosis'. This syndrome was derived from the notion of a relationship between genital activities (including sexual intercourse and menstruation) and changes in the nasal mucous membranes: in addition Fliess emphasized the bisexuality of all human beings and elaborated an almost mystical numerology built round the numbers 28 and 23, derived from menstruation and from the interval between menstrual periods.

It is of course perfectly true that the nose and the genitalia share certain physical characteristics, including the presence of erectile tissue, mucous membranes and secretions. Lest it should be thought that this connection, with all its potential for symbolic elaboration, has escaped the notice of contemporary anthropologists, attention should be drawn to Colin Murray's brilliant (1975) essay on snuff among the Sotho. Freud's interest in this topic was aroused (and maintained) because there was in his thinking throughout his professional life what has been described as 'a "hidden thesis"', presented by him in quasi-biological terms. According to this there was postulated 'an "organic repression" derived from man's adoption of the upright posture and his consequent devaluation of olfactory stimuli and the replacement of smell by sight as the dominant sense. Anal eroticism, and later, by association with the excretory zone, genital sexuality, fall victims to civilized man's increasing distaste for bodily odours' (Wollheim: 1971:146).[4]

There is little evidence that Freud's hypothesis did much more than reinforce for a few selected individuals long held and widespread assumptions about the link between civilization and smell. Savages are closer to animals than we are, hence they have keener noses than we have and, at the same time, smell worse than we do. This view is reflected in some of the ordinary technical literature of western medicine. For example, synoptic accounts of diseases of the nasopharynx, when dealing with the nature of presenting symptoms, use it as an explanation for the relative infrequency with which patients consult surgeons because they have either lost their sense of smell, or suffer from too great a sensibility to smells, or are aware of unpleasant smells; three conditions known, respectively, as anosmia, hyperosmia and cacosmia. In such accounts, anosmia and hyperosmia are taken to be 'objective' conditions, referring to smells which actually exist, whereas cacosmia, while sometimes due to long-standing atrophic or suppurative affections of the nose or adjacent structures, is described as being not uncommonly a 'subjective' condition, characterized by the perception of unpleasant smells which have no independent existence or no identifiable physical source of origin. Overtly hallucinatory cacosmia, frequently in association with so-called uncinate epileptiform seizures, is a striking diagnostic feature of cerebral tumours located in the region of the hippocampus; but 'subjective' cacosmia is more commonly encountered in what used to be classified as delusional insanity, where it was often described as the outward expression of profound guilt feelings on the part of the patient and, to those caring for him, as a danger signal of considerable prognostic gravity. 'Subjective' cacosmia, whether accompanyiny space-occupying lesions in the brain or presenting as a psychiatric symptom without any apparent organic basis, is therefore not without general interest as the mirror-image (as it were) of psycho-physiological responses to external stimuli involving an association of unpleasant smells with an awareness of danger.

An extreme example is that of the response to danger induced in most people by the sight and smell of a corpse in a moderately advanced stage of decomposition. In this case fear or horror, instances of what Lazarus (1971), in order to distinguish a psychological response from purely physiological stress, has termed 'dystress', may operate through internal secretions of the adrenal glands to produce nausea if not actual vomiting. This is an illustration of the complex reaction of disgust aroused by a range of sensory stimuli. In many societies the smell of putrefying organic matter, particularly that of corpses and faeces, seems to serve as an archetype, an almost primordial olfactory reaction. This may be reflected linguistically by the use of a term which, although of relatively restricted primary reference, is also widely employed as a metaphor; in English the words 'stink' and 'stench' are obvious examples. It is a matter of common observation that among individuals accustomed to the smells of putrefaction, such as those involved in specialised occupations, conditioning modifies or suppresses a response which may well have a biochemical basis, even though reinforced by socio-psychological factors. Certainly the opposite is true: a process of social conditioning is involved in the identification of those smells which may be categorized as disgusting in particular cultures. As in other aspects of classification, such arbitrary features as role, time and place apply; and here we are on the boundaries of familiar anthropological territory.

The rubric for the 1975 conference included a reference to the argument put forward by Durkheim that sequences of periods of greater and lesser social interaction generate concepts of time and space, and hence conceptual thought itself. As far as one can tell there are no human societies where the act of excretion and its products are not subject to public and private arrangements, to expectations involving time and space, regularity and appropriateness. There is a link here with the view put forward by Hall, who in *The Hidden Dimension* (1966) maintains that the living arrangements existing in a society determine to a great extent the needs of the people living in those arrangements, that any group of human beings becomes culturally adapted to the pattern of building arrangements which become conventional for that group. Hall maintains that any actual physical building represents an adaptation of the small-scale social system to the constraints of its environment, an adaptation mediated by a particular pattern of building conventions. These conventions are themselves, of course, related to a wide range of social and environmental factors, including the climate, the type of terrain, the density and distribution of the population, the variety of available raw materials and the techniques employed in dealing with them. But at the heart of them is the module provided by the human body. Its dimensions and its needs are the basic determinants of the use of space and of the physical structures erected to give shelter, protection and privacy. Consider bricks, for example. Men have made them, sun-baked or fired, for at least six thousand years. When made in moulds, their dimensions have been subject to periodic official regulation. But their sizes have in fact varied very little for centuries because the determining factor has usually been the size of a man's hand.

The allocation of space for domestic activities is more complex. Module and ideology combine to produce divisions which may reflect key aspects of the social structure; boundaries may, for example, be drawn between those parts used by old and young, male and female, married and unmarried. Reserved areas within or around a dwelling may thus be set aside for particular persons or for particular activities such as sleeping, cooking, eating and receiving visitors. In all societies some such demarcations seem to apply to excretion; an important consideration in most cases seems to be the smell of human faeces. The importance attached to it may be only relative, however. Although among the Tallensi the homestead is usually kept fairly clean indoors, the people defaecate just outside it. Fortes remarks that 'the stench around a Tale homestead in the rainy season is enough to shake the hardiest anthropologist' (1945:8). Similarly Meggitt reports that the Walbiri people of Central Australia, while saying that 'human excreta are physically offensive or "stinking" and should be covered with sand if deposited too close to the camp' (1965:271), in fact often take little care to do so. It may be noted that among both the Tallensi and the Walbiri little part is played by exuviae in witchcraft beliefs and practices. Among the Tallensi, however, human faeces mixed with animal manure forms an essential fertilizer for their farms. Here it should be noted that Panoff (1970) has shown how important for the Maenge is the 'ambivalent category' of the rotten for a horticultural people, who nevertheless exhibit feelings of revulsion for 'dirty substances' (including human faeces) which they use as fertilizer.

Malinowski (1929) had an axe to grind when writing of excretion among the Trobriand people. He is at pains to demonstrate that, 'in the matter of excretory functions, (the Trobriander) shows far more delicacy that most Europeans of the lower classes, and certain "sanitary" arrangements current in the south of France and other Mediterranean countries would horrify and disgust him'. Malinowski notes that the Trobrianders had separate reserved areas in the bush at some distance from the village, one for each sex. He says that they would never go to these reserves together, or ever defaecate near one another, though at sea a man would defaecate in the water, below the surface, held up by companions in the canoe. In villages between the sea and the lagoons, where there was a lack of dry land away from the settlement, Trobrianders allocated different parts of the beach to men and women. Other islanders from inland villages regarded these places as evil, holding their noses against the smell as they passed; though it is clear that the local inhabitants did their best to defaecate where their excrement would be covered by the tide. Human excreta were regarded as disgusting in particular because of the smell; the Trobrianders believed that malignant 'magical' influences achieved their greatest potency by entering through the nose and that 'the smell of ordure and decomposing matter' is 'noxious to human Health' (*op.cit.*:378).

The Trobrianders were not alone in holding such beliefs. High court judges in Britain started carrying prophylactic nosegays into court long before the bacteriological basis for the connection between endemic diseases and evil-smelling drains was demonstrated in the nineteenth century; indeed, the nosegays were aimed much more at warding off infection from what Shakespeare makes Coriolanus refer to as the stinking breaths and greasy caps of the populace. Even in the nineteenth century improved sanitation may well have loomed larger in the minds of the newly risen urban middle class, as an element in the process of distancing themselves from the great unwashed out of which they had themselves in many cases so recently emerged, than among the still predominantly rural landed gentry and old aristocracy. The latter, more securely established, may have been less fussy and less discriminating, as is suggested by an aunt of Bertrand Russell: writing of her childhood in the early years of Queen Victoria's reign, she remembered that 'in those old careless days... bad drains were considered rather a joke. If they smelt people considered it a sign of bad weather approaching and were rather pleased to have the warning' (Young 1934:85). As for Queen Victoria herself, the royal family's private apartments in Buckingham Palace in the 1840s were ventilated through the common sewer.

From top to bottom, therefore, members of the population of mid-nineteenth century Britain were much more familiar with the smell of raw sewage than their descendants are today. The valve water-closet was invented in England in the late sixteenth century but, although a number had been installed in (for example) some of the hospitals in London in the eighteenth century, the development of this amenity on any scale had to wait until after the introduction of cast-iron water mains, round about 1820, made possible a continuous piped water supply to ordinary houses. Within a few years there were published designs of water closets for cottages in order to 'prevent the rising of air, and consequently of smell', from cesspits; one ingenious arrangement of doors, valves and pulleys was presented to property developers, designed so as to prevent smell and to ensure flushing 'independently of any care of the person using it', a contrivance particularly recommended for water-closets in public buildings and infirmaries (Loudon 1833:701).

Mostly these improvements were confined to the larger towns. Elsewhere the disposal of excreta was as uncomplicated as ever. In medieval Britain the evidence from those buildings which survive shows that only major ones occupied by special kinds of people had relatively sophisticated indoor facilities for dealing with sewage. The most elaborate were the communal reredorters, with carefully designed drainage systems, found in monastic establishments. Royal residences, castles and larger fortified houses were usually provided with garderobes projecting from, or built into, the thick walls of such buildings, and drainage into the moat or into special pits. Ordinary citizens seem to have resorted for defaecation to areas enclosed by bushes, where privies, if there were any, were situated, and where the contents of any chamber pots used indoors were also deposited. Apart from the fact that this is how their descendants managed until quite recently, there is contemporary literary evidence to suggest that these were the usual arrangements among the general population of this country at that time. For example, in *The Owl and the Nightingale*, written in the first half of the thirteenth century, the relative merits of religious poetry and the poetry of love are debated in verse; the owl rebukes the nightingale because the latter's song is profanely contaminated by his habitual abode deep in the thickets behind dwelling where human beings went at night in order to defaecate. By contrast, a very different literary source indicates some of the difficulties arising more than six hundred years later for visitors to those parts of the British Isles less well provided with cover than those parts of the south-east visited by nightingales. Richard Best relates of J.M. Synge on the Aran Islands in 1900 that 'in the daytime, when he had to obey a call of nature - there were no up-to-date lavatories - he told me rather shyly that he used to go out into a field, overlooking the wild Atlantic, and just when he would be about to perform the major operation he would look round to see if there was anybody nearby; and nearly always, he said, some young woman would put her head over the stone wall. So I said, "What did you do?" "Well," he said, "I used to postpone that part of my daily routine until night. Then when the inmates of the house were all asleep, I would open my door and steal out over their bunks into the darkness of night"'(Rodgers 1972: 102).

Some farms in Wales to-day still have the remains of two, three or four-holed communal earth-closets 'out the back', relics of the days of large households. But nowhere in the delightful house and farmstead plans by Rees (1950) are privies marked as such. In one or two plans an un-named small building is shown at the end of the garden, its anonymity in marked and eloquent contrast to all the others, each of which is carefully labelled 'stable', 'calves', 'pig-stye', 'dairy' and so forth; even ovens, bee-hives, coal-sheds and fruit bushes are indicated. Rees goes to some length in examining the reasons for the nearness of the stable to the dwelling house; he rejects the utilitarian explanations of his informants, that horses used to be the most valuable animals on the farm and needed to be within earshot at night, in favour of patriotic arguments about the nobility of the horse in Celtic tradition since the Bronze Age. But nowhere does he mention the whereabouts of a privy.[5]

Most anthropologists readily admit that there are glaring gaps in their field data. Since few of us are fireproof in this respect, pots and kettles are free to call each other black. My own material regarding the Zulu-speaking people on farms in South Africa, among whom I worked for nearly two years, is woefully inadequate concerning such seemingly straightforward matters as where and when they usually defaecated, and what they said they thought about these things. A private act no doubt, and difficult to observe; but certainly a good example of those 'drab, everyday, minor events' which Malinowski (1935:241) blamed himself for not treating 'with the same love and interest as sensational large-scale happenings'. African informants told me that people are generally secretive about such actions and that as far as possible a clear distinction is maintained between the places used by men and those reserved for women and girls. I was also told that adults are expected, and children taught from an early age, to bury their faeces. In the old days it is said that members of the Zulu royal house were attended by honoured servants known as 'sod-turners' when they went to defaecate, the latter's task being to dig the hole and afterwards to cover up the excrement deposited in it.

In the course of routine clinical work I certainly found among African patients considerable emphasis on defaecation and on what were presented as supposed malfunctions of excretion, particularly - but not exclusively - among men in relation to themselves and among women in relation to the excreta of their children and grandchildren. The frequency with which my clinical notes refer to cases of bowel complaints among small children is not surprising; gastro-enteritis constituted (and still constitutes) what is probably the most common killer of black infants and toddlers in South African rural areas, and it is also an example of those conditions which are generally regarded as appropriate for treatment by white physicians as opposed to indigenous healers. Whether this is true for bowel complaint among adults is less clear. Men consulted me in many instances in order to obtain purges, the more powerful the better, not simply for the relief of a wide variety of complaints but as preventatives and because of the generally good effect on health (in the widest sense of that term) which they said they believed lay in regular purgation. Indeed I got the clear impression from many, and explicit information from a few, that the principal effect it was hoped to achieve was broadly speaking aphrodisiac.[6] In some cases this was sought as an indirect side effect of the treatment of *umnyama* (literally 'blackness'); this is not only a state of lowered resistance and heightened vulnerability to disease (see Ngubane 1976) but also associated with feelings best described as guilt, anxiety and depression, which are believed to be related to lack of popularity with members of the opposite sex and lack of success in amorous adventures. I was also informed that many if not most Zulu men, particularly those past middle age, administered regular enemata to themselves, usually at weekends, using a variety of medicaments and with the same general aims as purgation. Small children, including new-born infants, are similarly given regular enemata, in addition to a process designed to draw a small amount of blood from the child's anal orifice and widely believed to be essential to the future physical, mental and moral well-being of the individual (see Loudon 1970). But I missed chances of exploring further what no doubt exists, namely, a close association between bowel habits and ideas about them, and wider aspects of status and morality. Nor did I consider collecting direct concrete evidence about the mundane minutiae of excretion behaviour.[7]

In the course of one of the most detailed and perceptive series of studies ever undertaken of the link between body products and ritual, especially rituals of mourning and kinship, Wilson writes that she was 'driven to the conclusion that [the Nyakyusa] are dogged by a sense of guilt, for this is the only hypothesis on which the linking of misfortune and sin, the preoccupation with "filth", and the recurring acts of purification can be explained. One of the sources of this sense of guilt', she suggests, 'is awareness of anger within, for it is believed that such anger injures others' (Wilson 1959:161). She demonstrates elsewhere how these pre-occupations first make themselves plain in the Nyakyusa emphasis on 'cleanliness in person, in house and courtyard, in cooking and in disposal of excrement'; for them cleanliness is next to dignity, which is in turn associated with other cardinal virtues such as geniality, urbanity and hospitality. But here again it is not clear where the Nyakyusa go to defaecate nor whether the familiar boundaries based on sex and age in this respect are as important as those based on smell. The only specific reference to how the Nyakyusa dispose of excrement is when Wilson mentions that pchildren are trained to take a hoe with them when they go to defaecate and to bury the faeces' (Wilson 1957:208).

A well-known instance of explicit connection between the smells and substances of death and excretion is that of the Trobrianders, among whom the sons of a dead man are described as sucking the putrefying flesh from the exhumed radius, ulna and tibia, and as saying that it is right and proper that a child should suck those particular bones, 'for the father has held out his hand to its excrement and allowed it to make water onto his knee' (Malinowski 1932:133). But Monica Wilson's account of mortuary rituals among the Nyakyusa is a more systematic demonstration of the links between filth, particularly the smells of corpses and of human excreta, and the well-being of individuals and society. We repeatedly read of the importance of ritual, 'especially the death ritual, as a protection against going mad, against the disintegration of the personality' (Wilson 1957:232). Madness, the disintegration of the personality, among the Nyakyusa is apparently most commonly expressed in terms of eating faeces; and an essential expressive element in their mortuary rituals is the preparation of what are described by Wilson as 'sausages' of millet porridge, which are sprinkled with powdered medicine and half-buried in the ground, all the participants being expected to bend down and, without using their hands, to bite off and consume a mouthful of what quite explicitly represents a turd. This miming of the actions of a madman in eating excrement is regarded by Wilson's informants as a prophylactic and protection against madness. Furthermore the corpse and faeces are identified with one another; 'the corpse is filth', said her informants, 'it is excrement. And so, when a madman for whom they have not performed the ritual eats filth, that is the corpse' (*op.cit.*:49). It should also be noted that one result of neglecting the third main event of Nyakyusa mourning ritual, the so-called ritual of the gasping cough, is believed to be diarrhoea; and diarrhoea is associated with filth in general and (interestingly enough) may also be prevented by biting the 'sausages' of millet porridge, the other end of which is allowed to remain in the ground. This means, according to the Nyakyusa, that one's faeces will thereafter be 'hard (not watery)' (*op.cit.*:59); presumably the word hard used here may be taken to mean firm, while the meaning of the word translated as diarrhoea, defined in terms of consistence rather than in terms of frequency, is no doubt important in relation to the ways in which some variations in form and consistence of faeces are deemed to be closer to the category 'filth' that others.

A corresponding but opposite notion to that linking body

products and madness is that between body products and the integration of relationships. Among the Nyakyusa the most vivid instance is that by which semen (and other secretions) is powerful because it is seen as embodying the 'shades' which control conception and indeed all fertility. The part played by sexual secretions in the establishment of primary relationships is exemplified by the Ceŵa ritual of *kutenga mwana* ('taking the child') described by Marwick (1965). The mingled sexual fluids of the child's parents are rubbed onto the child's back; this is said to be done in order to strengthen it, though this is surely a metaphor for recognition of the child as a person and as the legitimate offspring of the couple. The ritual is not performed until the child is old enough to smile for the first time, at the age of four or five weeks; at the same time the child is said to move from a state of susceptibility to danger to one of security. A comparable example of the part played by sexual fluids and their smell in cementing relationships is reported by Turner (1968) for the Ndembu. A link between a newly married couple, the bride's mother and the unborn children of the union is shown by what happens to the water in which the couple wash themselves after their first official act of intercourse. Turner reports (*op.cit.*:260-61) that the bride's instructress visits them in the morning and takes the water in which the couple have washed and throws it under the bed of the bride's mother, being forbidden to throw it away in the bush; she also gives the mother the white beads which the bride has worn in her hair 'to keep until her daughter's first pregnancy rites'. The explanation, in Muchona's indigenous exegesis, is that the bride's hands 'are bad and smelling after sleeping with her husband and after wiping his penis'. But clearly there is an important sense in which the smell of the bride's hands is good, as representing the quintessence of a newly established series of relationships of the utmost importance. Here the biological basis of social relationships is expressed through the medium of body products and their smells.

FOOTNOTES

[1]An earlier version of this paper, substantially as spoken at the ASA conference in Belfast, has been published elsewhere under a different title (Loudon 1975). Parts of that version are incorporated here. I ought to add that I am very grateful to a large number of friends, colleagues and correspondents for providing helpful criticism and for offering abundant further material and comments, little of which I have been able to use here.

[2]See, for example, Firth's (1973) discussion of hair as private asset and public symbol, together with the works cited therein. Amputation of the terminal phalanx of the fifth finger, known as the *iNdiki* custom, continues to be practised among certain groups of Zulu-speaking people in Natal in the 1970s. For a discussion of excision of the vulva see Kennedy (1970).

[3]There is obviously something special about pumpkins. Wilson (1957:132), in a discussion of taboos of pregnancy among the Nyakyusa, reports the avoidance of pumpkins by menstruating women; she reports that the growth of a small tumour in the vagina (perhaps a cystocele) is attributed to contact with pumpkins - 'the menstrual blood resembles the pumpkin and they fight together.' And Junod (1927 (Vol. I): 184) reports for the Thonga that 'it is taboo for a girl to walk amongst pumpkin plants' because if she does so 'she would emit a bad odour, and no boy would court her.'

[4]Wollheim (1971) provides the original sources of the 'hidden thesis'. Given the prominence of rodents among macrosmatic mammals, it is not without interest to find that Freud's notions on this topic appear in a footnote to the case of the Rat Man, as well as in his letters to Fliess and in his *Civilization and its Discontents*.

[5]The same sad delicate omission occurs in the comparable study by Williams (1955) of farmhouses in Cumbria.

[6]This is analogous with the approach of ratings in the Royal Navy which is encapsulated in the phrase 'A shit, a shave and a shove ashore'.

[7]A recent report from India might be said to whet the appetite for such a detailed study. Kochar *et al.* (1976) give an account of a two year field study in rural West Bengal, designed as part of a project concerning the ecology of interacting dense populations of hookworms and human beings. This involved close observation, with a stop watch, of defaecation and ablution behaviour among many hundreds of villagers. Although the emphasis is on the 'biological' aspects of excretion as an action subject to cultural rules, what is particularly well demonstrated in this study is the significance for public health of the contrast between ideal behaviour and minutely observed reality.

REFERENCES

Barnes, S.B. (1969). Paradigms - scientific and social. *Man (N.S.) 4*, 94-102.

Bryant, A.T. (1949). *The Zulu people*. Pietermaritzburg: Shuter and Shooter.

Douglas, M. (1975). *Implicit meanings.* London: Routledge & Kegan Paul.
Firth, R. (1973). *Symbols: public and private.* London: George Allen & Unwin.
Fortes, M. (1945). *The dynamics of clanship among the Tallensi.* London: O.U.P.
Fortes, M. (1975). On the concept of the person among the Tallensi. In: *La notion de personne en Afrique noire.* Paris: Colloques Internationaux du Centre National de la Recherche Scientifique.
Hall, E.T. (1966). *The hidden dimension.* New York: Doubleday.
Johnston, H. (1908). *George Grenfell and the Congo.* London: Hutchinson.
Jones, E. (1964). *The life and work of Sigmund Freud.* (Abridged edition). Harmondsworth: Penguin Books.
Junod, H.A. (1927). *The life of a South African tribe.* (Second edition). London: Macmillan.
Kennedy, J.G. (1970). Circumcision and excision in Egyptian Nubia. *Man (N.S.) 5,* 175-91.
Kochar, V.K., Schad, G.A., Chowdhury, A.B., Dean, C.G. and Nawalinski, T. (1976). Human factors in the regulation of parasitic infections: Cultural ecology of hookworm populations in rural West Bengal. In: Grollig, F.X. and Haley, H.B. (eds.) *Medical anthropology.* The Hague: Mouton.
Lazarus, R.S. (1971). The concepts of stress and disease. In: Levi, L. (ed.) *Society, stress and disease.* London: O.U.P.
Le Gros Clark, W.E. (1939). *The tissues of the body.* Oxford: Clarendon Press.
Loudon, J.B. (1970). *White farmers and black labour-tenants.* Cambridge/Leiden: African Studies Centre.
Loudon, J.B. (1975). Stools, mansions and syndromes. *Royal Anthropological Institute News,* 10, 1-5.
Loudon, J.C. (1833). *Encyclopaedia of cottage, farm and villa architecture and furniture.* London: Longman, Rees.
Malinowski, B. (1929). *The sexual life of savages.* London: George Allen & Unwin.
Malinowski, B. (1935). *Coral gardens and their magic.* London: George Allen & Unwin.
Marwick, M. (1965). *Sorcery in its social setting.* Manchester: University Press.
Murray, C. (1975). Sex, smoking and the shades: a Sotho symbolic idiom. In: Whisson, M.G. and West, M. (eds.) *Religion and social change in Southern Africa.* Cape Town: David Philip.
Ngubane, H. (1976). Some aspects of treatment among the Zulu. In Loudon, J.B. (ed.) *Social anthropology and medicine.* London: Academic Press.

Panoff, F. (1970). Food and faeces: a Melanesian rite. *Man (N.S.) 5*, 237-52.
Reading, P. (1950). Nose and nasopharynx diseases. In Horder, T. (ed.) *British Encyclopaedia of Medical Practice*. London: Butterworth.
Rees, A.D. (1950). *Life in a Welsh countryside*. Cardiff: University of Wales Press.
Rodgers, W.R. (comp.) (1972). *Irish literary portraits*. London: British Broadcasting Corporation.
Turner, V.W. (1968). *The drums of affliction*. Oxford: Clarendon Press.
Williams, W.M. (1955). The farmhouses of south-west Cumberland. *Transactions Cumberland and Westmorland Antiquarians and Archaeological Society*, 54.
Wilson, M. (1957). *Rituals of kinship among the Nyakyusa*. London: O.U.P.
Wilson, M. (1959). *Communal rituals of the Nyakyusa*. London: O.U.P.
Wollheim, R. (1971). *Freud*. London: Fontana.
Young, G.M. (1934). *Early Victorian England 1830-1865*. London: O.U.P.

THE INTRODUCTION OF POSES TO A PERUVIAN BROTHEL AND CHANGING IMAGES OF MALE AND FEMALE

Katherine Arnold

This paper is concerned with the way in which changing images of men and women relate to the changing use of the prostitute's body within El Descanso, one of the two licensed brothels in Cuzco.

Prostitution is the hiring of a body for sexual purposes. Sexual behaviour is one of the most imtimate areas of a person's life, and it is correspondingly difficult to investigate. Prostitution in Peru is legal, however, and it was easier to collect verifiable and systematic information in a legal brothel than has proved possible in countries where organised prostitution is illegal (Sandford 1975, Winn 1974).

Moreover, among the *mestizo* population of Cuzco (the Spanish-speaking majority of the people), it is accepted that men need extra-marital sex, and assumed that if they are not going to the brothel they must be having affairs. The brothel is regarded by most wives as the preferred area for a man to fulfil his sexual needs. It is therefore possible to discuss the brothel relatively openly, and indeed, prostitution is a topic of great interest for many wives, who speculate on how much these exotic women earn, and how they live in general.

Bataille points to the relationship between the family and the prostitute when he says: 'Prostitution provides a complement to marriage'. (Bataille 1962:13). It allows other women to be elusive. He says that a prostitute, 'even if she is not ashamed of it', lives in a profane world of transgression which allies her with the animal world (*op. cit.*:135).

The hiring of the body of the prostitute is of great relevance in understanding the nuclear complex, notions of masculinity and femininity, and recent changes in Cuzco society. *Machismo* is the culture complex that orders relations between the sexes in Cuzco. It is concerned with ordering men along a continuum of lesser or greater honour, which Salvador Reyes Navares has described as follows: 'The principal of honour in men is manliness, and women is shame.... Manliness implies strength of body and spirit. Furthermore, it involves the notion of effectiveness. The man must be self-sufficient and able to solve his problem in life without hesitation or confusion (Navares 1970:14).

Within the brothel, the particular elements of *machismo* that come to the fore are, violating women; showing off to other, preferably weaker, men; and 'completing' (*completarse*) or fulfilling oneself. How a man does this is a central problem of the paper.

The image of women which is required to complement her *macho* counterpart is also well described by Navares: 'The characteristics of femininity are normally passive - patience, faithfulness, resistance in the face of misfortune, gentleness.... The "ideal woman" is the sum of these virtues, but raised to such a height that they are ultimately ridiculous and counter-productive' (Navares 1970:17).

The myth, of which *machismo* is a dominant element, creates images of both men and women which are based upon pretence. Octavio Paz writes: Whether as prostitute, goddess or *grande dame* or mistress, woman transmits or preserves but does not believe in the values and energies entrusted to her by nature or society' (Paz 1961:35). She is 'a domesticated wild animal, lecherous and sinful from birth who must be subdued with a stick and guided by the "reins of religion"' (*op. cit.*:36). To maintain the myth, women must be controlled and prevented from undermining the belief that they are passive, asexual beings, and that men are omnipotent and independent.

The prostitute challenges these myths by her independent existence, and is therefore feared, controlled and excluded from Cuzco society as far as possible. She is a *mala mujer* (bad woman) as Paz describes her, 'hard and impious and independent, like the *macho*' (*op. cit.*:39). She is active sexually and independent economically. She threatens the *macho* by her strength. At the same time, she is similar to the *macho* and demonstrates the frailty and brittleness of his display, and, as we shall see, knows and shares what he fears are his weaknesses.

THE NUCLEAR COMPLEX

Men are regarded by women with a mixture of fear and indulgence. The husband's jealousy of his wife, which can lead to her being constantly chaperoned, is seen by women as a compliment, an annoyance, and something which must be tolerated. Meanwhile his own infidelities are accepted and blamed more upon the woman involved than the husband himself. He is treated like a dangerous small boy: both hard and cruel, but also weak and at the mercy of the evil desires of women.

The situation in one family illustrates these attitudes very well. They were upper-class *cuzquenos*, and the expression of *machismo* was generally more subtle than in other *mestizo* homes. The husband's mistress, a nurse, persistently telephoned his home to contact him. Finally, in exasperation, his wife told her that if she was so desperate for a man, she (the wife) would send the chauffeur round. The wife's mother commented that she could not understand why nurses were always so desperate for men. The husband, explaining why the wife must nevertheless be faithful, said that a woman is like a glass of water: one drop of ink makes it all filthy. At home, he was treated like a small boy, or a king, by his wife and mother-in-law, who would peel his oranges at meals, and so on.

The Peruvian sociologists, Castro, Silva, Majluf and Estrada have examined the dynamics of the Peruvian family. They investigated the families of workers, employees and Indians (or *campesinos*). I am concerned only with the first two categories, as Indians do not enter the brothel either as clients or prostitutes. They found that two types of family were most important: the *despotico* and the *patriarcal*. The 'despotic' father provided for the family, made decisions and expected absolute submission: 'The father demands admiration and love as his rights, without any inclination to give anything in exchange. The mother wants security and the opportunity to give herself emotionally. The children hope for love and security. The mother provides the only real contact between members of the family' (Castro *et al.* 1964). The 'patriarchal' family is also regarded as a possession of the father, but he is more involved with and inclined to give affection to his family. Even in this family, 'There is internal strife, there are strong currents of hate, the family is divided into two fronts (the mother and children on one side, and the father alone on the other) and insecurity prevails' (*ibid.*).

There is a tendency to seek a spouse who will be a good parent, probably not only to the children but to one's self also. There is further cross-generational confusion expressed in the use by a father to his daughter, of endearments such as *mamita* (mummy), and to his wife, *hijita* (little daughter). The term *mamacita* (mummy) has strong sexual overtones, but reflects a similar confusion of a female image as both a mother and a sexual object.

The ideas of *machismo* have a profound effect upon the dynamics of the *mestizo* family and its relations with the outside world, and also determine what is acceptable behaviour for a man or woman in Cuzco. It is not, at present, accepted that a woman may be economically independent, nor may she obviously enjoy sex or be the active partner in sexual relations with her husband.

The experience of economic insecurity, which is often given as a major reason for entering prostitution (*negocio*), is similar to the insecurity of a woman within her family. But I suggest that economic factors are not in themselves as important as the system of ideas related to *machismo* in precipitating the recruitment of women to prostitution (cf. Henriques 1965:308-309). I do not deny that the ideology of *machismo* may itself be a product of socio-economic conditions, but in direct relation to prostitution, it is of greater significance than the search for employment.

ATTEMPTS TO CONTROL THE PROSTITUTE BY THE *MACHO*

In Peru, prostitution is legal within certain licensed brothels, which are required to be at a certain distance from the town. A girl can take out her *carnet* and enter the profession. Most girls take a work name at this stage, perhaps symbolising their entry into a new profane world, but their real name is always known to the police. They are then subject to various checks and supervisions, as is the owner of the brothel, who by law is a woman. On entering the house, the girl gives her papers to the administrator, who takes them to the P.I.P. (police) office, where she is registered and her card stamped. This is 'control', and she has to attend every Thursday afternoon for another stamp on her card. She can then proceed to the next check, which is the visit to the hospital. There she is checked for signs of disease, and whether or not they are found, given a large dose of penicillin. If signs of disease are found, she is not allowed to work during that week. The visit to the hospital is at 8.00 a.m. on Saturday, which is a very awkward time for people who work from 6.00 p.m. to the early hours of the morning and whose busiest days are Friday and Saturday.

There are also *mariposas* (butterflies), non-registered prostitutes who work in the street without any of these checks and are looked down upon by the registered workers. Most girls have worked as 'butterflies' at one time or another, and then graduated to the house. It seems that a balance is struck: a woman who becomes a legal prostitute is strengthened by the law, she is protected from pimps and works for another woman, whose position is also secured and controlled by the law. In return she submits herself to various checks which are designed to protect the clients from the powerful forces that she represents.

The landlady is similarly under the eye of the police. She has to pay for licences for each different part of the evening: for *callejon*, the hours between 6.00 and 10.00 when the girls are available for clients but there is no dancing or bar service; and for *salon*, when the bar and the juke box (also licensed) operate. She also has to insure her employees.

Thus the whole business of hiring a body for a few minutes is hedged around with constraints and controls which are apparently applied arbitrarily by the State, as well as by clients, families and prostitutes.

SALON AND *CALLEJON*

Limits put upon the meetings of prostitutes and clients are expressed in the use of time and space. Within El Descanso there are two distinct parts of the evening, involving different parts of the building and allowing different expressions of *machismo*.

The house was purpose-built in 1971 (see Fig. 1). There are rooms for nineteen girls and the administrator around the dance hall, with a narrow, open passageway linking the numbered rooms. There is a courtyard surrounded by the kitchen, lavatories and store rooms. The whole complex is situated on a hillside outside Cuzco, at the legally required distance from the town. The owner also keeps pigs and chickens on her land there, as if to emphasize that the *negocio* and the *mala mujer* (bad woman) must be kept away from the city and allied with rural life. The rooms are of bare concrete, and contain a bed, a chair and a chest of drawers with a mirror. The girls have to provide their own primus stoves and buckets.

From 6.00-10.00 p.m. is the time of *Callejon*. *Callejon* means 'alleyway', and its use draws attention to the distinction Cuzco culture makes between a woman inside or outside her home. During the hours of *callejon* the girls stand in the doorways of their rooms, which occupy a particularly strategic intermediary position in the courtship complex, while men arrive on the bus, in taxis, or in cars. The distance of the house from Cuzco may protect society from the 'bad woman', but at the cost of inconvenience to the *macho*. They walk around until they choose a girl who is not working. The usual price is fifty *soles* (50p), but varies with the service and the girl required. Most men want five minutes, but according to the girls even this is too long for many of them. During *callejon*, the men who come are those anxious to live up to the *macho* ideal and to fulfil themselves, but in reality unable to achieve the wider social requirements of *machismo*: they are men who do not have sufficient money to indulge in the display of the *salon* hours. Some may stay in the kitchen drinking coffee or having supper, but most come only for a few moments. These relatively inadequate men are divided in time from the men with greater pretensions to *machismo*. Some girls refused to work at this time, as the value would increase if their bodies

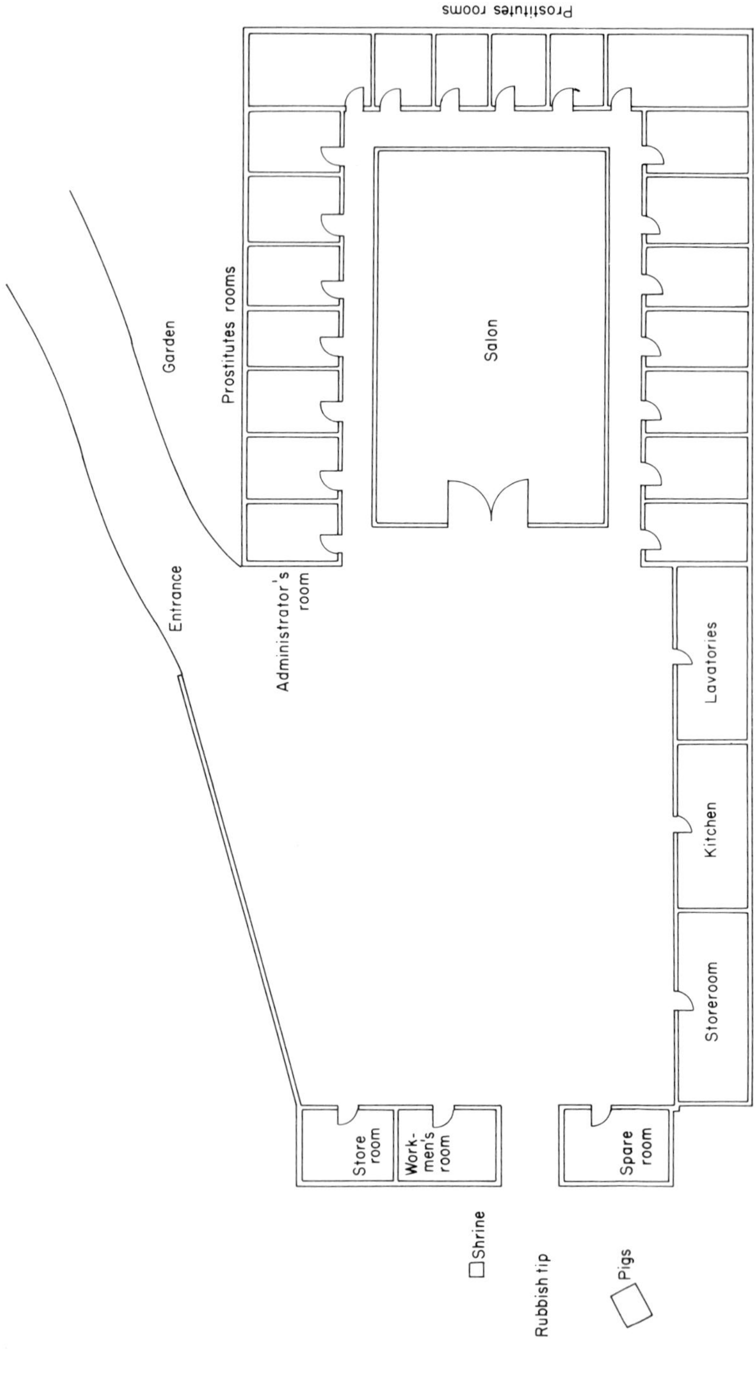
Prostitutes rooms
Garden
Prostitutes rooms
Salon
Prostitutes rooms
Entrance
Administrator's room
Lavatories
Kitchen
Storeroom
Store room
Work-men's room
Spare room
Shrine
Rubbish tip
Pigs

Figure 1

were undefiled by contact with these non-*machos*.

During *horas de salon*, from 11.00 p.m. onwards, the brothel offers far more. The girls come into the hall in pairs or small groups, wearing smart, sometimes specially made, short or transparent dresses, make-up, and, occasionally, wigs. Men stand at the bar talking amongst themselves or with the administrator. A group of men may approach a girl with the offer of a drink. There is both hostility and distance between the prostitutes and her potential clients. The men tend to use the women as props in verbal demonstrations of their masculinity. Many girls refuse to take part in this and remain behind the bar watching.

Later, couples start to dance, and this again provides an opportunity for the *macho* to show off in front of his friends. Often only one girl from each group will join in, and will be passed around a group of men. Unlike a 'respectable' woman, a prostitute moves her hips and body when she dances in a way that reminds everyone of the purpose of the visit to El Descanso. More aggressive girls go to find men to *cachar* (screw). This is a battle which no one can win without simultaneously sustaining a loss: the woman gains money but loses self-respect and that of society; the man gains *machismo*, but there is always a risk that the prostitute will put him down in front of his friends. Also, the position used during sexual intercourse may sometimes undermine his sense of domination of the female.

To drink much is also an important part of the *macho* self-image. Throughout the evening men and women drink, many of them becoming very drunk. Prostitutes try to avoid going to their rooms with a drunkard, but for the man it becomes increasingly necessary to demonstrate that he does not care what she thinks, and he will come anyway. The tension increases all the time. Navares has written about the drunkenness of the *macho*: 'it awakens various reactions in women - it terrifies them, it scandalises them, it bores them; whichever of these it evokes, what is at the bottom of the *macho* conduct is a phrase "So they (i.e. other men) see that it doesn't matter what she wants. I do whatever suits me"' (Navares 1970:16). The prostitute is forced to conform to the submissive, passive image of women which fits with the ideals of *machismo*.

INTERCOURSE - '*NATURAL*' or '*POSE*'

Before a couple leave the *salon* for the girl's room they decide what is wanted and agree upon a price, which is paid in advance. The negotiations are usually private, since the danger of humiliation for both parties is great. This also allows for the possibility of subsequent exaggeration.

Intercourse takes place either *natural* or using a *pose*.

Natural involves the woman removing her pants while the man remains clothed and lies on top of the woman. Until recently (I have the impression that this means since about 1970), this was the only mode of sexual intercourse offered in Cuzco brothels, although in Lima and the North *poses* have been a major part of the attraction of the brothels for some years.

Cuzco is regarded by prostitutes, and clients, as a relatively peaceful, undemanding, slightly dull place to work. The recent introduction and increasing frequency of *poses*, or non-normal positions for sexual intercourse, is beginning to change this way of life, in prostitution and marriage alike. On my second visit to Cuzco in 1974, I found that more girls were giving *poses* than in 1972-73. By 1974, prostitutes in Cuzco provided fellatio (*chupe*, *bambino*) and various positions for intercourse which a Cuzceno wife apparently would be unwilling to perform: *la secretaria*, in which the woman sits astride the man's lap; *patas al hombro*, in which the woman lies with her feet upon the man's shoulders (he sometimes performs cunnilingus); *el perrito*, in which the man enters from behind; and *filo de catre*, in which the man kneels and the woman lies on the edge of the bed. There are other *poses* which involve the woman being on top of the man; *el caballito*, in which the woman sits astride the man who lies on his back; and *el caballito mexicano*, in which the woman lies on top of the man.

The fact that a woman allows her body to be used in *poses* provides the major source of shame for the prostitutes in Cuzco. In Cuzco, in contrast with Lima where *poses* are *older*, it is still thought that some prostitutes will 'make love for love, not just for money'. *Poses* are believed by some girls to damage the ovaries, and reasons such as this are given to avoid doing them. Only a few girls were willing to tell me about *poses* that they themselves did: most would describe *poses* and attribute them to someone else. The administrator told me that no one wanted to pay the extra price (20-50 *soles*), and so no one did them, but on another occasion she said that if a girl offered *poses* there was always a queue outside her door. *Poses* seemed to evoke strong and contrasting images of masculine and feminine.

Poses generate strongly ambivalent feelings for the prostitute and client alike. I was told about the more exaggerated *poses* (e.g. *el caballito*) after some obvious hesitation, at the end of a list of other *poses*, accompanied by a lot of giggling. One girl said that she left her husband because he was unable to enjoy such a position: he was *frio* (cold), whereas she was *caliente* (hot, meaning lustful).

Many men want such *poses* but, like the prostitutes, their reaction is one of laughter when confronted with this anxiety-provoking need which reverses the dominant position of the *macho* and they are acutely aware of a woman who is already in a position to challenge and threaten the male self-image. The steady introduction of such *poses*, which reverse the usual male/female hierarchy, makes it possible for the prostitute and client to relate in two different ways: one which reinforces the male domination of the woman by using *poses* that make his power more obvious, and one in which the *macho* is able to relax his demonstration of *machismo*. There remains the question of why such *poses* are now appearing in Cuzco, and what exactly their introduction means in terms of *cuzqueno* society. On an economic and pragmatic level, *poses* brings the prostitute more money, which she needs to pay the doubled cost of her room, but on the level of symbolisation the *pose* relates to changes in images of men and women.

THE *PARRANDA MACHISTA* (ORGY OF MASCULINITY) PROVIDED BY THE BROTHEL

The situation of being in a brothel is one in which all concerned are vulnerable. Although in Peru one is legally obliged to carry identification papers all the time, they are rarely asked for, and, in fact, the only time I was asked for mine was at the brothel, where it is a fairly regular occurrence. The administrator said that the P.I.P. are looking for boys too young to be there, that is, under 16, and for criminals from Lima and other places. Although prostitution is to an extent controlled and is part of the accepted Cuzqueno way of life, it is still associated with bad elements of society, and the young must be protected from the forces of the 'bad woman'.

In addition to the threat of police interference, the vulnerability of the *macho* is further heightened by the presence of one of the few Peruvian women who is economically independent. Both the girls and the house as a whole do extremely well. Individual prostitutes vary in their earnings, but most girls in their early twenties are able to earn as much as a university lecturer, which puts them among the economically well-off in Cuzco.

The fundamental cause of tension in the brothel, with its corresponding vulnerability, lies in the fact that it is here that the *macho* attempts to demonstrate his qualities in front of a male audience who are all competing with him in an attempt to live up to the myth of the *macho* ideal. It is a *parranda machista*, an orgy of masculinity. He who laughs loudest, makes the best jokes and drinks the most, is doing well but, as Navares says:

> out of all the techniques we have invented to make those near us understand that we exist, he, (the *macho*), has raised up one in particular - his importance depends upon his sexual prowess. He can admit that he's not up to much intellectually, he can agree that he has no great ability for work, that he is mediocre at whatever you care to mention, except in his masculine role' (Navares 1970:19).

The visit to the brothel is correspondingly of crucial importance for the *macho*. It is his chance to come as close as he can to displaying publicly his sexual prowess. Many young men say that they have to go to the brothel to *completarse* (fulfil themselves).

There is a continual hostility between the sexes, both inside and outside the brothel, a feeling that either sex can only benefit at the expense of the other. In the house, it receives expression in that the prostitute wants her client to 'get on and get off'. To encourage this, some girls first wash the penis in warm water. At the same time the man who finishes quickly is called *gallo* (rooster), and mocked by men and women. The prostitute's interest in limiting the time spent with each man so that she can earn more money, is directly opposed to that of the man, who wants to demonstrate his sexual prowess. A delicate balance of the two has to be maintained.

The continual conflict of interests between male and female, prostitute and client, readily erupts into violence: the prostitute is always at risk from men who want to 'cut her face'. Her only real protection comes from the administrator, who is another woman. The men who work as waiters and cleaners cannot be relied upon, in their marginal and complex roles, to protect the prostitute.

CONCLUSIONS

A further use of the word *natural* is given by Pierre van den Berghe in his discussion of ethnicity. He defines it as 'a term used to refer to a local person as distinguished from a stranger' (Van den Berghe 1975:22). The *natural* position in sexual intercourse has similar overtones of something which belongs within the domestic sphere of life and is therefore both safe and familiar. In contrast, the prostitutes say that *poses* have been introduced to the brothel by foreigners, or by men who bring foreign magazines which illustrate such exotica. Thus the introduction of *poses* adds a further dimension to the prostitute's position *outside* the home, within which the passive woman required by *machismo* exists. A woman who is not contained domestically, challenges and questions the categories and dichotomies which have ordered relations between the sexes for centuries, certainly since the Spanish conquest.

To discuss this further it is necessary to consider:

1. The anthropological theories that consider the relations between images of men and women. I shall refer particularly to Nadel's study of the Nupe, as he provides a full and interesting ethnography of the women, including the prostitutes.

1a. The relevance of these theories to societies in which notions of honour and *machismo* are important. These are generally defined as Mediterranean or Latin societies.

2. Two sets of polarities that require elaboration:
   (a) *macho* and *manso* as applied to men
   (b) whore and virgin as the elements into which the image of women is frequently split in such societies.

When these points have been clarified, the particular significance of the changing use of the prostitute's body should become more apparent.

1 *Images of Men and Women*

The distinction of sex is made on the basis of body form, but there have been many debates as to how far the biological sex differences influence the cultural expectations of each sex. Recently, many anthropologists have turned their attention once again to the 'problem' of women. The literature has been summarised in Rosaldo and Lamphere's book, and there are points that are relevant to my field data.

According to Ortner, women have been universally defined as inferior to men. She proposes that this is because society is concerned to demarcate itself from nature, and women are on three levels closer to nature than men, and therefore despised:

i The woman's *body* is more involved with 'species life', in contrast with man's physiology, which allows him to devote himself to culture rather than nature.

ii The woman's body forces her to adopt *social roles*, such as child rearing, which are culturally defined as of a lower order than men's.

iii Women are deemed to have or require a different *psychic structure* from men, in order to fulfil their functions at the first levels described. This psychic structure is defined as inferior to men's.

Rosaldo has conceptualised some women as being confined to a domestic sphere of life, from which they have to escape in order to achieve anything that will be culturally valued. In Cuzco, women find themselves in a cultural double bind. The only escape is into prostitution, which is despised. This is because 'male, as opposed to female activities, are always recognised as predominantly important, and cultural systems give authority and value to the roles and activities of men' (Rosaldo and Lamphere 1974:19). Hertz says, 'To the right hand go honours, flattering designations, prerogatives: it acts, orders and *takes*. The left hand, on the contrary, is despised and reduced to the role of humble ancilliary, by itself it can do nothing; it helps, it supports, it *holds*' (Hertz 1973:3). He specifically links male with the right hand, and female to the left hand, as have many other writers. Women are excluded from sacred rites - 'the profane and the impure are essentially weakening and deadly; the baleful influences which oppress, diminish and harm individuals come from this side' (Hertz 1973:8). Women are feared, as they are associated with evil forces. The prostitute is structurally the left hand of the women: she represents the extreme of this categorisation system, which separates male and female, and associates them with different qualities and powers.

Hertz describes both the physical and cultural constraints imposed upon mankind by our structural models, when he says,

> if the constraint of a mystical ideal has for centuries been able to make man into a unilateral being, physiologically mutilated, a liberated and far-sighted society will strive to develop the energies dominant in our *left* side and *right* cerebral hemisphere and to ensure by an appropriate training a more harmonious development of the organism' (Hertz 1973: 31).

The concepts of masculinity and femininity, as opposed to male and female, enable different models to be constructed to describe the prostitute's position in society. Masculinity refers to a group of qualities, including vigorousness, decisiveness, capacity for objectivity, and action, which may be more expected of men. Femininity is a complex of qualities associated with women, particularly necessary in their role of caring for children. Notable amongst these are sensitivity and awareness of others' feelings, with a resulting ability to relate to other individuals in a deeply personal way. In Western societies it has become common place to recognise that men have feminine feelings and talents, as women have masculine elements of their makeup. Our images of both men and women have become more complex and closer to reality than they were, or still are in many other societies, and even in areas of our own. Where there exist models of male and female that attempt to ignore or transcend reality, there is, inevitably, a great deal of tension.

The wider the gap between the ideal and reality, the greater the 'problem' of women is likely to be.

A similar lack of fit, or pretence, leads to tensions in the Nupe images of male and female. The model of male behaviour assumes men to be able to control their wives' sexual affairs by economic sanctions, which the majority of men are in fact unable to do. This fact, and other problematic areas of male and female identity, are dealt with in the Gunnu rituals, which represent ideal or inverted behaviour. The major elements of the ritual are:

1 the exclusion of women from the esoteric rites, which provides a wish-fulfilling demonstration of male dominance and independence of women, which is far from being the position in reality.
2 Childless women are ostracised and punished, when in reality many women live happily in this state, especially those who trade and prostitute themselves.
3 Abandoned behaviour, which is not usually permissible in a society that is prudish about sex.
4 Boys dress as girls, which callenges the view of male existence as always preferable to that of the females.

Nadel describes a nostalgia among the Nupe for a time when women knew their place, but the ideal is always out of step with reality and women are made responsible for the problems of this mis-matching. Women are 'the witches of the society made to carry its fault or failure' (Nadel 1954: 176), when the dissonance between ideal and reality becomes inescapably apparent.

1a In Pitt Rivers's words, quoted by J. Davis, honour is 'the value of a man in his own eyes, but also in the eyes of his society' (1965:21). In the brothel the elements of self-knowledge and appraisal are neglected, and the emphasis lies instead upon display and competition with other men. Honour relates to a man's political and economic power, in comparison with other men. For the purpose of this paper, the most important aspect of honour is that it also rests upon the sexual behaviour of the man's female relatives. (His ability to control them is, in fact, intimately related to economic power.)

Davis describes an honourable man as 'one who is seen to be a good husband, an efficient producer, able to support his wife and children without sending them into the fields, and whose wife is faithful, whose daughters are chaste.' A man, in fact, who is able to be effective, as *machismo* requires. In Cuzco such a man is not stigmatised for going to the brothel, where he becomes involved in the more extreme displays of *machismo* which occur there. The men who use the prostitute, especially during *callejon*, are precisely those who have difficulty in fulfilling the more positive requirements of honour - for instance, in providing adequately for their families. They have to resort instead to the humiliation of the prostitute to 'complete' themselves.

The passivity and submissiveness of women required to co,-plement the *macho* have been described. This goes along with a great fear of women's powers, and especially of their sexuality. Pitt Rivers has described the beliefs about women held by the Spanish people of the Sierra '... supernatural power from grace belongs to women' (Pitt Rivers 1954:195). 'During the time of their period women possess certain involuntary destructive powers' (*op. cit.*:197). Menstruating women are said to be *en celos* (in season). The uncertainty for men over when a woman is menstruating, and the undefined nature of her evil powers at that time, leads to a continual fear of women amongst men. Women who are said to experience sexual desire are described as *calio* (hot). They are feared, Pitt-Rivers says, for their possession of the 'potentially anti-social force of female sexuality'. This fear of female sexuality relates to the degree of unreality of the images of male and female which the society attempts to maintain as fact. Women are said to be submissive and asexual, but the lie is given to the fiction that this can be true of all women, all of the time, by men's openly expressed fear of women's powers related to their sexuality.

The images of male and female among the Sarakatsani, described by Campbell, and Neapolitan slumdwellers, studied by Parsons, point up interesting parallels and contradictions within these images in two societies, both concerned with honour and *machismo*. These illustrate the hypotheses above, and help to clarify the discussion in the previous section about the 'mutedness' or otherwise of women's models of society.

The Sarakatsani are a Greek shepherding people, who divide their universe into men and sheep, who are placed in opposition to women and goats. Women are feared - female demons are described as 'ladies who bring good luck' (Campbell 1964:332), a propitiatory euphemism which indicates their power. 'Women are not, of course, simply creatures of the devil, but the nature of their sexuality, which continually threatens the honour of men, makes them willingly or unwillingly, agents of his will' (*ibid.*:31). Fear of female sexuality is shared by women themselves, who feel 'an instinctive revulsion from sexual activity, an attempt in dress, movement and attitude to disguise the fact that she possesses the physical attributes of her sex' (*ibid.*: 270). Denial and self-hatred, which is so strong a part of the prostitute's self-image, is common to all women whose view of themselves is determined by *machismo*, which sees female sexuality as potentially dangerous and disruptive, notably to the male's self-image as powerful and able to control himself and his environment. As Campbell says, honour for the Sarakatsani involves a 'struggle of self-discipline over cowardice and sensuality', which requires a man to have 'boundless self-confidence in his own powers' because few others will help him (*op. cit.*:269).

A further interesting point which Campbell makes clear, is that the seduction of a woman involves a *loss* of prestige for her and her family. Perhaps it is in relation to this that a client 'completes' himself by using the prostitutes body in El Descanso.

Campbell says that honour is dependent upon the opinion of others. It requires an outward conformity to a code of sexual shame for both men and women, but it does not relate to a person's own estimation or knowledge of themselves. (This echoes Paz's description of women as not embracing the cultural values to which they are forced to conform.)

This hollow, outward conformity to a code of conduct is the common theme running through the images of men and women in societies concerned with honour. The use of the prostitute's body in the brothel provides a means for the near-display of sexual prowess, which the Cuzqueno concept of honour requires, without actually having to investigate for oneself or others the reality behind the image. As John Davis says, 'the control of women's sexuality, and the resulting combination of sex and self-importance makes a unique contribution to the human comedy in life and literature' (Davis 1969:69).

Anne Parsons discusses Neapolitan families. She quotes an informant who says that, of course, women have honour (which is here equated with virginity): after all 'how would anyone ever know if a man had it or not?' (Parsons 1969:349). From a woman's perspective honour belongs to women, but 'there is, nevertheless, a sense in which the idea of honour is relevant to masculine identity, as well as feminine' (*ibid.*:349). There are interesting points of variance between models built up from a male or female perspective. Parsons describes a situation in which men tend to join peer groups, who take the streets as their territory rather than remain within the home. The reason for abandoning the home is economic. Men often cannot support their families as the ideals of *machismo* require. They therefore avoid the area in which this harsh reality challenges the male self-image. The exchange of gifts and food is an important element of the Neapolitan culture, and is associated with women. The central importance of the Madonna in the cultural and religious life of the city provides further evidence for Parsons's view that 'cultural values centre on the feminine image' (Parsons 1969:367).

In many respects this is also true of Cuzco, where, for instance, families are held together by the mother. But the feminine image is a myth, as is the masculine one, and the problems arising in the attempt to live the myth as if it were reality are expressed, and perhaps resolved, in the resort to the prostitute's body.

2a *Macho* has been discussed enough already, but there is a term that can be related to it, which sheds further light upon its meaning. This is the word *Manso*, which means a tame, meek, domesticated animal. Lois Paul describes how, in the Guatemalan village that she studied, it is also used as slang to refer to an impotent man.

2b The roughly equivalent dichotomy expressed along a female dimension, is the split between virgin and whore. Whore and Virgin can be most readily distinguished as being sexual and asexual images of women. As described, women's sexuality is feared as anti-social and potentially disruptive, and so it is not surprising that whores tend to be feared for their powers. I have described the ways in which society tries to control the prostitutes of Cuzco, especially by distancing the brothel from the city. A further expression of space as a dimension that divides the virgin from the whore is the domestic or extra-domestic area of their existence. There are many societies in which women of the streets are scorned, feared, and perhaps also admired, for the ability to leave the safe confines of the home, particularly since, as Nadel says, the 'economic independence of the women connotes the inadequacy of the man' (Nadel 1954:175). In Naples, 'a woman of the streets is one who had violated the taboos and in a sense taken over masculine prerogatives' (Parsons 1969: 347). Perhaps in societies concerned with honour, one can also relate these four concepts - *macho*, *manso*, whore and virgin - along the continuum of masculinity/femininity.

In Cuzco the prostitute is *macho* in that she is able to look after herself, and her children efficiently, and make adequate financial provision for them. The sexuality of the prostitute is equated with the desired quality of *machismo*, so that there is a sense in which a client can recharge himself from the prostitute's body. The world view of women's roles is categorised as follows:

Whore/sexual	Virgin/asexual
extra-domestic	domestic
exotic	familiar

dangerous	safe
macho	*manso*
masculine	feminine

How does this relate to the new *poses* required of Cuzco prostitutes?

NATURAL AND *POSE*

The need for prostitutes is expressed by men of Cuzco as a need to complete themselves. In the jungle oil camps, where men have to live for months isolated from town-life and women, the oil companies consider it essential to fly prostitutes in to service the camp at regular intervals. If the weather prevents this, after about six weeks the men stop work, saying that they *cannot* continue until they have had the use of a woman's body. The need is seen as fundamental. Once a prostitute has arrived and moved quickly around the dormitories, work continues as usual. These visits are a high priority for the company and for the girls, who can earn a great deal of money very quickly. Here again the man is seen as very dependent upon the woman: masculine pride forbids masturbation, and he cannot continue to function without access to a woman's body. The *macho's* facade of independence is extremely brittle.

Men aim to gain *machismo*, and using a prostitute's body fulfils an important, and at times crucial, part of this aim. If a man seduces a woman from outside the brothel, he gains *machismo* by humiliating her male relatives, who lose honour. A prostitute has no male protectors. She is dependent upon female strength and solidarity. No men lose their honour when a man takes a prostitute to bed. The loss involved when a man 'completes' himself in this way is therefore not of the same order as a family's loss of honour. In the context of the *parranda machista*, the man is looking for *machismo*, and this is what he gains. Because sexuality is equated with masculinity, and asexuality with femininity, the *parranda machista* is an orgy of sexuality which therefore becomes an orgy of masculinity, and from participating in this orgy, the elusive quality of *machismo* can be gained. The display which takes place in the brothel derives its cultural significance from a confused equation of masculinity with both sexuality and dominance. When sexual acts are viewed as acts of communication, the position *natural* and the *poses*, which make the man the active, sexually dominant partner, reinforce this oversimplified condensation.

Poses which require the woman to dominate the man express an opposite conceptual system, as they explode the myth of women as asexual, and submissive. Prostitutes, at least, are recognised as sexual, active beings. At the same time an area appears in which it is culturally acceptable for a man to be passive and to capitulate to a woman. The repertoire of concepts of male and female is enlarged. Instead of one dichotomy between the sexual, dominating man of *machismo* and the sexually pure, passive woman, these new *poses* create several polarities, active/passive, dominant/submissive, sexual/asexual, - which interrelate in different ways in different situations.

Such fundamental changes in world view necessitate and set in motion profound shifts in many areas of society, in this case particularly in marriage. At the time of my study the brothel was an arena in which it was possible to observe and understand the generation of new cultural forms and symbols as they received their first tentative expression in the embrace of prostitute and client.

ACKNOWLEDGEMENTS

I visited El Descanso from December 1972 to July 1973. I was working with Dr George Primov, who has generously allowed me to use those parts of our material collected together. We were very lucky to find the owner, and especially the administrator, of the house extremely helpful and understanding. Without their friendship, the study would not have been possible. When I returned for a brief visit in 1974 they again enabled me to visit the house and talk with the prostitutes whenever I wished. I am very grateful to Dr John Davis, Sean Conlin, W.D. Wilder, and Elizabeth Long, who read earlier drafts of the paper and helped me enormously in re-writing and re-ordering my ideas.

REFERENCES

Bataille, G. (1962). *Eroticism.* London: Calder.
Van den Berghe, P.L. (1975). *Class and Ethnicity in Peru.* Leiden: E.G. Brill.
Campbell, F.K. (1964). *Honour, Family and Patronage.* Oxford: Clarendon.
Castro, Silva, Majluf and Estrada (1964). *Dinamica de la familia Peruuna* Mimeo: Lima.
Davis, J. (1969). Honour and Politics in Pisticci. *Proceedings of the Royal Anthropological Institute,* 69-81.
Davis, J. (1974). *Choice and Change* (ed.) L.S.E. monograph No. 50. London: Athlone Press.
Hertz, R. (1973). The Pre-eminence of the Right Hand. In R. Needham (ed.) *Right and Left.* Chicago: University of Chicago Press. 3-31.
Nadel, S.F. (1942). *A Black Byzantium.* Oxford: Oxford

University Press.
Nadel, S.F. (1954). *Nupe Religion.* London: Routledge and Kegan Paul.
Navares, S.R. (1970). El Machismo en Mexico. *Mundo Nuevo.* Argentina: Instituto Montevideo Latino-Americano de relacienes internacionales, *46*, 14-19.
Ortner, S.B. (1974). Is female to male as nature is to culture? In Rosaldo, H.R. and L. Lamphere (eds.) *Women, Culture and Society.* California: Stanford University Press. 67-88.
Parson, A. (1969). Is the Oedipus Complex universal? In W. Muensterberger (ed) *Man and his Culture.* London: Rapp and Whiting, 331-384.
Paul, L. (1974). The mastery of work and the mystery of sex in a Guatemalan Village. In Rosaldo, H.R. and L. Lamphere (eds) *Women, Culture and Society.* California: Stanford University Press. 281-299.
Paz, O. (1961). *The Labyrinth of Solitude.* New York: Grove Press.
Pitt-Rivers, J. (1954). *The People of the Sierras.* London: Weidenfeld and Nicholson.
Rosaldo, H.R. and L. Lamphere. (1974). *Women, Culture and Society.* California: University of Stanford Press.
Sandford, J. (1975). *Prostitutes.* London: Secker and Warburg.
Winick, C. (1968). *The New People.* New York: Pegasus.
Winn, D. (1974). *Prostitutes.* London: Hutchinson and Company.

THE DEFINITION OF FEMALE SEXUALITY AND THE STATUS OF WOMEN AMONG THE GUJERATI-SPEAKING INDIANS OF JOHANNESBURG

Leora N. Rosenthal

In recent years a few writers have followed up de Beauvoir's argument that a society's definition of female sexuality is related to the inferior status of women. Women are said to be regarded as sex objects - playthings whose primary function is to satisfy men's sexual urges (Greer 1971:58, Firestone 1973:139-147, Mitchell 1974:110, de Beauvoir 1975:543). Women's own sexual feelings are said to be weak, passive and insignificant, and women's enjoyment and need for sex are believed to be less than that of males. Greer maintains that this definition of female sexuality serves to deflect and dissipate women's energy, so that their human potential is drastically reduced (Greer 1971: 68-69). Firestone suggests that the stereotype of women as love objects reinforces female dependence on males by preserving direct sex pleasure for the latter, and allowing women satisfaction only through vicarious identification with men (1973:141).

My aim in this paper is to examine the definition of female sexuality amongst the Gujerati-speaking Indians of Johannesburg, and to see how this is related to the status of their women. I shall first define what I mean by 'female sexuality'. The concept comprises female sexual physiology, the stereotype of femininity, and freedom of female sexual expression.

(1) *Female sexual physiology.* This concerns the ways in which society perceives the female sexual response, whether it believes women to be capable of sexual desire, orgasm etc. These beliefs may be based on the study of books such as Masters and Johnson's study, *The Human Sexual Response*, and the *Kamasutra*, or on oral tradition.

(2) *The Stereotype of femininity.* This concerns those characteristics that society idealises in women and which constitute what may be termed 'the Feminine Ideal'. The stereotype is likely to be closely associated with society's perception of female sexual physiology.

(3) *Freedom of female sexual expression.* This concerns the

degree to which women's need for sexual expression is taken into account, and the ways in which society views female sexual enjoyment.

Before I examine definitions of these three aspects of female sexuality among the Gujerati-speaking Indians of Johannesburg, a brief introduction to their society and culture is necessary.

HISTORICAL BACKGROUND

Indian immigration to South Africa began in 1860, following an agreement between the governments of Britain, India and Natal, which made Indian immigrant indentured labour available to sugar farmers in Natal (Pachai 1971:60). Those Indians recruited to work under the five-year labour contract were mainly Tamil-Telegu-and Hindi-speaking Hindus. In 1870 a new wave of immigration began from North India. These immigrants were mainly Gujerati-speaking Hindus and Moslems, and were known as 'free' or 'passenger' immigrants, because they paid their own passages to South Africa and were not bound by any contract. The Gujerati-speaking Indians initially made their livings as traders among the indentured labourers, providing them with spices and essential commodities. Later they expanded their markets and began to trade among other ethnic groups in South Africa (Meer 1969: 10-21). Many traders left Natal, trekking northwards into the Transvaal and later into Rhodesia (Dotson and Dotson 1968:35-36). When Johannesburg was founded in 1880, a few Gujerati Indian traders were already in the area and ready to establish their businesses (Randall and Desai 1967:1-2).

The present Indian population of Johannesburg is about 40,000, of whom approximately 50% are Gujerati-speaking. However, only 18,000 Indians live in the city of Johannesburg itself. The rest live in a satellite township, Lenasia, 22 miles from the city centre. They have been compelled to move there under the Group Areas Act, which sets aside specific areas for members of each 'racial' group for the purposes of business and residence. This transfer of Indians to their own 'group areas' also involves the eviction of business owners from their premises in Johannesburg, which usually means the loss of an important source of livelihood. The Government made a concession to Indian business owners by building the Oriental Plaza, a large shopping complex providing premises for about 400 Indian shops in Johannesburg. Only a small number of evicted businessmen, however, are able to obtain premises at the Plaza.

Two years' fieldwork was carried out between July 1972 and March 1975 among the Gujerati-speaking Hindus and Moslems living in the city of Johannesburg.

GUJERATI INDIAN FAMILIES IN JOHANNESBURG: ECONOMIC BACKGROUND

The section of the Indian population that lives in Johannesburg is mostly in Vrededorp, Fordsburg and Fereirastown, which were among the first suburbs to be established in the city. Dwellings in these areas are about sixty to eighty years old, and comprise small semi-detached houses and blocks of flats, many of which are in dilapidated condition.

Overcrowding is common, and it is quite usual to find a family of six or more sharing two small rooms. Despite these conditions, the traditional extended family system has survived. In a survey of 204 Gujerati households which I carried out in these areas, 44.6% of the sample was found to comprise various types of extended family, while 55.4% comprised nuclear families. The most common type of extended family, comprising father, mother, their married sons, daughters-in-law, grandchildren and unmarried daughters, constituted 17% of the sample.

Many informants said that in spite of overcrowding they preferred to live in Johannesburg rather than in the smarter township of Lenasia. The main reason for this preference was that in Johannesburg they were closer to their places of work, particularly in the case of business owners, who live above their shops. Most businesses are owned by Gujerati-speaking Hindus and Moslems. Half of the gainfully employed males in my sample were full or partial owners of businesses. Of the remainder, the majority were in white collar occupations, with only 5% in skilled or semi-skilled work, and no unskilled labourers.

Gujerati Indian businesses are essentially family businesses: various members of the family are to be found serving behind the shop counter, and shareholders are generally family members (Kharsany 1970). The Gujerati shop owner typically brings his sons into the business, perhaps initially as assistants, but later making them partners, and eventually leaves the business to them on his retirement or death. Brothers may continue in the same business for a while after the death of the father, but usually each one attempts to branch out on his own.

STATUS AND ROLE IN THE FAMILY

Status in Gujerati families is determined by age and sex. The old are superior to the young and men are superior to women. Younger members of the family are expected to behave respectfully towards older members, although siblings, particularly if the age gap is small, tend to treat each other as equals. 'Respectful' behaviour means refraining from such activities as drinking or smoking or engaging in light-hearted conversation. It also means obeying the requests of the respected person and not 'talking back' to him or her. The most respected person is the father, who wields considerable authority over junior family members. Although grown-up children are permitted in most cases to voice their disagreement with his decisions, he nevertheless has the final word. As sons grow older their own authority increases, yet they retain their respect for an elderly father, and would not act without consulting him.

Respectful behaviour is also due to the mother. However, whereas the father/child relationship is characterized by authority and submission, the mother/child relationship is characterized by mutual affection. Indian mothers pamper, cosset and spoil their sons. Daughters are expected to help their mothers in seeing to the comforts of their brothers. This exceptionally close bond between mother and son leads to a situation of potential conflict when the latter marries and brings his wife to live in his parental home, for the mother-in-law and daughter-in-law now both compete for the affection and attention of the same man.

Traditionally, Indian marriages were arranged by the parents of the couple, who were not supposed to see each other until the wedding day. In my survey, 57% out of 283 married women of all ages had never seen their husbands until the wedding day. The rest had been permitted at least one or two interviews with their prospective spouses before the proposals were accepted. After marriage the bride moves from the family in which she has grown up to the home of her husband's parents, where she is a stranger. She is expected to plunge immediately into a round of chores, and comes under the authority of her mother-in-law, whom she is expected to obey without complaint, and who is depicted in the traditional stereotype as a harsh, exacting slave-driver. The husband usually sides with his mother in situations of conflict, and the wife is expected to bear all insults and ill-treatment with a good grace, for she has been taught since early childhood that it is her prime duty to serve her husband and his family regardless of the burdens and hardships which might be imposed upon her.

During the course of my fieldwork, however, I did not encounter this extreme situation. While young couples were expected to spend at least a few years in their husbands' parents' homes, some were permitted to set up independent households immediately. These were considered extremely fortunate. Most of the daughters-in-law whom I interviewed appeared to get on amicably with their mothers-in-law, and were not placed under severe restrictions. Informants attributed this to westernization and some maintained: 'Nowadays the daughters-in-law are telling the mothers-in-law what to do'. This however, is an exaggeration, for the daughter-in-law living in her mother-in-law's house is still in a subordinate position. She cannot come and go as she pleases, and requires her mother-in-law's permission before she may visit her own mother. When visitors come to the house, she is the one who is expected to serve them. She may not even dress as she pleases, and among the Hindus she is expected always to wear a sari. Unmarried women living in the same house wear more comfortable clothing such as dresses or slacks, as do young married women living in their own independent households.

Many Gujerati Indians employ African domestic servants to help with the cleaning, but certain chores, notably cooking and child care, are never assigned to servants. People explained that servants could not be trusted with these tasks, and that Indian cooking requires skill that can only be developed over a period of years. My own view is that these tasks constitute the most important aspects of the woman's traditional role, and it is therefore unfitting that they should be taken over by servants. For example, the wife's obligation to serve her husband usually takes the form of her preparing his food and serving it to him. In Gujerati homes, men are usually served their meals first by women, who eat afterwards. It is unacceptable for a man to serve his wife. Divorce is rare. There were only eight divorces out of 520 married people in my survey. It is said that regardless of the circumstances of the divorce, the women is always blamed, because she is expected to remain loyal to her husband no matter how badly he may treat her. While many conceded the unfairness of this rule, they maintained that if a woman found herself married to a scoundrel she should not divorce him, but rather try to reform him through an example of moral rectitude.

SOCIAL CHANGE AND THE PROBLEM OF FEMALE SEXUALITY

The rising cost of living and the displacement of Indians from family businesses has made it increasingly necessary for women to seek employment outside the home. 344 out of 578 women in my survey were classified as 'potentially employable', that is, over fifteen years of age, not studying, and not too old to work. But of this number only 60, or 17%, were gainfully employed outside the home, and most of these fell into the younger age categories. The numbers of employed Moslem and Hindu women were roughly equal, but Hindu women showed greater achievement than Moslem women in the field of education. No Moslem woman over 30 years old had achieved matric education, but 32.6% of those between 15 and 29 had done so. 27.6% of the Hindus between 15 and 29 had achieved matric education, compared with 16.6% of those between 30 and 44, and 10% of those between 44 and 54. No Hindu woman over 54 had achieved matric education. I found, in addition, that many parents are sending their daughters to secretarial colleges. The result of these changes is that more young girls, who had previously led cloistered existences in the home, are now exposed to the influences of western education and mass media. Many are beginning to copy western ways, and are beginning to question the traditional restrictions which their elders still seek to impose on them.

Those areas in which change is most desired are in household organization and in the choice of spouse. The young women expressed the desire to set up their own independent households at marriage, and also to select their own husbands and to get to know them before marriage. Already certain changes are detectable in the system of arranged marriage. Thus, only 26% of married women between 17 and 29 had not seen their husbands before the wedding day, compared with 43.5% of women between 30 and 39, 85.3% of women between 40 and 49, and 93.5% of women over 50. On the other hand, 18% of women under 40 had had a limited say in the choice of spouse, compared with 5% of women over 40, while only 11.5% of women under 40 had had complete freedom in the choice of spouse. No women over 40 had had such freedom. There was no significant difference between Hindu and Moslem women in this respect. I also found that norms and values pertaining to courtship and marriage were strikingly similar in both groups, and that one group was not more westernized that the other, despite the greater educational attainments of the Hindu women in the survey.

In spite of these changes, courtship and dating are strictly forbidden among the Gujerati Indians, and young girls who wish to date boys must do so in secret. Parents seemed to assume that unless girls are conscientiously restricted in their movements and closely watched, sexual promiscuity would be the outcome. It is believed that if girls were allowed to date boys, they would naturally go to bed with them. Indeed, girls who are known to have been alone in the company of the opposite sex are automatically suspected of being unchaste. Since Indian schools are co-educational, they are looked upon by parents as potential breeding grounds of vice, and a schoolgirl who wishes to complete her education must be extremely circumspect in her behaviour. This means, for example, that she must never be seen talking to boys in public. Girls who work come under similar suspicion, and married women who work must keep their distance from men. In one case, a jealous young Moslem stabbed and injured his wife for coming home from work fifteen minutes late. An elderly member of the community commented that the husband was to blame. If he wants his wife to be beyond reproach he should keep her at home and not send her out to work.

Most middle-aged informants maintained that if women were given the freedom to come and go as they please, it would lead to sexual promiscuity, divorce, adultery and illegitimacy, and they pointed to western society as one ravaged with sexual immorality as a result of the freedom of women to associate with men.

It is widely believed that it is in the nature of women to be unable to control their sexual desires. Further evidence of this emerged in discussions that my husband, Dr N.E. Rosenthal, held with a group of Gujerati Hindu men between the ages of twenty and thirty, and a group of Gujerati Moslem men of approximately the same ages. They considered that Indian women, in their normal restricted environment, manifest little, if any, interest in sex. If, however, the restrictions are lifted and they are permitted to associate freely with men, their sexual desires emerge strongly. Thus, one informant maintained: 'The Indian woman has been in a cage. When she is let loose she is twice as desirous as the man. She tries to make the best of every hour.' Another said: 'If a woman is not kept under lock and key she will usually give in to the man's sexual advances. Once she gets used to going out, her desire will become equal to or even greater than that of the male.' A similar belief is expressed by the South African Indian Moslem religious writer, Adam Peerbhai, who writes:

> It needs no great imagination to know of the mystery of women and the dynamic story of their amazing secret life. Women who hold high positions in society, appear refined, cultured, educated, virtuous and innocent by day, but represent a totally different picture by nightfall. By night, under "neurotic compulsion" they commit the immoral act, bringing with them human degradation, destruction and sometimes death. Women are thus pruned out of their peaceful existence.
>
> A woman is not drunk when she commits the act. She gets befuddled by the powerful "sex currents" and "sex waves" that are ever present in the atmosphere. This is the first indication of her desire - her sex desire - which presses upon her loins and sweeps upwards into her body, filling the mind with one impelling thought. Probably the woman, after years of caution, of devoutness to religion, of self control and self denial, gets release of her heated blood, through the compulsion of her uncontrolled emotions and her long repressed passions (Peerbhai 1968:8).

Manu, the Hindu lawgiver of the 4th century A.D. wrote about women in a similar vein:

> Women do not care for beauty, nor is their attention fixed on age; they give themselves to the handsome as well as to the ugly, just for the fact that he is a man. If she be chaste, it is because she has not found the proper man, place or

> opportunity.... It is the nature of woman to seduce man in this world. She is able to lead astray not only the ignorant but even the learned man and make him a slave of lust (Kapadia 1966:254).

Most of the informants with whom I discussed Peerbhai and Manu seemed to regard them as fanatics. Nevertheless, these writers express, albeit in exaggerated form, a sentiment that seemed to prevail among Gujerati Moslem and Hindu men - namely a mistrust of the unshackled female libido. This sentiment can only be properly understood, however, in terms of a fundamental Indian value which I have hitherto not mentioned: the value of female chastity.

THE GUJERATI INDIAN IDEAL OF WOMANHOOD

Karve maintains that there was a time in the history of ancient India when female chastity was not greatly valued. Unmarried girls in their father's homes were permitted to have lovers. At the same time impotent husbands could beget children by engaging strangers to sleep with their wives, and the custom of leviratic marriage was practised. Around the beginning of the Christian era, however, these customs fell into disrepute and were even condemned, while at the same time female chastity came to be greatly valued (Karve 1965:82-85). Karve also notes that the value of female chastity seemed to emerge at the same time as laws which stress the importance of physiological paternity in determining the legitimacy of children (*ibid.*:82).

Koranic injunctions require premarital chastity and postmarital fidelity of both men and women, although men are permitted to have four wives at any given time, and similar standards of sexual morality are expected of Hindu men and women. But these standards are enforced only for women, while sexual licence among men is tolerated and even accepted as normal.

A girl is taught that virginity is her most prized possession, and one attribute that she must have in order to qualify for wifehood. A great mystique surrounds the first night of marriage, when the bride's virginity is physically put to the test. If a husband should find that she is not a virgin, he may annull the marriage there and then. In order to avoid this possibility a man is unlikely to accept as a wife a girl with a tarnished reputation. The stigma of a bad reputation may be acquired without any direct proof of misconduct. Vague rumours are sufficient to secure a girl's condemnation by the community. Even chaperoned courtship is disapproved of by Gujerati Indians, not because anything could go wrong in the presence of a chaperone, but because young women are expected to preserve their emotional as well as their physical virginity. If a man should win a girl's heart and then not marry her, she loses her marriageability owing to the belief that 'a woman only gives her heart once'.

Young men, on the other hand, are expected to gain considerably premarital sexual experience with 'bad' women. In South Africa these are often women of other 'racial' groups who do not fall under Indian codes of sexual morality. They may also be Indian women who, having lost their reputation, became 'good time girls'.

The value on virginity must also be seen within the context of westernization and changing value systems. Thus, while all parents were convinced that a bad reputation would ruin their daughters' marriage chances, I came across only one case in which a potential suitor did not propose because the girl was known to have had previous boyfriends. The same girl, however, subsequently married another man, who did not seem to mind about her past 'misconduct'. Informants were unable to point out to me any women who had remained spinsters owing to loss of reputation, although some maintained that the odd spinster in the community was probably a woman who had lost her virginity and subsequently refused to marry for fear of being rejected by her husband after the wedding night.

Most of the young unmarried men who were interviewed, said that they would be prepared to forgive their wives for loss of virginity. The following is an extract from an interview with a young male university student:

Question: Would you marry a girl whom you knew had been out with other boys?

Answer: This would depend on the girl and the boys she had been out with. If she was a serious girl who had dated respectable boys and been let down, it would not matter to me. If she had just gone out for kicks and had run around with bad types it would matter to me. On the whole, though, if I were in love with a girl who had been out previously I would marry her.

Girls, on the other hand, feel less confident that boys will accept them as wives if they are known to have dated others. Most of the girls whom I interviewed had dated boys at some time or another, and many of them suffered strong guilt feelings on that account.

'We feel', said one girl, 'that if we go out with a boy and he does not marry us, then we are being unfaithful to the man whom we eventually marry'. She continued by way of explanation: 'It's the men who want it that way. They want us to be like from the cradle when they marry us.'

In spite of considerable westernization, then, the orthodox Gujerati Indian feminine ideal remains that of a pure virgin who becomes a chaste, loyal wife. The woman's body, and particularly her genital tract, are regarded by both men and women as the property of the man who marries her. Indeed the very social environment of an Indian woman is designed to preserve her premarital and postmarital chastity for the man who owns her. Even young westernized women regard their bodies as the property of their future husbands.

GUJERATI INDIAN ATTITUDES TO FEMALE SEXUAL EXPRESSION

A society's attitude towards female sexual expression must be seen in the context of its views on female sexual physiology. For example, in Victorian society, since women were believed to be devoid of sexual feeling, their need for sexual expression was not taken into account. Indeed, sex for women was considered to be an unpleasant marital duty. Among the Gujerati-speaking Indians, since the sexual needs of women are recognized, some satisfaction of these needs is considered appropriate.

My informants held that Indian women are entitled to sexual satisfaction in marriage. Some even referred to the ancient Hindu treatise on sex, the *Kamasutras*, as proof that the sexual lives of Indians are uninhibited, and one middle-aged man maintained: 'It is the husband's duty to teach his wife about sex and to initiate her into the joys of lovemaking'. Another suggested that early marriage was a solution to the problem of sexual desire in a young girl.

There are, however, factors that may possibly inhibit female sexual expression. Many young female informants claimed that the poor sex education which the Indian girl receives in her parents' home does not prepare her psychologically for a successful sexual union with her husband. A twenty-year-old unmarried girl told me:

> Our mothers do not teach us about sex. When a girl starts to menstruate at the age of about twelve, her mother tells her to stop talking to boys and not to go near men. The girl does not know what will happen to her if she goes near a man. She thinks that something terrible will happen to her. She grows up to think of sex as something disgusting.

Aileen Ross describes similar findings in her study of modern urban Hindu families in India (Ross 1960:266).

Older women, on the other hand, did not think that it was necessary for girls to learn about sex from their mothers, and maintained that it was the husband's duty to teach a young bride.

A number of young women also believed that the arrangement of marriages by parents was responsible for poor sexual adjustment between the spouses.

> In arranged marriage the man is a relative stranger. The wedding night is traumatic because the girl is pounced upon by a strange man. Men do not try to be gentle on their wedding nights. They practically rape their wives. This leaves a scar, and after that the girl feels that sex is repulsive. She can't give herself to her husband, and there is no close relationship between husband and wife. In a love match, on the other hand, you know your guy inside out, and you are not scared to sleep with him.

There was by no means unanimous agreement on this point. Most older women felt that falling in love and courtship could take place after marriage, while some younger women even suggested that it might be exciting to go to bed with a stranger.

Owing to the great value on female chastity, however, any sexual expression of women outside marriage is severely condemned by both men and women. A woman whose husband has extramarital affairs while she is totally neglected is expected to suffer in silence. She cannot divorce him, nor can she take a lover, without being ostracized by the community. Such behaviour would be antithetical to the feminine ideal of the pure, loyal, self-sacrificing and devoted servant.

DISCUSSION

I would like to return to the contention of de Beauvoir and other writers, who have maintained that the subordinate status of women in European and American society is related to a view of their sexuality as being weak and passive, and the belief that their sexual needs are less than those of men. Among the Gujerati-speaking Indians of Johannesburg, as I have shown, women are regarded as sensual creatures and their need for sexual satisfaction is recognized. Yet their status is unequivocally subordinate to that of men.

What should perhaps be emphasized is that while in most societies women have inferior statuses to men, the definition of female sexuality is not necessarily uniform, although different definitions may be equally demeaning to women, each in its own way. I also suggested in my introduction that female sexuality comprises at least three separate aspects, each of which may be considered separately in relation to status. Among Gujerati Indians it is chiefly the feminine ideal which demeans women, in that it defines them as the sexual property of men. The fact that sexual feelings in women are recognized, simply leads to their being kept under stricter supervision in order to prevent them straying, and to preserve their sexual purity for their husbands. Indeed, while the Indian view of female sexual physiology is the opposite of the 'western' view described by Greer, de Beauvoir, Firestone and others, its effects are to surround the woman in a web of restrictions and prohibitions, and to cast suspicion on her every movement. While it may be true, as some writers have suggested, that female sexuality is defined by men to benefit men, it is nevertheless possible that men in different cultures have different goals and needs, and that their definitions of female sexuality may consequently differ. Further research in this field may reveal some interesting relationships between variables that have hitherto been studied mainly in western societies.

ACKNOWLEDGEMENTS

I would like to thank Mrs Sabitha Jithoo of the University of Durban Westville for her criticism of an earlier draft of this paper; and my husband, who conducted the group discussions on sex with Indian men in the course of his medical work.

REFERENCES

de Beauvoir, S. (1975). *The Second Sex.* Penguin. (Orig. 1949).
Dotson, L. and Dotson, F. (1968). *The Indian Minority of Zambia, Rhodesia and Malawi.*
Firestone, S. (1975). *The Dialectic of Sex.* Paladin.
Greer, G. (1971). *The Female Eunuch.* Paladin.
Kapadia, K.M. (1966). *Marriage and the Family in India.* London: Oxford University Press.
Karve, I. (1953). *Kinship Organization in India.* Poona: Astra.
Kharsany, E.E. (1971). *A Profile of the South African Indian Business Community, Reasons for Success and Failure.* Unpublished dissertation for the degree of Master of Business Administration, University of the Witwatersrand, Johannesburg.
Meer, F. (1969). *Portrait of Indian South Africans.* Durban: Avon House.
Mitchell, J. (1974). *Women's Estate.* Pelican.
Pachai, B. (1971). *The South African Indian Question 1860-1971.* Cape Town: Verry.
Peerbhai, A. (1968). *Women of Evil and Islam.* Durban: Islamic Institute.
Randall, P. and Desai, Y. (1967). *From Coolie Location to Group Area.* Johannesburg: South African Institute of Race Relations.
Ross, A.D. (1960). *The Hindu Family in its Urban Setting.* Toronto: University of Toronto Press.

TO DANCE IS HUMAN

Some Psychobiological Bases of an Expressive Form

Judith Lynne Hanna

Anthropologists have generally had a limited view of dance, although it is a nearly universal and often complex behaviour. 'Then they danced' is the common treatment *ad nauseam*. Ethnocentrism has reigned, and false dichotomies have been drawn between 'primitive' and 'nonprimitive' dance. Dance and similar motor behaviours have not been distinguished. Dance as emotional behaviour has been overemphasized. Focus on the promotion of social harmony (Radcliffe-Brown 1922) has neglected dance's disharmonious consequences (Evans-Pritchard 1965:74), its performance as duty (Reay 1959) and work, and its impact on sociocultural change (Hanna 1975). When dance has been shown to play a critical part in sustaining social institutions, the properties that enable it to do so are rarely discussed.

This essay is an attempt to sketch a fuller picture of dance possibilities, to show that the configuration of human behaviour that is called dance is significantly different from the behaviour of other animals, including that which has also been labelled dance, but that human dance has its roots in phylogenetic and ontogenetic evolution, firstly in predisposing psychobiological processes and secondly in social experience. The essay seeks to identify both the basic psychobiological factors that may generate or modulate the many diverse dance forms found through time and across sociocultural areas, and certain processes that dance has in common with other forms of nonverbal communication, motor activity, play, 'artistic' behaviour, and altered states of consciousness. This is done in the hope that more attention will be directed to the properties of dance that may help to explain why it is a common attribute of temporal and transcendental behaviour in the religious and secular worlds, why it is a component of sociopolitical manipulations, and why it has potential for individual and group therapy.

Because the word 'dance' has numerous referents, an operating definition is called for. The derivation and conceptualization are discussed elsewhere (Hanna 1977b).

Dance is defined as (1) human behaviour composed, from the dancer's perspective, of (2) purposeful, (3) intentionally rhythmical, and (4) culturally patterned sequences of (5a) nonverbal body movement and gesture which are (5b) not ordinary motor activities, (5c) the motion having inherent and 'aesthetic' value. Within this conceptualization, behaviour must meet each of the four criteria in order to be classified as 'dance'. That is to say, each behavioural characteristic is necessary, and the set constitutes sufficiency; the combination of all these factors must exist.

In the following sections, I will discuss the interrelated psychobiological bases for each property of dance. My argument draws from a variety of perspectives. Cumulatively, these may amplify our understanding of dance as thought and action.

DANCE AS HUMAN BEHAVIOUR

Human dance behaviour emerges from an evolutionary process lasting millions of years; therefore, we look for presages in other animals and in infants. On the basis of recent studies, a relatively thin behavioural barrier between humans and other primates is acknowledged. Yet although nonhuman primates carry the germ of visual, motor, auditory, kinesthetic, affective, and cognitive patterns that are more fully developed in humans, certain incipient capacities in nonhumans are realized only in humans. The problem is to find out to what extent humans have a biologically based predisposition for dance as, for example, there seems to be a genetic programme for language (Lenneberg 1967).

It is true that a human can dance mechanically or perform a dance pattern conceptualized and created by someone else, in the same way that a nonhuman can be trained to perform a dance by a human. We have all seen 'dancing' chimpanzees, horses, dogs, bears, parrots, or elephants. However, differences between humans and other animals occur in symbolization, emotional expression, and motor-gesture patterns. The distinction within a species between fixed action patterns and selected action patterns is critical. Bases for these contrasts lie in the development of the human brain, as well as the increased dependency and socialization of human offspring.

Symbolic capability. Evolutionary changes in the human brain are generally accepted to have had relevance for toolmaking and language, and to those we must add dance. The limbic region makes up most of the cerebral cortex in the most primitive mammals, and is apparently associated with emotional behaviour necessary for species survival. In higher primates, areas of the cortex known as association areas have differentially expanded. The development of an 'association area of association areas' in the parietal region, which has connections only with other association areas and not with primary sensory or motor regions or with the limbic system, is greatly increased in humans. The capacity for neurological cross-modal cortical connections allows relatively unlimited expansion of symbolization.

What other animals communicate in motor patterns ethnologists call ritual performances (or dances). These seem to involve immediate emotion and drive (fear, hunger, well-being, aggression, submission, sexual arousal, etc.) and autonomic rhythms. Humans share these behaviours, but they can also communicate abstract concepts removed from an immediate stimulus, distanced in time and space, and in deliberately chosen rhythms. In what is called the 'bee dance', bees communicate through motor patterns to members of the colony, the direction, distance, and richness of the pollen source discovered minutes earlier. It is perhaps less an immediate reaction to the environment than a relating of certain aspects of past experience. Humans have the capacity to project experience extrinsic to themselves, to alter feeling and thought, and to transform it symbolically through varieties of sensori-motor-aesthetic patterns, using different devices and spheres of encoding messages (see later section on aesthetics). They can deliberately express or withhold an experience. Humans can embed symbols within one another, use opposites and inversions, situational qualifiers, and interrelated symbol systems.

Non-humans appear to lack the human level of synesthesia, the capacity to perceive and transmit simultaneously stimuli in several senses. In dance there is the sight of performers moving in time and space; the sounds of breathing, the impact of feet upon the ground, and other physical movements; the odours of bodily exertion; the feeling of kinesthetic activity (the dancing) or empathy (observing); the touch of body to body and/or performing area; and the proxemic sense. Humans can combine media in the dance. They often interweave music, song, and movement, extend or otherwise sculpt the body through costuming and accoutrements (e.g., elongating it through stilts or masks, creating illusions of foreshortening through body paint). They can 'fossilize' or capture the motor-gesture patterns through objectification in lithic, ceramic, or painted form (Kurath and Marti 1964; Hanna 1975), or, more recently, in notation, film, or videotape.

There is, of course, a difference between cognition or

knowledge about dance, and behaviour. One can know the grammatical rules for dance and appreciate its denotations and connotations with notions of appropriateness and not have the skills or opportunity for performance.

Related to the human's greater range and complexity of symbolic behaviour is the fact that no other mammal has the dominance of the left half of the brain, and the dependency upon it for learning certain kinds of knowledge. The left hemisphere appears to be involved in abstract symbolic functions, sequential information-processing, and complex patterns of movement. The right hemisphere seems to involve elementary perceptual tasks, nonverbal processing of spatial information and music, and emotional reactivity. Although the Broca and Wernicke areas, located in the left hemisphere, have been associated with language expression and comprehension, they may be equally important for creating, performing, and understanding complex patterns of movement.

A comparison of the design features of dance with those of verbal language described by Hockett and Ascher (1964) emphasizes the symbolic potential of dance. Dance, like language, has directional reception, interchangeability (someone can be both a transmitter and receiver, arbitrariness (many characteristics cannot be predicted), discreteness (bound to the time continuum), displacement (reference can be to something not immediately present), productivity (messages never seen before can be sent and understood within a set of structural principles), duality of patterning (a system of action and a system of meaning), cultural transmission, ambiguity, and affectivity (expression of an internal state with potential for changing moods and a sense of situation).

There are, however, important differences between dance and verbal language. In dance, motor/visual-kinesthetic-auditory-olfactory-proxemic-tactile channels predominate, instead of the vocal/auditory channels. Whereas language exists primarily in a temporal dimension, dance involves both temporal and spatial dimensions. The language performer's ability to perceive everything relevant to his performance is not generally possible in the multisensory dance. Whereas in language someone need not be totally involved in the act of communication, this is less applicable to dance.

Motor-gesture potential. Both humans and other animals have gentically programmed and learned motor-gesture patterns. Humans have achieved a greater potential for generalization, which allows them to combine or isolate various locomotor and gesture patterns in different contexts and using different parts of the body.

Humans seem to have more control over innate motor-gesture

patterns, possibly through greater attention span and learning capability. Nonhumans tend to move on the basis of urges, irritability, and external stimuli, i.e., movement is symptomatic or drive-motivated. Their dance-like behaviours are displays, stereotyped, usually repetitive, though not entirely arbitrary behaviours, which are ritualized in the biological sense that they have been subject to selection during evolution for their communicative values (cf. Jolly 1972:140, 144-45). In human evolution, it seems that programmed action sequences gradually became replaced by those in which social learning and individual choice played a more important role.

Humans have a greater potential not only for motor-gesture variety, control, and learning, but also for creative manipulation of patterns within rules (Huxley 1966), so that different dance styles and structures can be chosen. Dance is an open-ended system that is relational rather than just sequential, and tends to be cumulative within a culture. The choreographer generates and expands themes and structural/stylistic patterns to transform what is known and produce new forms by such techniques as adding, deleting, and rearranging different dance elements (cf. Williams in press).

Although nonhumans may be trained to do human-like dances, there is no evidence that they create understanding and transmit to other animals dance sequences that are physically, affectively, and symbolically complex, or leave something representing dance behaviour that lingers beyond their life-time.

PURPOSE IN DANCE

Having focused on dance as human behaviour, we will now explore the psychobiological bases of the other necessary and sufficient criteria of dance as specified in this paper. Purpose will be discussed in terms of adaptation, and of cognitive and affective communication.

Adaptation. The phenomenological movement and gesture of dance may be its primary end and reward (i.e., autotelic). However, the pleasure principle does not negate the instrumental purpose but subserves it. Dance may have been, and may still be, an adaptive pattern spreading and differentiating at the expense of less efficient precursors. Yet dance need not be adaptive to exist. One of the mechanisms of adaptation to an environment is natural selection, and dance may be the result of processes that have been selected. Alland (1973), in discussing art, refers to exploratory behaviour, a sense of rhythm, metaphorisation, and the ability of the brain to make fine distinctions. In some societies natural selection could even favour individuals with dance skills. For example, in cultures where choosing a mate is influenced by the exhibition of skill in a dance that incorporates qualities which are deemed predictive of success in life, the incompetent may be excluded from procreation. This occurs in much war dance behaviour (discussed in Hanna 1977a) which might also confer a selective advantage if it prepares a childless warrior for battle success both physically and emotionally. Alternatively, patterns of behaviour originally developed to serve directly adaptive ends may recur through habituation, even though they cease to be adaptive (Darwin 1965:28). They may be so intrinsically rewarding that they become used for other purposes.

Purely exploratory behaviour, pretending, the anticipatory working out of imminent problems, or experimenting with something new in dance, may in themselves be adaptive (cf. Kreitler and Kreitler 1972:330). There is evidence that the absence of such exploratory behaviour has serious effects (cf. Jolly 1972: 231-235), though this has not yet been shown to apply to dance.

Darwin, with whom the modern scientific study of movement behaviour may be said to have begun, regarded motor behaviour as subject to the same laws of natural selection, adaptation, modification, and extinction as are bodily structures (Darwin 1965:xii).

Cognitive communication. Survival of a species depends on accommodating to the environment. Dance is part of the cultural system in which adaptively valuable information is communicated to oneself and others. Communication includes both the performer's intentional communication and the inadvertent transfer of information. Dance can communicate information purposefully and provide an open channel that could be used. Like other cultural codes and patterned interaction, it is a way of ordering and categorizing. It is possible that some statements made in the dance form cannot be made in another. Shared knowledge about a form, experience in its use, and notions about when, where, how, and why messages are sent, as well as information sufficiently lucid to be perceived through surrounding distractions or impediments, are conditions for effectiveness. Dance may support or refute linguistic, paralinguistic, and other forms of communication. Its presentation may be an interpersonal dialogue or a monologue in the presence of others. The power of dance lies in its cognitive-sensori-motor and aesthetic capability to create moods and a sense of situation for performer and spectator alike.

A people's basic cultural assumptions and orientations, procriptions and prescriptions, as well as the seeds for their destruction or alteration, may become locked into this mode of communication. Dance may mirror or refract social and political structures and techno-environmental factors. However, it may also be a generative force, a processual agent, reflecting 'anti-structure' (Turner 1974), even going beyond what Peckham (1965) refers to as the human's 'rage for chaos', the need to experiment with novelty within the safety of an artistic, 'pretend' situation.

Clues to the basis of nonverbal communication and the use of parts of the body to create patterns that command attention are found in ontogenetic and evolutionary perspectives. Body language is as basic a means of expression as the cry, and it can be used meaningfully without verbal language, although it is generally analyzed in relation to verbal communication.

Hewes (1973) even argues that gestural communication was the first form of human predicative or propositional communication. Lyons (1972:76-77), however, argues that there is no evidence from language to support the belief that language evolved from nonverbal communication, while Fouts (1973) suggests that vocal and gestural modes developed simultaneously. What matters is not whether language or gesture is antecedent, but that both are conceptual vehicles and can reinforce and often substitute for each other.

In accounts of human learning during an individual's early life written by such diverse specialists as Jean Piaget, Arnold Gessel, Maria Montessori, Gordon Allport, and Erik Erikson, the acquisition of communicative motor skills is viewed as critical to a young child's ability to cope with the world of persons, things, and ideas.

Dance is used to communicate a variety of ideas and emotions, but in this respect its meaning cannot be said to be intrinsic, i.e., self-evident: it depends primarily on the meanings that are assigned or associated with it in different societies. Norms, values, and cooperative skills involved in procuring food may be transmitted through the dance. It can teach coordination, concepts, problem-solving, and the discrimination of elements which build up skills necessary for hunting, agriculture, or war (Hanna 1977a). Dominance-submission hierarchies and other differential statuses may be communicated through dance patterns homologous with such statuses (Kurath 1964; Hanna 1975). Assertion of territorial possession occurs through dance as a conceptual boundary marker. And sometimes it is a means of asserting ethnic or class identity or aspiration in a heterogeneous situation (cf. Mitchell 1956).

There are, of course, instances where the values and meanings that are assigned to dance may coincide with a significance that is biological. For example, dances designed to exhibit the physical being may be functionally, as well as stylistically, similar to the display behaviours of other animals.

Affect. In dance affective and cognitive communication are intertwined. Spiegel and Machotka (1974:24-25) argue that in body communication the contrast between emotion and cognition is not sharp. Physical movements associated with affect may stimulate or sublimate a range of feelings (cf. Clynes 1969), and may be elicited for pleasure, or for coping with problematic aspects of social involvement.

Immigrants to town, for example, may find stability and solace in a heterogeneous transient milieu, through familiar dance performances.

Alternatively, dance may provide and communicate the kind of excitement that Turner (1974:46, 270) describes as liminality, the suspension of usual rules and conditions outside or on the peripheries of everyday life in rites of passage which mark changes in the individual's social status or cultural or psychological state; that Berlyne (1971:141-161) calls arousal, that is, novelty, counter expectations, complexity, conflict, ambiguity, and instability; or Ludwig (1969:13-17) refers to as altered states of consciousness. In most cases, the individual clearly feels a qualitative shift in his pattern of mental functioning through disturbed time sense, loss of control, body image change, perceptual distortion, change in meaning, sense of the ineffable, feeling of rejuvenation, or hypersuggestibility.

Altered states of consciousness may be induced by socialized responses to the contextual situation, by auto-suggestion, and by physical behaviour such as energetic dancing, which may alter brain wave frequencies, adrenalin, and blood-sugar content, and induce giddiness through high speed or sensory rhythmic stimulation in more than one sensory mode. Kinesthetic stress, overexertion, and fatigue, also increase susceptibility to rhythm.

Peckham (1965) argues that art forms provide protected situations categorized by high walls of excited insulation, in which 'disorientation' or 'discontinuity of experience' can be savoured. Man's drive for order, he believes, prevents him from changing his orientation when conditions within which it worked change. Art thus permits humans to experience chaos symbolically and without danger; it provides novelty which functions to break up old orientations.

Many psychologists, psychiatrists, and dance therapists recognize that dance may serve as a psychological defense mechanism embodying psychologically or socially unacceptable impulses, and so gratifying or deflecting needs. What is made sensorily perceptible in dance, such as anxiety and fear, is thereby accessibly to purposive action by individuals or groups. Dance may therefore function like play, exploratory behaviour, rituals of rebellion, or cathartic outlets for deviance, in which a segment of the psyche or world is represented in a non-threatening manner in order to understand or cope with it. Mead (1928: 311), for example, notes the reversal of the pattern of conformity of Samoan society in the dance, where the individual is given a chance to exhibit his own individuality and skill. Keil (1967:33) describes a high degree of role differentiation and organizational complexity in Tiv dance groups, which run counter to other institutions.

DANCE IS CULTURALLY PATTERNED

Birdwhistell (1970) hypothesizes that expressive body movement is a culturally determined channel of communication learned through social experience, so that patterns of movement are commensurable with social roles, age, sex, and class. Other scholars have argued that some expressive movements are universal (cf. Clynes 1969; Eibl-Eibesfeldt 1972:307). However, in dance these patterns may be manipulated and combined, and may be assigned particular significance in culture.

Movement styles, structures, and purposes, reflect patterns of group interaction. There is also great variability in the creative potential and output of individuals who have an impact on a culture's dance patterns, since an individual's dance is shaped by both cultural involvement and a unique set of personal experiences within the patterns of group behaviour.

There are psychobiological bases and propensities which act as constraints to cultural patterning. These interdigitate with historical-cultural and environmental factors to create conditions for a particular dance pattern. They are:-

1. *Unconscious mental structures.* As is the case with much linguistic behaviour, in which an individual creates sentences and responds to them without being conscious of how he does it, dance processes sometimes operate without awareness of them. Entirely novel sequences of movement and gesture may be performed without the creator being conscious of how the creation occurred, and an audience may understand these sequences.

Such dance movements are not to be confused with symptomatic behaviour, which includes epileptic or hysterical fits, a child jumping or trembling from excessive nervous excitement, and similar spontaneous movements. These are called 'response' behaviours by Spiegel and Machotka (1974), in contrast to the 'presentational' behaviour of dance.

2. *Maturation.* Dance is learned through social interaction within the constraints of general psychobiological maturation. Psychologists recognize that our primordial experiences with the body have an impact on later motor behaviour. Body image, rooted in sensory experience and proprioception, refers to the

memory of experiences we have had with our bodies and the attitudes toward or expectations about the personal body.

Gardner (1973:101-109), a proponent of neoteny, builds upon Erik Erikson's theory of human development. Three early developmental stages in which the individual resolves the conflicts that pervade a young child's functioning, form a substratum upon which all experience is initially assimilated. In the first stage, which focuses on the oral sensory zone, the characteristic mode of approach is getting and taking in, or biting and grasping. Some vectoral properties include quick versus slow speed, regular versus irregular time, and narrow versus wide space. The predominant mode during the anal period is retention and expulsion. Associated vectoral properties are ease versus strain, hollow versus full, and thick versus thin. The modes characterizing the genital state of intrusion or inception-inclusion are associated with the vectoral properties of open versus closed, and directionality, force, depth, comfort, and texture. The resolution of these developmental stages (with fixations and regressions sometimes occurring) reflect common elements discerned in the symbolic media of the arts.

3. *Physical equipment*. The human body works through the complex interactions of the skeletal, muscular, and nervous systems, which impose constraints on culturally patterned dance. Bodybuild affects speed, power, elevation, the ballet 'turn out', and 'double-jointedness'. Differential body size, strength, pelvic structure, and fat deposit are usually sex-related. Co-ordination, balance, endurance, flexibility, speed of reaction and movement, sensory acuity, level of excitation, all affect dance performance. The human body's bilateral symmetry and sensory frontal plane equipment lead to innate forward, upright, and bipedal movement. Hunt (1968:62) claims that the ability to discriminate the side space of the body must be learned.

The capabilities of the sensory system limit the orientation mechanisms of the body in relation to features of external space: vision is important in maintaining a stance, information from the inner ear, ankles and feet, also contribute. The ratio of complex non-visually to visually based choreography, which includes imitation within and between dance cultures and individuals, may be related to the proportion of innate or learned reliance on the left or right hemisphere of the brain (cf. Diamond and Beaumont 1974). Choreographing set dances and improvisation both require knowledge (usually tacit) of relational rules for using a motor lexicon, and these are left hemisphere functions that emphasize digital, analytical, and sequential information processing. Imitation merely depends on learning a set pattern, which is a right hemisphere function that involves analogic and spatial abilities.

Although genetic programmes, body structure, and early life experiences impose limits on cultural patterning in dance, humans often attempt to defy these with lengthy training. Examples of specialized dance training to counter the body's usual performance are Western ballet, Indian Kathakali, and Balinese ritual dance.

INTENTIONAL RHYTHM IN DANCE

Intentional rhythm refers to the choice of time pattern, of alternationa of quiescence and activity. It is the organized flow of energy in time and space which contrasts with autonomic-interactional rhythms of other human and nonhuman motor patterns.

Dance can be viewed within three time perspectives: the duration of the performance, the duration of the intervals portrayed in the dance, and the duration of the intervals during which the audience perceives, understands, and/or reacts to the dance. Rhythm in dance may be created by the transformation of time itself (as by foreshortening) within these perspectives, or by content.

In assessing dance rhythm, *accent* is the significant stress, the greater relative force or intensity with which energy is released. *Duration* is the relative length, the amount of time of movements, phrases, patterns, and performances. *Metre* is the underlying, consistent numerical grouping of beats and accents. *Tempo* is the rate or speed at which movements follow one another. These elements may be simple or complex, uniform or variable, within a dance. And in some dance forms metre is absent.

Intentional rhythm in the creation or performance of dance seems to be rooted in the alternations of quiescence and activity which are immanent in our experience (cf. Luce 1971, Doob 1971). Humans can harmonize with the rhythms they recognize, or they can attempt to counter them. Psychobiological bases of intentional rhythm in dance may be found in sensorial reaction to natural time, which refers to ecological variables such as the fluctuation of the seasons, climatic variations, and diurnal or circadian rhythms. Luce concludes, 'Invisible rhythms underlie most of what we assume to be constant in ourselves and the world around us' (Luce 1971:1). 'Rhythmicity', she suggests, 'may have been one of the first forces of natural selection, since organisms that timed their activity and life processes in accord with the changing light, temperature, humidity and other factors of the environment would have had an edge in survival' (*ibid.*:145). Time structure may be crucial to health. There is evidence that inconsistent light-dark circadian rhythms seriously perturb certain diurnal psychological rhythms in man (Foulks 1972:83). Virtually all physiological functions in all living creatures show periodical fluctuations. These affect mood, memory, and sensory and motor capacities.

Biological time centres on rhythm in the human organism. All autonomic functions of the body, such as brain waves, muscle contractions, heart and respiratory rates, hormonal tides, and peristalsis, are rhythmic. The human foetus experiences intrauterine rhythms, the maternal heartbeat, its own in synchrony or counterpoint, and the mother's rapid eye movement cycle (Luce 1971:36).

The three time dimensions of dance (duration of performance, of intervals portrayed in the dance, and dance motor time) which affect the perception of the dance can be related to natural and biological rhythms and so manipulated to provide security or novelty and to arrest attention, or some combination of these. We could hypothesize that intentional rhythm in dance is homologous to the biological rhythms that all human experience and/or to individual congenital rhythms when the purpose of dance is to provide affective security and consistency. Alternatively, these may be modified in dances that attempt to manipulate the environment for some other purposes.

DANCE IS EXTRAORDINARY NONVERBAL BODY MOVEMENT AND GESTURE, THE MOTION HAVING INHERENT VALUE

What is ordinary in one society may be extraordinary in another. The distinguishing characteristic that sets extraordinary nonverbal body and movement and gesture in dance apart from other activities is the manipulation of ordinary motor activities within an aesthetic domain, with an emphasis on the importance of bodily movement, the illusion and residual action resulting from the kind of movement produced. Ordinary movements and gestures are transformed or incorporated into dance configurations. According to the Kreitlers, (1972:163, 173) remoteness from the habitual seems to be more important in dance than in other expressive forms as a means of arousing interest and generating aesthetic experience. Arnheim (1954:361) explains this as a response to a change in the environment which may provoke a reaction. Motion is related to the experience of arousal and motivation. The Kreitlers (*op. cit.*), using a refined model of motivational dynamics, argue that the focal point of the arts is pleasure, which includes altered states of consciousness. These states can be induced through the driving technique involved in repetitive rhythmic movements and central nervous system tuning. Couched in social situations generating energy and releasing inhibitory mechanisms, dance may autonomically affect the human body (cf. Lex 1975).

AESTHETIC DIMENSION IN DANCE

The concept 'aesthetic' refers to notions of order, appropriateness, quality, or skill intrinsic and extrinsic to dance. It encompasses the specific movements used and their manipulation in time and space with energy (see Table of Dance Movement categories). Aesthetic parameters specify how dance should achieve its purpose within a particular culture or subculture. The dancer's aesthetic is related to his or her reference group (one to which he or she belongs or aspires to belong).

TABLE I

Dance Movement Data Categories

Movement definition: the visual result of energy release in time and space through muscular response to a stimulus; it is the essence of dance. Its structure (interrelation of parts) and style (characteristic mode and quality of all the contributing elements) can be analyzed in these terms:

Space (design):
- *direction* (path the moving body cuts through space)
- *level* (high--weight on ball of foot, low--body lowered through flexing knees, middle--normal stand, elevated, kneel, sit, lie, curvilinear, diagonal)
- *amplitude* (size of movement, relative amount of distance covered or space enclosed by the body in action)
- *focus* (direction of eyes and body)
- *grouping* (overall spatial pattern of movement in relation to dancer, created dance space):
  - free form or organized pattern (individual, couple, small group, team--linear or circular, symmetrical or asymmetrical)
  - physical link (none, parts of body, length of contact)
- *shape* (physical contour of movement design--includes direction, level, amplitude)

Rhythm (time, flow):
- *tempo* (rate at which movements follow one another)
- *duration* (relative length of movements, patterns, performance)
- *accent* (rhythmically significant stress)
- *metre* (basic recurrent pattern of tempo, duration, and accent)

Dynamics (force, relative amount of energy, effort--tension and relaxation released by the body to accomplish movement):
- *space* (indulgence--minimum or maximum use ('direct' straight

lines or 'flexible' curves and deviations)
flow (control, continuous transfer of energy which qualifies movement--free, unimpeded or bound, hampered)
locomotion (means of moving from one place to another)
projectional quality (texture produced by combination of elements--relative quickness or slowness of energy release in space)

Characteristic use of body (instrument of dance):
posture (axial movement)
parts of body (hinge, rotation, flexion, extension, vibration) [gesture--a movement in space that does not carry weight]
locomotion (walk, run, leap, hop, jump, skip, slide, gallop)

Although some movements may have universally shared meanings, such as approaching, fleeing, and attacking, jumping with joy, or drooping with sorrow, most are semantically culture-specific (Kreitler and Kreitler 1972:181). Mimesis or conventional stylization within specific movements and gestures have been identified as the chief means of symbolization in dance.

There are at least six modes of signification that may be applied to dance at the etic level, and they may be conventional or idiosyncratic.

(1) A *concrete representation* produces the outward aspect of a thing, event, or condition, such as a mimetic portrayal of an animal.

(2) An *icon* represents the properties or formal characteristics of a thing, event or condition, and is responded to as if it were what it represents, such as a danced deity revered as the deity.

(3) A *stylization* encompasses arbitrary gestures or movements, such as pointing to the heart as a gesture of love, performing specific movements or dance styles as an emblem, or dancing abstract images within particular parameters.

(4) A *metonym* constitutes a conceptualization of one thing for that of another of which it is an attribute or with which it is associated or contiguous; for example, the Ubakala dance for the birth of a child performed for a bridge-launching is metonymical to the creative process.

(5) A *metaphor* expresses one thought, experience, or phenomenon by another which resembles the former and is somehow analogous to it, such as dancing a leopard to refer to the power of death.

(6) *Actuality* is an individual dancing in terms of one or several of his or her usual statuses and roles, such as Louis XIV dancing as a king and being so treated.

These modes are signs that may function as signals when they are directly related to the action they signify, such as a war dance to herald a battle.

Each dance culture may use a device singly or in combination with others within either a dance genre or an individual dance. Devices may be interchangeable according to context, have many referents that may change over time, or become paramount at different phases of a dance performance.

Meaning depends on context. In a study of Nigerian dance I suggested that the devices may operate within one or more of seven spheres: dance as (1) sociocultural event, (2) total human body in action, (3) entire performance, (4) discoursive performance (unfolding of a sequence of configurations), (5) specific movements or gestures, (6) intermeshed with other communication modes, and (7) a medium for song, music, costume, accoutrements, and speech (Hanna 1976).

Thus, the psychobiological bases of aesthetics lie in the human capability for symbolic thought and action. Cultural patterning contributes to the selection of mechanisms that transform affect and cognition into meaningful patterns of body-movement, which arrest attention because they are intentional and extraordinary.

CONCLUSION

I have suggested some psychobiological factors that seem to generate dance, and have directed attention to properties that may help to explain why dance is a component of temporal and transcendental behaviour, sociopolitical manipulation, and mental health therapy. Dance was viewed as the result of processes that have had a selective advantage and as being potentially adaptive. Phylogenetic and ontogenetic perspectives focused on the distinction between human dance and other animal nonverbal communication. Table II summarizes the distinctions that were drawn. Human dance is a form of thought and action which, compared with other animal nonverbal communication forms, serves a wider variety of purposes with greater complexity in open, lexical, semantic, and syntactic systems. Unfortunately, there is a dearth of systematic research on the impact of dance or the relative salience of what is universally or culturally determined.

Implicit in the foregoing discussion is a processual model of dance performance in a sociocultural context. This will be briefly summarized to suggest the interaction between human endowment and sociocultural context and the dynamics involved in what is dance and what is assigned to dance. This perspective is schematically presented in Figure 1 and illustrative

TABLE II

Comparison of Dance and Nonverbal Communication of other Animals

Dance	Nonhuman Nonverbal Communication
Purpose	
Affective and cognitive motivation - - - }	Mostly displays
Social bonding, territorial possession - - - }	Mostly displays
Courtship and mating - - - }	Mostly displays
Dominance submission - - - }	Mostly displays
Child care - - - -	
Exploratory behaviour - - - -	
* Coordinate actions from immediate and distant context, assert continuity in defense of threat of mortality.	Coordinate approach, withdraw, follow in relatively immediate or limited context
* Express, evoke and transform emotion	Express and evoke emotion.
* Multipurpose	Food, fear, and sex motivational states mostly energize and drive organism
Intentional Rhythm	
* Choice of rhythm in harmony or counter to psychobiologically based rhythms	No deliberate choice
Culturally Patterned	
* Open (productive) system of movement, semantic, syntax	Limited intraspecies variation
* Complex movement, semantic, syntactic systems; modifiers and context important	Relatively simple systems; response to relatively few simple perceptual cues
* Learned arbitrary forms	Learning, but not of deliberately chosen forms
Extraordinary Nonverbal Movement and Gesture, Motion Having Inherent Aesthetic Value	
Extraordinary movement and gesture - - - -	

TABLE II (cont.)

Motion has inherent value	- - - -
* Aesthetic values, notions of culturally based rules of performance	Recognition of naturally based performance

- - - refers to continuity across species

commentary will be drawn from a study of Nigeria's Ubakala.

Effective communication depends upon shared knowledge and the interplay between skilful dance expression and sensitive perception. For every performance there is a catalyst determining who dances, why, where, and how: a village market, family birth or death, or a festival, virtually compels Ubakala performers and spectators to participate in a dance-play; the dry season or a moonlit night provide the memory and desire for the pleasure of a dance-play, the enjoyment of the performance *per se* and its related sociability.

Selective perception is affected by cognition and emotion. For the Ubakala, motivations, incentives, and desired rewards to participate in a dance-play include fear of being shamed, pleasure (aesthetic involvement, alleviation of boredom and desire for arousal, and opportunity for catharsis), social approval, money received from a praiseworthy performance, and/or the need to be aware of what is occurring in order to pursue strategically individual and group goals. Furthermore, once a dance-play is successfully under way, it commands attention.

Interaction in the dance-play involves the performers amongst themselves, particularly in practice sessions and leader-group exchanges. Audience and performer interaction in general may involve individuals and/or groups whose responses are not necessarily homogeneous. Assuming no interference, messages of intrinsic form, which Armstrong (1975) calls 'affecting presence', and/or extrinsic reference are decoded. These may have an effect that is or is not recognized, or no effect; functions, in turn, may have implications for the dancer, the observer, and/or the structure of the event, the catalytic environment.

Feedback refers to information about a performance being introduced into the performance system. Feedback loops involve multiple relationships or elements linked by reciprocal influences. The feedback may be positive, serving to reinforce performance in the same current pattern, or negative, serving as a disincentive for the current pattern. When Ubakala performers receive feedback from their own kinesthetic proprioceptors

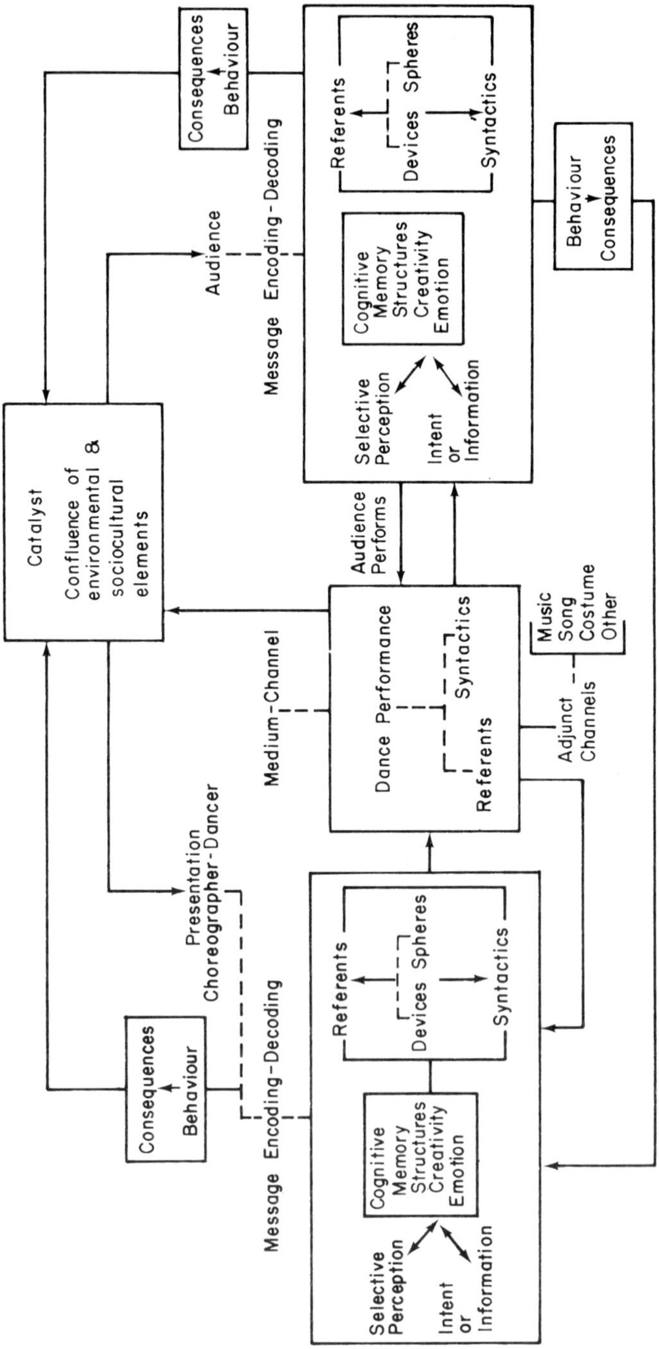

Fig. 1 Dance Semiotics: A Processual model.

and other internal sources, amongst themselves, they respond to the unfolding of movement style and structure and other channels. Audience praise may stimulate further excellence or a continuation of the performance, while their denigration may lead to a change or termination.

Feedback related to the outcome of the performance may occur through the performers' or spectators' thought and actions. Among the Ubakala, interference may be caused by natural events such as rain, by a loose animal or a fight, or the presence of an outside government official.

The dancer makes movement choices and performs with the intention of informing, evaluating, prescribing, or effecting, or without a specific intention but with the possibility that information transfer can occur. Intent becomes transformed into performance by means of rules for semantic and syntactic encoding that are related to the values, beliefs, and norms of Ubakala society.

If dance is rooted in human capabilities for comprehension, choreography, and performance, how do we account for the varieties of dance in similar environments and the absence of dance in some times and places? It appears that a natural tendency to dance may become repressed, inverted, or otherwise culturally modified. Kern (1975), for example, has discussed the influence of culture on the uses of the human body. In the U.S.A., where verbal modes of expression have predominated because of a bourgeois, puritanical, and technological heritage, a trend toward greater involvement in nonverbal communication is reflected in increased dance study, performance, and observation, partially to counter disillusionment with an impersonal society (cf. Theeman 1973).

The persistence of dance from the eras of cave paintings and archaeological artifacts to the twentieth century suggests its versatility. It also poses the need for anthropologists and other social scientists to give it greater attention as a form of human thought and action.

ACKNOWLEDGEMENTS

I am most appreciative of the useful comments of Adina Armelagos, Eliot D. Chapple, Ralph Holloway, Carlos Medina, James M. Murphy, Pamela Squires, Suzanne Youngerman, and my most critical readers William John Hanna, long-time supporter and colleague, and Alexander Alland, Jr., recent mentor, on earlier drafts of this paper. To James Fernandez I owe the encouragement to examine symbolic dimensions in dance and to John Blacking, the provocation and opportunity to explore the psychobiological bases of dance.

REFERENCES

Alland, A., Jr. (1973). The evolution of art. Paper prepared for Burg Wartenstein Symposium No. 59, *Ritual: Reconciliation in Change*, July 21-29.

Armstrong, R.P. (1975). *Wellspring*. Berkeley: University of California Press.

Arnheim, R. (1954). *Art and visual perception: a psychology of the creative eye*. Berkeley: University of California.

Berlyne, D.E. (1971). *Aesthetics and psychobiology*. New York: Appleton-Century Crofts.

Birdwhistell, R.L. (1970). *Kinesics and context: essays on body motion communication*. Philadelphia: University of Pennsylvania.

Clynes, M. (1969). Toward a theory of man: precision of essentic form in living communication. In N. Leibovic and J.C. Eccles (eds.), *Information processing in the nervous system*, 177-206.

Darwin, C. (1965). (1872). *The expression of the emotions in man and animals*. Chicago: University of Chicago.

Diamond, Stuart and Graham Beaumont (eds.) (1974). *Hemispheric function in the human brain*. London: Elek.

Doob, L.W. (1971). *Patterning of time*. New Haven: Yale University.

Eibl-Eibesfeldt, I. Similarities and differences between cultures in expressive movements. In Robert A. Hinde, (ed.) *Non-verbal communication*, 297-311.

Evans-Pritchard, E.E. (1965). *Theories of primitive religion*. Oxford: Clarendon Press.

Foulks, E.F. (1972). *The arctic hysterias of the North Alaskan Eskimo*. Anthropological Studies No. 10. Washington, D.C.: American Anthropological Association.

Fouts, R.S. (1973). Capacities for language in great apes. Paper prepared for the IX International Congress of Anthropological and Ethnological Sciences, Chicago.

Gardner, H. (1973). *The arts and human development: a psychological study of the artistic process*. New York: John Wiley.

Hanna, J.L. (1975). Dances of Anáhuac - for god or man? An alternate way of thinking about prehistory. *Dance Research Journal 7*, 13-27.

Hanna, J.L. (1976). *The anthropology of dance ritual: Nigeria's Ubakala nkwa di iche iche*. Unpublished doctoral dissertation, Columbia University.

Hanna, J.L. (1977a). African dance and the warrior tradition. In Ali A. Mazrui (ed.), *The warrior tradition in modern Africa*. Special Issue of Journal of Asian and African

Studies, *12* (1-2), forthcoming.
Hanna, J.L. (1977b). Toward a cross-cultural conceptualization of dance and some correlate considerations, in John Blacking (ed.), *The Performing arts: music, dance, theatre.* The Hague: Mouton (forthcoming).
Hewes, G.W. (1973). Primate communication and the gestural origin of language. *Current Anthropology 14* (1-2), 5-24.
Hinde, R.A. (ed.) (1972). *Non-verbal communication.* Cambridge; University Press.
Hockett, C.F. and Ascher, R. (1964). The human revolution. *Current Anthropology 5,* 135-168.
Hunt, V. (1968). The biological organization of man to move. Impulse 1968: dance, a projection for the future, 51-63.
Huxley, Sir J. (ed.) (1966). A discussion on ritualization of behaviour in man and animals (June 10-12, 1965). *Philosophical Transactions of the Royal Society of London 251,* 337-349.
Jolly, A. (1972). *The evolution of primate behaviour.* New York: Macmillan.
Keil, C. (1967). Tiv dance: a first assessment. *African Notes 4* (2), 32-35.
Kern, S. (1975). *Anatomy and destiny: a cultural history of the human body.* New York: Bobbs-Merrill.
Kreitler, H. and Kreitler, S. (1972). *Psychology of the arts.* Durham: Duke University.
Kurath, G. and Martí, S. (1964). *Dances of Anáhuac: the choreography and music of precortesian dances.* Chicago: Aldine.
Lenneberg, E.M. (1967). *Biological foundations of language.* New York: John Wiley and Sons.
Lex, B.W. (1975). Physiological Aspects of Ritual Trance. *Journal of Altered States of Consciousness 2* (2).
Luce, G.G. (1971). *Biological rhythms in human and animal physiology.* New York: Dover.
Ludwig, A.M. (1969). Altered states of Consciousness. In Charles T. Tart (ed.), *Altered states of consciousness.* New York: John Wiley, 9-22.
Lyons, J. (1972). Human language. In Robert A. Hinde (ed.), *Non-verbal communication,* 49-85.
Mead, M. (1928). *Coming of age in Samoa.* New York: W. Morrow.
Miller, G.A. (ed.), (1973). *Communication, language, and meaning: psychological perspectives.* New York: Basic Books.
Mitchell, J.C. (1956). *The Kalela dance.* (Paper No. 27) Manchester: University for the Rhodes-Livingstone Institute.
Peckham, M. (1965). *Man's rage for chaos: biology, behavior and the arts.* Philadelphia: Chilton

Radcliffe-Brown, A.R. (1922). *The Andaman Islanders.* Glencoe: Free Press (1964).

Reay, M. (1959). *The Kuma: freedom and conformity in the New Guinea Highlands.* Melbourne: University Press.

Spiegel, J.P. and Machotka, P. (1974). *Messages of the body.* New York: Free Press.

Theeman, M. (1973). *Rhythms of community: the sociology of expressive body movement.* Unpublished doctoral dissertation, Harvard University.

Turner, V.W. (1974). *Dramas, fields, and metaphors: symbolic action in human society.* Ithaca: Cornell University.

Williams, D. In press. *Deep structures of the dance.* In E. Schwimmer (ed.), *Yearbook of Symbolic Anthropology.* London: C. Hurst.

MUSIC AND POSSESSION TRANCE

Gilbert Rouget

Possession trance is indisputably to be seen as a behaviour making use of a special technique of the body (c.f. Marcel Mauss's *technique du corps*) in order to realise a certain kind of religious experience. Everywhere, from Bali to Brazil, through Vietnam, India, Tibet, the Middle East, the Mediterranean, East and West Africa, possession trance is associated at some point or another with music and dance. Very often, music is closely associated with the onset of trance. From this comes the widespread idea that certain kinds of music are endowed with the particular power of inducing or precipitating trance. Two questions then arise:

(1) Is it really so? and in that case, from where do those musics get such power?

(2) If it is not so, of what order are the relations between music and trance?

Referring not to trance but to therapy, which is so often associated with possession, the programme for the conference cites two possibilities: 'Are music and dance therapy effective because certain patterns of sound and movement may automatically affect the human body, or because the intensity of the associated social situation generates energy and/or releases inhibitory mechanisms?' Before trying to answer these questions, let us first ask what data are available to discuss them.

Apart from the observations made by Maya Deren about the effects of music during her personal experiences of trance, no introspective data on trance in relation to music are, as far as I know, available. For many reasons, introspective data concerning lived experiences of trance are very scarce, and I consider Deren's observations suspect, though Sheila Walker (1972:22) considers them very reliable and quotes her at length, precisely about music. We do not have available, either, experimental data resulting from the observation of trance *en situation* seen in its relation to music.

Finally, there has not been any convincing laboratory experiment on the possible effects of music on trance. Despite the fact that they have sometimes been taken very seriously, Neher's experiments must be considered as devoid of scientific value. Sheila Walker relies heavily on Neher in her chapter, 'The neurophysiology of possession', and in his remarks on Rodney Needham's paper (1967) on percussion and transition which itself proposes a biological interpretation of the effects of sound, W.C. Sturtevant (1968) considers that Neher's theory provides the real answer to Needham's question.

A. Jackson (1968) agrees. My principal reasons for dismissing Neher's theory are: (1) The parameters of his laboratory experiences, as he has described them (Neher 1961), are in no way comparable to those involved in actual possession ceremonies: (2) the physiological results obtained (*ibid.*) are in no way comparable to what can be observed during actual ceremonies; (3) if the 'driving' effect he is referring to can be produced by beating a drum at a speed which can vary from 8 to 13 beats per second (Neher 1962:154), and if that speed can eventually be as slow as 4 beats/sec (Neher 1961:449), then the whole of sub-Saharan Africa should be in trance from the beginning to the end of the year.

This being so, data relevant to our problem are to be found exclusively in the ethnographic literature. The following considerations will be grounded both on a fairly large and, it is hoped, representative sample of this literature, together with a considerable number of films and recordings, and on personal observation of possession cults mainly made in the Republique populaire du Benin (formerly Dahomey) among Fon, Gun and Yoruba. Specific references will not be given here, since they form a major part of a forthcoming book (Rouget in press). Ethnographic data show that:

(1) Although trance is generally associated with music, in any given cult it can sometimes as well take place without any music at all.

(2) Considered from the point of view of the sound, the variety of music associated with possession trance all over the world is so great that the effect of music on trance cannot generally be of a physical order, except perhaps in a few cases: signals are so widely different from one another acoustically that they cannot in themselves have the same sort of neurophysiological effect.

(3) The effects of music associated with trance vary so widely according to circumstances both with different persons and for the same person, that it is unlikely that this effect may be of the order of a conditioned reflex (Pavlov) to a particular music. The theory that the onset of trance may be considered

the result of a conditioned reflex was first put forward by Herskovits (1943:25), and has since been picked up by several authors.

(4) The effects of music on trance are contradictory: on the one hand it is most often supposed to induce trance while on the other, it is considered to cool it off.

(5) Shamans are always musicians, and in shamanism, the shaman is the musician who induces his own trance. On the other hand, in possession cults (for a distinction between shamanism and possession, see Heusch 1964) music is made by specially appointed people, and as a rule possessed people dance but neither play instruments nor sing. The exceptions can easily be accounted for, and so I am limiting myself here to the general rule. The relation that the person in trance has with music is thus of two very different kinds: active when it is of the shamanistic type, and passive in the case of possession. The effects of music are necessarily different according to what is being considered.

(6) The effects of music (or of dance) evidently differ in kind according to the type of music (or of dance) that is performed. In the case of what I should call 'inspiration trance', such as the one shown during the *dhikr* performed by certain Moslem sects, where people are inspired by Allah but not possessed, music and dance fulfil complementarily a programme of overstimulation and overventilation which is doubtlessly to be held as having direct neurophysiological effects. The same applies to certain Christian sects (Chlüstü) of old Russia, or for the curing dance of the Bushmen. Trances of that type are characterized by their frenetic aspect, and many possession trances show the same character: this was well illustrated in Jean Rouch's film on the Songhay, *Les maîtres fous*. But there exist possession cults where trance is highly sublimated, and where dance and music are of the most quiet type. For example, the Vietnamese trances described by Pierre Simon and Ida Simon-Barouh (1973) are quiet, and a most striking example from Africa is that of the Moundang of Chad, which has been filmed by Zempléni, Adler, and Pineau. These two types of trance belong to two different techniques of the body. Music is performed in both cases, but in relation to two different behaviours.

In this respect, ecstasy (*wajd*) such as is observed among the Sufis and results from the repetition of the name of God (*dhikr*) from an intense emotion (*tarab*) due to 'listening to music and singing' (*samā*' being accompanied by dancing, but silently) raises another problem (see, in particular, Macdonald 1901 and 1902, and Gardet 1952).

(7) Clear differences are to be made among the diversity of

trance states. In French as well as in English, the words 'trance' and 'ecstasy' are most often used as synonyms in the ethnographic literature or at least are frequently interchanged. Shamanism has been labelled by M. Eliade (1968) 'technique archaïque de l'extase'. Possession cults and shamanism have been grouped by I.M. Lewis (1971) under the general heading of 'ecstatic religions'. Both these authors (and many others) constantly speak of trance to designate these 'ecstatic' states. It is, however, obvious that if the mystic state described by Santa Teresa de Avila is to be called 'ecstasy', the one commonly known as characteristic of the Shakers (to take another example from Christianity) must be labelled differently. One is attained in silence, solitude and immobility. The other requires music and dancing and is to be lived collectively. Ecstasy should be viewed, I think, as a particular case of that wide range of states of consciousness more and more generally called trance by anthropologists, as shown by recent works such as Beattie and Middleton (1969), Bourguignon (1973) and Walker (1972). From one extreme represented by Santa Teresa's ecstasy (or by Indian *samādhi*, which is totally different) to the other one, represented, for instance, by the frenzy exhibited by *ndöp* dances among Wolof and so well described by Zempléni (1966), there exists a continuous chain of intermediary states. All of them have to be considered in their particular relations to music.

(8) If we now limit ourselves to possession trance in the narrow sense of the term, *i.e.* as different from mediumship trance, inspiration trance, shamanism trance (or, in Plato's terms and in a different perspective, telestic *mania* as different from the mantic, the poetic, and the erotic: see Brisson 1974), we see that the musics associated with the onset of trance offer two 'universals' which are:
(a) to be meant to identify with (most commonly by glorifying him/her) the divinity responsible for the trance, (b) to be dance music.

Having made these very general observations, we now come to the interpretation of the data.

Two centuries ago, J.J. Rousseau (1970:165) wrote:

> To illustrate the physical power of sound one cites the curing of tarantula bites. This instance proves quite the contrary. There need be neither specific sounds nor the same airs to cure those who have been bitten by this insect. Each needs the air of a melody which he knows and sentences which he understands. The Italian needs Italian airs, the Turk, Turkish airs. Each is affected only by accents with which he is familiar; his nerves respond to them only so

> far as his soul predisposes him: he must understand the language spoken to him in order that what he is told might set him moving. Bernier's cantatas are said to have cured from fever a French Musician; they would have given it to a musician of any other nation.

Ethnographic data from all regions of the world largely confirm these views. Enlarged and expressed in a modern way, they could perhaps be presented in the following manner:
(a) Despite what Rousseau's old enemy, Rameau, thought (and so many others after him), the effects of music on possession trance are not to be looked for at the 'natural' level, but at the 'cultural' level, with the difference, though, that in tarantulism the music (the *tarantella*) 'cures' not because of a particular virtue, but because it is music for dance. Moreover, as a dance, tarantella 'cures' not only because of its psychophysiological effects but because it allows the tarantulated to identify himself with the tarantula which bit him or her. This process of identification is the fundamental principle of possession cults.
(b) What has just been said about tarantulism holds good for possession cults in general, whether Indonesian, Mediterranean, or African. Possession cults are to be seen as aiming fundamentally to realise a relation of alliance with a divinity (god, spirit, ancestor, etc). Despite appearances, exorcism is to be seen as a secondary and, from the present point of view, misleading (or misunderstood) aspect of possession.
(c) Possession music must not be seen as a musical therapy but as the means to put the person possessed in communication with the divinity who possesses him/her and at the same time with the society where the cult functions. It is the communication and the identification which create the therapeutic situation.
(d) As a multichannel language, possession music is the unifying factor around which possession ceremonies are built. It unifies the society of devotees with the society as a whole, the society of the living with the society of the dead, men with their gods, man with himself. Thanks to it, the change of personality is lived simultaneously at the conceptual, emotional, and motor levels.

If a Yoruba peasant can behave publicly as Shango, the thunder god (and he behaves in this way only publicly and in certain circumstances), it is because music enables him to do so. Music dictates his behaviour, and without music he could not afford to be that personage or play that role. Fundamentally, possession trance is theatrical, as M. Leiris has shown.
(e) Since each possession cult is a system, obeying its own internal logic within the general logic of possession, the role

of music varies from one cult to another. To understand its role precisely in each case, one has to place the music within the general structure of the cult.
(f) Ultimately, if possession trance is to be considered as a *technique du corps* defining a certain type of religious behaviour, possession music is to be considered as a technique - and perhaps as the principal one - to socialize trance (Cf. Bastide 1972:92).

In order to say exactly what one means by 'to socialize trance' one has to say first what 'trance' means. But this is another problem.

REFERENCES

Bastide, Roger, 1972 *Le rêve, la transe et la folie.* Paris: Flammarion.

Beattie, John and John Middleton (ed.) 1969 *Spirit mediumship and society in Africa.* London: Routledge and Kegan Paul.

Bourguignon, Erika (ed.) 1973 *Religion, altered states of consciousness and social change.* Columbus, Ohio: University Press.

Brisson, Luc, 1974 Du bon usage du *dérèglement.* In J.P. Vernant (ed.) *Divination et rationalité.* Paris: Sueil, 220-248.

Deren, Maya, 1953 *Divine Horsemen, the Living Gods of Haiti.* New York: Thames and Hudson.

Eliade, Mircea 1968 *Le chamanisme et les techniques archaïques de l'extase.* Paris: Payot. 2nd edition.

Gardet, Louis 1952-3 Un problème de mystique comparée: la mention du nom divin (*dhikr*) dans la mystique musulmane, *Revue thomiste* LXème année - *LII* no. 1, 642-679, and LXIème année - *LII* no. 1, 197-216.

Herskovits, Melville J. 1943 *Pesquisas Etnologicas na Bahia.* Bahia: Museo d'Estado.

Heusch, Luc de 1964 Possession et chamanisme. Reprinted in *Pourquoi l'épouser? et autres essais.* Paris: Gallimard, 1971, 226-244.

Jackson, Anthony 1968 Sound and ritual. *Man* (n.s.) *3,* 293-299.

Leiris, Michel 1958 *La possession et ses aspects théâtraux chez les Ethiopiens de Gondar.* L'Homme. Cahiers d'ethnologie. Paris: Plon.

Lewis, I.M. 1971 *Ecstatic Religion. An Anthropological Study of Spirit Possession and Shamanism.* Penguin Books.

Macdonald, Duncan B. 1901-2 Emotional religion in Islam as affected by Music and Singing. Being a Translation of a Book of the *Ihyā 'Ulūm ad-Dīn* of al-Ghazzāli with Analysis,

Annotation and Appendices. *Journal of the Royal Asiatic Society* 1901: 195-252 and 705-748; 1902, 1-28.

Needham, Rodney 1967 Percussion and Transition. *Man* (n.s.) *2*, 606-614.

Neher, Andrew 1961 Auditory driving observed with scalp electrodes in normal subjects, *Electroencephalography and Clinical Neurophysiology (Jl. Internat.* Soc. for Electro-enceph. and Cl. Neuro.) *13*, 449-451. Amsterdam: Elsevier.

Neher, Andrew 1962 A physiological explanation of unusual behaviour in ceremonies involving drums, in *Human biology, IV*, 151-160.

Rouget, Gilbert In press *La musique et la transe. Eléments pour une théorie générale des relations de la musique et de la possession.* Paris: Gallimard.

Rousseau, Jean-Jacques 1970 *Essai sur l'origine des langues. Où il est parlé de la mélodie et de l'imitation musicale* (1781). Charles Porset (ed.) Bordeaux: Ducros.

Simon, Pierre J. et Ida Simon-Barouh 1973 *Hâù Bóng. Un culte viêtnamien de possession transplanté en France.* Paris: Mouton (Cahiers de l'homme).

Sturtevant, William C. 1968 Categories, percussion and physiology. *Man* (n.s.) 133-134.

Walker, Sheila 1968 *Ceremonial spirit possession in Africa and Afro-America. Forms, meanings and functional significance for individual and social groups.* Leiden: E.J. Brill.

Zempléni, Andras 1966 La dimension thérapeutique du culte des rab, ndöp, tuuru et samp. Rites de possession chez les Lébou et les Wolof. *Psychopathologie Africaine II*, 295-439.

SOME NOTES ON THE ANTHROPOLOGY OF DANCE

Roderyk Lange

Dance Anthropology is only now becoming an independent area of studies. For a very long time, there were no methods of investigation established, no theory of dance available, and not even any appropriate means of movement notation used. Dance anthropology, therefore, lags far behind modern progress in the human sciences.

Since dance is an activity very intimately connected with the human condition, it reveals many aspects of human development. In dance, the only instrument used is the body itself: the dancer is at the same time the creator and the bearer of the dance activity. The texture of dance is the movement of the dancer, and no other media are necessary to reveal expression, symbolism, and eventually poetry, non-verbally. Perhaps nowhere else has man ever expressed himself so directly and completely as through dance.

DANCE AS AN INDEX OF SOCIAL STRATIFICATION

Dance also mirrors sensitively social conditions, and throughout human history appropriate forms have been adopted according to social needs. Human dance cultures have been roughly divided into two different streams: the dances of the upper and lower classes (Sachs 1937:281-2). This is by no means a rigid division. Indeed, these two streams of dance culture have influenced each other at different periods and in varying degrees.

To illustrate contrasting dance cultures as a reflection of the interaction of social groups, I have chosen the dance known as the 'polonaise'. Originally of Polish origin, it spread in a wave of fashion towards the end of the eighteenth century.

Fig. 1A (Kinetogram R. Lange) illustrates the version that is still being executed in Polish villages. In this version of

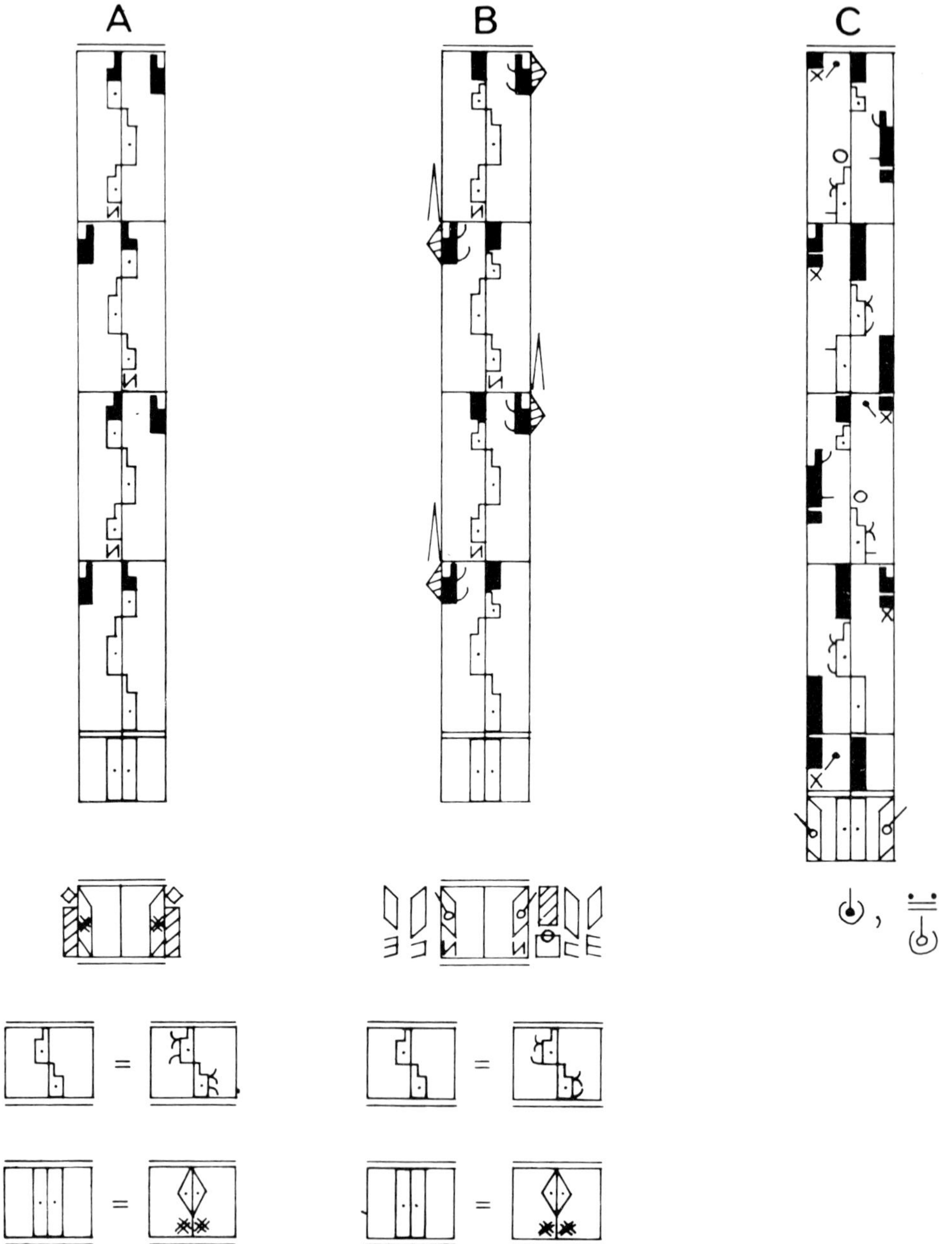

Fig. 1. *Dance as an index of social stratification: contrasts between styles of dancing, illustrated with Labanotation.*

(A) *Village version of the* 'polonaise'. (B) *The* polonaise, *as danced by the landed gentry and in the royal court.* (C) *A step of the* courante, *a court dance of the sixteenth and seventeenth centuries.*

---

the dance step, the progression mostly stressed is the main characteristic of the dance, whose name is actually *chodzony*, the 'walking dance'. In this peasant version, it is indeed a dignified way of walking, with an extended step on each first beat. No other adornments are included. The body is kept upright, whilst the dancer steps from the heel onto the whole foot, with relaxed knees. This style stresses the self-contained way of dancing, and so reflects the social situation. There are no alien onlookers, no audience, no outward exhibition is necessary.

Choreotechnically, however, this simple structure contains a number of features that offer the performer stylised experience of walking. It is absolutely different in form and function from ordinary walking. It has become a poetic transformation of it, and acquired unusual features if compared with utilitarian movement; it becomes dance, it becomes art. Accordingly, the movement texture of ordinary walking has changed. In this danced 'walk', the steps are designed not to carry the body to a particular spot, but in such a way as to supply the performer with the sensation of 'being carried'.

This is the essence of any dancing. Continuity has to be maintained to an unusual degree to enable the dance activity to last (Sheets 1966). There is no other basically utilitarian purpose to motivate dancing action (Sachs 1937:6).

Fig. 1B (kinetogram R. Lange) represents the same dance as adopted by the landed gentry and successively transferred to the big manors and to the royal court. It is already being called *polonez*, 'The Polish Dance', raising the anonymous status of the village to the status of national representation. Here the self-contained body movement of the peasant version acquires quite different features. There is a leaning out of the upper part of the body on each third beat, allowing the preparing foot to glide on the floor, whilst the other leg bends. The whole dance is intended to be spectacular, to display elegance. Dignity turns here into pride. Choreotechnically, it demands from the performer a display of gracious skill. In this version of the dance, performers expect to be seen.

This is a good example of adaptation in dance: a social group importing an alien element, translates it into its own idiom.

To compare with the 'polonaise', Fig. 1C (kinetographic transcription by M. Drabecka from G. Taubert's *Rechtschaffener Tantzmeister*, 1717), shows a recorded 'step' of the *courante*, a typical European court dance. Originating from the sixteenth century, the *courante* survived till the end of the seventeenth century and was still remembered in the eighteenth. Here, the gliding gesture of the foot in preparation for the next step of the *courante*, is similar to that in the 'polonaise', Fig. 1B. This was a typical feature of European upper class dancing. It is therefore not surprising that the peasant *chodzony* was translated into this style, allowing the expected moment of outward-directed display to appear in the 'polonaise'.

DANCE AS AN INDEX OF RURAL/URBAN CONTRASTS

Differences occurring between rural and urban traits in culture are even more explicit in dance. Fieldwork in Eastern Europe has revealed astonishing features of dance culture surviving unexpectedly close to neighbouring industrialised regions.

The tendency in anthropological research has been to turn to cultures where social bonds are still closely knit, and as a result non-European peoples and tribal societies have been a main interest in Western anthropology. But there are still areas of Europe where peasant culture has retained highly integrated patterns of life. This has been stressed only during the last few decades (Shanin 1971, Gunda 1963). In these conditions, dance is an integral part of life, whereas in urbanised conditions it is only a pastime.

From my fieldwork in rural Poland, I have chosen some examples to show the differences presented by the same type of dance when used in rural and urbanized environments. The intricate rhythm of the rural version is abandoned in the urban interpretation, and thus the whole form of the dance changes. In the rural situation, improvisation is based not on accidental chioce but on a range of accepted possibilities which change slowly. Many variants are possible, but this freedom rests on well-known and generally accepted norms. There is close interaction between musicians and dancers and the musicians must follow the dancers (not vice versa), 'reading' from their bodies the required changes of rhythms and tempo. All this is lost in urban conditions, where we meet set dance forms, strict metre, instruction from outside, etc.

These features are met all over the world and through the whole history of dance. At the same time there is a constant interchange noted between the 'rural' and the 'urban'. To illustrate this problem, I have chosen examples from Cuiavia

(Poland), where the most representative dance of the mazurka type is called *kujawiak* (see Lange 1960, 1964, and 1966). During 1954-1967, I studied in detail forty-five villages and settlements in that area.

Because of the importance of improvization in rural dancing, there is no possibility of finding one single way of dancing a *kujawiak* in the villages. Moreover, the same element is varied rhythmically in the course of dancing, even by the same person. As a result, during my research, I was compelled to establish a *scale* of possible variations of the same 'step'.

Now, the *kujawiak* was adopted in the early nineteenth century by townspeople and landed gentry and it spread all over the country to become one of the Polish 'national dances'. Taken away from its original surroundings, it changed considerably. Rhythmically, it keeps strictly to the 3-beat metre. Choreotechnically, it becomes a smoothed-out, simplified version of its peasant original, acquiring an entirely different character. It is this version that is being taught at dance schools in Poland and abroad.

The synoptic table summarises the result of my research into the problem of the *kujawiak*. (See Fig. 2).

1. The first line contains an example of music as interpreted by a town musician and notated metrically.

2. The second line shows the dance notated as it was performed by the townspeople and landed gentry. Again it is metrical. The choreotechnic units are equally measured, and there is a limited and quite regular movement of the body as the weight is transferred.

3. Line three contains the same music, but this time as performed by a village musician who has no idea of musical notation and who has never been schooled in music. It is played in the way in which is 'always has been played'. The musician has to follow the dancer; the song was first improvised by the dancer and then taken over by the musician, and there is constant interaction between the two. The $\hat{\circ}$ signs above the stave indicate moments when there is *rubato*, or flexibility of metre in performance.

4. Line four is a musicological reinterpretation of line three. In the course of the *rubato*, the time units inside the bars remain the same as in lines one and three. The inner structure, however, has been rhythmically remodelled. These changes are so far-reaching that one could even approximately transcribe the village version in a 5/16 beat instead of 3/8. This solution was proposed by the late Prof. M. Sobieski (1960) of Warsaw, and it has been used here as the musical version that agrees best with the peasant version of the dance, showing a predilection for 5 rather than for 3.

Fig. 2. *Dance as an index of rural/urban contrasts: versions of the* kujawiak, *a mazurka-like dance.* (A) *Music as performed by a town musician.* (B) *The dance as performed by townspeople and landed gentry.*

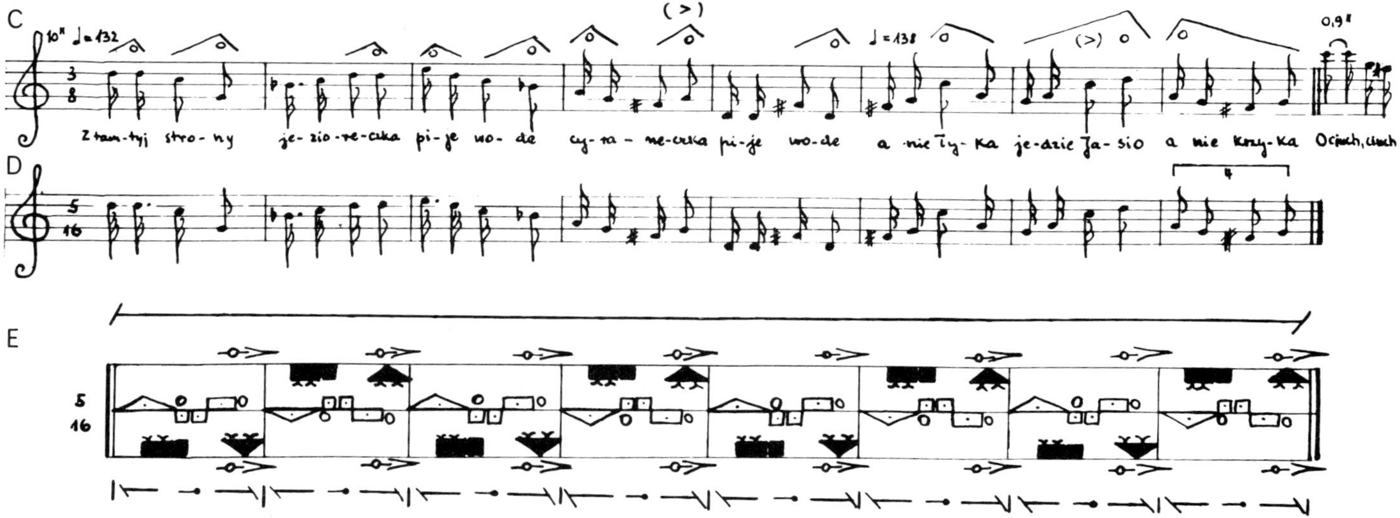

Fig. 2 (C) *The same music as in* (A), *performed by a village musician.* (D) *A more accurate transcription of* (C). (E) *The dance as performed by villagers.*

5. Line five contains the choreotechnic scheme of the same dance as interpreted by the villagers. Indeed, it can only be danced in conjunction with the music of line three; that is, nearly as a pattern of five beats with different stresses caused by the hesitant type of stride. The amplitude of the movements that occur with the transference of weight is much greater than in line two. There is also an inner vibration in the upper part of the body resulting from the intricate change of weight in the feet. The whole performance is entirely immersed in the flow of dancing.

In addition to this, when improvising the song prior to dancing, the performer indicates the rhythmic pattern with bodily movements, more or less on the spot, following the scheme contained in line five of the synoptic table. However, the arms and the upper part of the body of the dancer would move in counter-directions, and would follow an entirely different rhythmic pattern, making the whole movement picture even more complex. Here we come across polyrhythm in the dancer's body. Furthermore, to fill in the rhythmic phrase, which is longer than the text of the song itself, the dancer adds a series of vocables at the end of the phrase (line three).

The differences between the two versions of the same dance (2 and 5) signify changes conditioned by the different social and cultural backgrounds. The peasant version is decidedly confined to the rural situation, where there is close interaction between dancers and musicians and both dance and music are still improvised. This feature is lost in the urban situation.

In rural dancing, emotional tensions are heightened to that degree that it would be difficult or even impossible to apply any of the criteria that one uses in urbanised conditions. The rhythmic pattern has a pulsating character which follows the bodily processes of tension and relaxation. Strictly measured time-units were introduced only in later stages as a symptom of extreme organisation (Sachs 1953:65-67, Bowra 1962:85).

This is equally relevant with tribal and peasant peoples who are still not bound by the type of imposed and measured strictness as attained by 'civilisation' (urbanisation, Westernisation). In the former conditions of life, there is not yet a striving for mechanical discipline. There is still the possibility of manifesting and acknowledging the primary potential of human movement as revealed directly through the body. In these conditions it is still possible to live in agreement with the biological background. The many extensions of our capacities, among the specialisations and appliances of the urbanised world, are simply not *needed*.

The rhythmic patterns of machines of all kinds are generally alien to living beings, because they exclude the element of recuperation in movement, in which each effort must be balanced organically with an element of recovery.

However, rural inhabitants certainly explore rhythm in organising their movements to achieve the uniform actions necessary to perform various tasks. But this degree of automation is, on that level, still beneficial to the executor of the task, and the rhythm is not yet being dictated by a machine. A song or drum beat supplies a suitable rhythm, constructed according to the psychophysical capacity of the human being. The effort is in this way always well-balanced by the element of rest and recovery that follows the element of activity and strain (Bücher 1896).

It is, of course, not only the mechanisation of life that is responsible for the automation of our actions. Social restrictions imposed in the urbanised world have eliminated the direct use of the body in expressing ideas and symbols (Douglas 1970:12).

DANCE AS AN INDEX OF MAN'S BIOLOGICAL MAKE-UP

With this, we have arrived at a crucial problem: the use of movement as a primary form of expression and its limitations in an urban industrial environment. Movement as a means of expression and communication, is common to animals and to man. A baby, before it is born, moves in the womb and performs rhythmic patterns which are thought to have some communicative and expressive significance, such as discomfort or pleasure (Meerloo 1955). Primary expressions are largely confined to biological functions and they seem to be uniformly understood (for instance coughing, laughter, crying, yawning, itching, scratching, shivering, rocking). The urge of living beings to perform dance-like actions is in the first place the need to relieve, by means of movement, emotional tensions caused by joyful as well as disturbing events (cf. Darwin 1872). Similarly, the need to relieve the feeling of discomfort resulting from excessive use of particular joints in specialised, restricted everyday movements like sitting hunched over an office desk or before a conveyor belt in a factory, may provoke spontaneous, dance-like actions, thus providing a natural counterbalance (Gros 1899:471-3). During the mating period in animals, their communicative actions are exaggerated through exhibitionism (Meerloo 1952, Lorenz 1970:190, Ellis 1897:41-58, and especially 51). Although human dances are clearly phenomena of a very different order, the sexual aspect is often apparent.

These are examples of dance-like movements playing a physiological role in the life of human beings, and being connected with biological functions.

It is difficult to draw an absolute line indicating where 'dance' begins and ordinary movements end. Many spontaneous movements acquire a dance-like appearance quite imperceptibly, and the transition to 'dance' is blurred. Such 'dance' manifestations still continue to be executed spontaneously by children even in a technologically orientated culture. They may not belong to 'art', they may not be included in the investigations of aestheticians, but nevertheless they are an essential part of human development.

This ability fades away as people grow up; and it is particularly evident in man's general technological development; in the course of gaining mechanical substitutes, man is often confined to inactivity which excludes the 'formative or creative mental processes', the positive element in any productive action (cf. Read 1965:149, Fox 1970, and Steiner 1967).

Successively, the ability to use the body as an expressive medium has been limited or even lost in the urbanised world. This applies equally to the primary manifestations connected with formative processes as with the ability to express poetic formulations directly through the body in dance as an art.

But then art in urbanised conditions is not a commonly pursued activity. Instead it became one of our many specialisations executed by some people only, the remainder being the passive receivers.

Man, as a unique species, has to act in his own special way, but at the same time according to the laws of his biological make-up. As soon as this balance is disturbed, human beings become uprooted.

It is therefore evident that acknowledgement of the biological aspects of dance not only has significance as yet another element in the historical perspective of human development, but also has essential practical, contemporary value that has still not been explored or sufficiently appreciated.

The biological aspect of dance has primarily educational (especially for infants and younger children), recreational, and rehabilitative relevance.

At this point, a quotation of Sir Herbert Read might perhaps be relevant:

> Unless we discover a method of basing education on these primary biological processes, not only shall we fail to create a society united in love; we shall continue to sink deeper into insanity, mass neuroses and war. (Read 1959).

REFERENCES

Bowra, C.M. (1962). *Primitive Song,* London: Weidenfeld and Nicolson.
Bücher, K. (1896). *Arbeit und Rhythmus.* Leipzig: B.G. Tebner.
Darwin, C. (1872). *The Expression of the Emotions in Man and Animals.* London: J. Murray.
Douglas, M. (1970). (2nd ed. 1973). *Natural Symbols; Explorations in Cosmology.* London: Barrie & Rockliff.
Ellis, H. (1897). *Studies in the Psychology of Sex.* London and Philadelphia. Vol. I, Part 2.
Fox, R. (1970). The cultural animal. *Encounter.* July, 31-42.
Groos, K. (1899). *Die Spiele der Menschen.* Jena: Gustav Fischer Verlag.
Gunda, B. (1963). Die mitteleuropäischen Bauernkulturen und die Methode der 'Cultural Anthropology'. *VIe Congrès International des Sciences Anthropologiques et Ethnologiques, Paris, 1960, II,* No. 1. Musée de l'Homme.
Lange, R. (1960). *Taniec ludowy w pracach Muzeum Etnograficznego w Toruniu; Metoda pracy i kwestionariusz,* (Folk dance research at the Ethnographic Museum in Toruń; fieldwork methods and questionnaire). Toruń: Muzeum Ethnograficzne.
Lange, R. (1964). Tance kujawaskie (Dances from Cuiavia). *Literatura Ludowa, 4,* 13-20.
Lange, R. (1966). Kinetography Laban (Movement Notation) and Folk Dance Research in Poland. *Lud, L,* 378-91.
Lorenz, K. (1970). *Studies in Human and Animal Behaviour,* London, Methuen & Co. Ltd.
Meerloo, J.A.M. (1952). *Communication and Conversation.* New York: International Universities Press.
Meerloo, J.A.M. (1955). Archaic Behaviour and the Communicative Act. *Psychiatric Quarterly, XXIX,* 60-73.
Read, H. (1959). The Biological Significance of Art. *Saturday Evening Post,* Philadelphia, 26th September.
Read, H. (1965). *The Origins of Form in Art.* London: Thames and Hudson.
Sachs, C. (1937) (c. 1933). *World History of the Dance,*tr. B. Schönberg, New York: W.W. Norton.
Sachs, C. (1953). *Rhythm and Tempo; A Study in Music History.* New York: W.W. Norton.
Shanin, T. (1971). *Peasants and Peasant Societies.* Harmondsworth: Penguin Books.
Sheets, M. (1966). *The Phenomenology of Dance.* Madison and Milwaukee: University of Wisconsin Press.
Sobiescy, J. and M. (1960). Tempo rubato u Chopina i w polskiej muzyce ludowej, (Tempo rubato as used by Chopin and as evident in Polish folk music). *Muzyka, 3.*

Steiner, G. (1967). *Language and Silence*. London: Faber and Faber.

PATTERNS OF BODY MOVEMENT IN THE MUSIC OF BOYS' INITIATION IN SOUTH-EAST ANGOLA

Gerhard Kubik

INTRODUCTION

The *mukanda* institution is known by this name throughout a vast area of Central Africa comprising almost all east and south-east Angola, the northwestern parts of Zambia and portions of southern Zaire. With local variations it occurs among all the populations that Murdock (1967:9) includes as Cluster 13 under the designation Lunda, with his subdivisions Luvale (AC 11), Ndembu (AC 6), Chokwe (AC 12), Lwimbi (Luimbe) (AC 28), Luchazi (AC 28) and Luwa (AC 29).

Mukanda can be described as a boys' initiation and circumcision school and is the most important educational institution among Luvale, Chokwe, Ngangela, Mbwela, Mbunda and other groups, all of which are culturally and linguistically closely related and share common historical links. Western social anthropologists have been inclined to speak of *mukanda* as a 'circumcision lodge' or 'circumcision camp', terms which are often resented by members of the communities concerned and regarded as concepts of Western colonialism which deny the existence of schools to traditional African societies. People in liberated areas of Angola and also in Zambia actually refer to *mukanda* as a traditional school (*shikola*, from port. *escola*). This point of view is also shared by Mwondela, an outstanding Zambian educationist in his recent book (Mwondela 1972). I studied *mukanda* firstly in Angola, where I worked in the south-east among Mbwela, Ngangela and Chokwe groups living near the upper Kwitu river (Circunscricao do Cuito-Cuanavale) and its tributaries, and in the east near Kazombo (Cazombo) among the Luvale, from July to December 1965; secondly, in 1971 and 1973, when I was engaged in fieldwork in Northwestern Province, Zambia, mainly among Luchazi communities. The present paper is exclusively based on recordings, films and information collected in South-East Angola in 1965, where I worked in four different *mikanda* (see

TABLE I). Unless stated otherwise, data from my later work in nine Zambian *mikanda* are not included.

TABLE I
Mikanda documented in south-eastern Angola, 1965

Mukanda of village headman	Age and number of novices		Rites in which I assisted
Sakateke	6	ca. 5-10	1. *kukusha mpoko* (to wash the knife) 2. *kulundulula mukanda* (to remove the mukanda)
Limbwangu	4	ca. 6-10	*kulundulula mukanda*
Chintete	16	ca. 6-15	----
Sakapula	2	ca. 8-10	*kulundulula mukanda*

Elsewhere I have given detailed accounts of the social and cultural aspects of *mukanda* in south-east Angola (Kubik 1971), the meaning and cultural context of the associated *makisi* mask complex (Kubik 1969), music education (Kubik 1973), and songs of the initiates as expressive and didactic vehicles in the socialization process (Kubik 1974). Accounts by other authors describing *mukanda* as a ritual include Turner's works for the Lunda-Ndembu (Turner 1962, 1967), White (1961) for the Luvale, Baumann (1932) for the Chokwe, Koolwijk (1963) for the Ngangela, and other more scattered accounts.

My own preoccupation with *mukanda* developed in Angola, where I originally went as an ethnomusicologist. Very soon I discovered that any deeper study even of music alone was illusory without my being culturally assimilated by the communities concerned. I learned the Mbwela language and began to go through the rites of Mbwela youth. Besides *mukanda,* which is recognized by the people as the first stage in formal education I had to pass two further institutions: *vandumbu* (see Kubik 1971) and *mungongi*. (My material on *mungongi* initiation is still unpublished). What John Blacking has called an anthropology of the body implied, in my particular case, the involvement of my own body.

Mukanda is considered by Mbwela communities to be the first intervention in a male human's life to make him a complete person (*linga akapwe muntu*); *mukanda* does not make a child (*kanike*) to become a 'man', but a socially adapted person, and thus, a real human being (*muntu*). I see my preoccupation with *mukanda* partly as a reflection of my own process of integration and enculturation. As I grew into Mbwela culture I ceased to be regarded as *chindele* (European) and became *muntu*.

In this paper I consider one facet of Mbwela culture, namely the relation of the novices' body movement to some of the psycho-social processes in *mukanda* initiation. At the same time I also introduce a system of dance transcription.

The word *mukanda* denotes in Mbwela/Ngangela language a) the rectangular enclosure outside a village where the newly circumcised boys are kept in seclusion for between four to five months; b) the institution as such, with its associated activities. The season for *mukanda* coincides with the dry season, that is approximately from April or May to October/November. The age of the novices ranges from six to ten, though occasionally some boys may be older.

ORGANIZATION OF MUKANDA

The most important role inside the school is that of the *chilombola* (plural: *vilombola*), i.e. the guardian of the *tundanda* (novices). Usually a *chilombola* is recruited from the male relatives of a newly circumcised boy. He may be a cousin, elder brother, or a more distant relative of the boy. He is never the father or mother's brother. Every novice has his personal guardian, and in addition there are one or more assisting guardians called *tulombola-ntito*. The latter are youths, brothers, cousins, even friends, who are often just a few years older than the novice. Often a teacher for songs and dancing is employed and called *mulongesi* (teacher).

The mothers of the novices (*vanyatundanda*) are responsible for sending food to the boys secluded in the *mukanda*. Masked dancers, who are guardians in disguise, appear regularly in the village during *mukanda* and collect gifts from women to support the school. The women respond in chorus to the singing of the masked figures, cheer at them or run away in fear, according to what kind of mask it is, or mirror their dance movements (see Kubik 1969, 1971). When there is communal singing in the lodge during the night, the boys' mothers are expected to stay up in the villages, sometimes all night, and show their appreciation with *vingunda* (or *zingunda*), i.e. joyous ululations, produced while striking the mouth with the open hand.

In the first stage of *mukanda*, the novices are said to be 'still at the leaves' (*vachili hamefo*): the penis is wrapped in a soft leaf, and the main task of the guardian is to replace the leaf each morning after cleaning the wound with traditional medicine. As soon as the wound is healed, about two to three weeks after the operation, a period of intensive instruction begins. At this stage much emphasis is placed on discipline and cooperation. Knowledge about nature is acquired by the novices while roaming in groups near the *mukanda* during daytime, setting traps, shooting birds with bow and arrow, and fishing. In the *mukanda* itself some elements of craftsmanship are learned, such as plaiting and twisting of plants and strips of bark for making dance costumes. Traditional technology is emphasized more, however, in subsequent years when the initiates return to *mukanda* in new roles.

Nonverbal interaction is as important as verbal instruction in the *mukanda*. There is a specific category of songs referred to as *myaso yatundanda* (songs of the novices) which are performed only by the novices and which are accompanied by dance movements. An analysis of the content and symbolism of these songs (see Kubik 1974) shows concern with the internalization of new modes of behaviour that will be expected of graduates, especially towards the mother, but also to sisters, father, maternal uncle, guardian, and other members of the village community. 'Difficult' children may be given more verbal instruction than is usual by the guardians. This was especially marked in the case of a mentally retarded child in a *mukanda* that I studied at Mikula village (Katuva river), north-western Zambia, in 1971.

Another category of songs is performed by male visitors, guardians and novices, to the accompaniment of concussion sticks. These sessions of communal singing in the *mukanda* during the night are referred to as *kukuwa mingongi*. Some of the songs are didactic and recall important historical events, or they relate taboos (*vizila*, sing. *chizila*) to be observed by novices and others in the *mukanda*: for example, certain kinds of food are forbidden, and the novices may not enter through the gate of the guardians, but only through their own gate. Each Mbwela *mukanda* has two entrances, one for the novices and one for the guardians.

All these rules function as a set of behavioural prescriptions aimed at maintaining a precarious and rather fragile psychic balance among those involved in the *mukanda*, particularly in the first period which is experienced as dangerous. I was aware of the enormous tension when I stayed and lived in the lodge and assumed some of the roles. As the tension gradually diminishes with the successful healing of the novices' wounds and their 'getting better' in many ways, the taboos become less numerous and are less rigidly enforced. Some taboos extend, however, beyond the *mukanda* period until years after.

BODY MOVEMENT AND ITS FUNCTIONS

Though music and dance instruction in south-east Angolan societies is also in itself an important objective of the *mukanda*, and the results of the five months' instruction are lavishly displayed before the women on the day of the novices' return to the village, they are part of a general system of education. Movement is often an important means of expression. In one action drama during a rite of the second stage of *mukanda*, the boys are told that they will be killed with an arrow. They are told to line up outside the *mukanda* and then one by one move towards the 'tree of the dead' (*muti wa vakulu*) planted in the centre of every Mbwela *mukanda*. Arriving at the 'tree', the leader of the rite, with an arrow in his hand and red paint in a pot, whispers to the novice: 'You embrace the tree and fall down like a dead person.' The boy acts it out. The man puts red paint on his breast and the arrow near his body. When the next boy enters through the gate he sees his comrade in his 'blood' on the ground, believing that he was killed. The same is then repeated until all the boys are through. The movements of entering, embracing the 'tree' and falling down are acted out like a drama.

Some dance patterns are linked structurally with other patterns of body movement, such as beating bark cloth or winnowing. These links cross age and sex groups, in that a dance pattern of the novices may appear in a structurally similar form in women's work and this may be recognized by the people. Furthermore, the movement patterns do not stand in isolation but are part of a *movement style* which even transcends, in some aspects, ethnic and linguistic boundaries.

A movement pattern of novices called *kundekula*, for example, is known not only in south-east Angola, but also among the Luchazi and Luvale of north-western Zambia. There it is known as *Kawali* after the name of the most famous song that accompanies this body movement. There is a link between body movements and songs, in that certain songs always go with certain movements. Other movement patterns, such as the combination of strokes produced on the concussion sticks, transcend in their geographic distribution the area occupied by the *mukanda*-practising populations. The particular combination given in Fig. 4 below occurs in drum music of the Bemba, who are outside the *mukanda* area, and in many other areas, and the two-against-three combination is almost universal in sub-Saharan Africa.

Some dance patterns are specifically for the novices and do not occur in performances unconnected with the *mukanda* school. Dances from the village, such as *kachacha*, are not performed in the lodge, though they may be employed in the village during the chief rites of *mukanda* such as *kukusha mpoko* or *chikula*, in which the women may dance *kachacha* to the drums.

Since the movement patterns of the novices' dances are *mukanda*-specific in the sense that they are only seen performed when a *mukanda* is on, the possibility of a functional and structural relation to the novices' psychic states has to be taken into account. There is evidence from verbal testimony that when the novices perform various dance patterns in the *mukanda*, they experience relief from stress and sorrow caused by isolation from their mothers and seclusion in a camp in the forest. Dancing makes them 'not to think of their mothers.'

Such relief seems to come about in two ways. Firstly, stick or dance patterns associated with the status of novice are always performed in groups, never alone. Even in the rare case of a *mukanda* with only one novice, which I studied in Zambia at Sangombe's village in 1971, other boys who had already graduated from the *mukanda* came to help him perform in a group. Though there are many *mikanda* with two boys, particularly when the village is small and the area sparsely populated, two are not considered sufficient for a satisfactory performance.

The number *two* is emphasized during the first stage of the *mukanda*: the boys generally sleep in pairs and the cage-like sleeping devices are constructed in that way. The intense relation between the novices and guardian during the first period is another dual set-up that is later eased. After the feast *kulonda ntsimba* (see below), the paired sleeping devices are burnt, and all the boys sleep together and act more and more as a group.

The increasing emphasis on the group as soon as the novices 'get better', obviously absorbs a great deal of the same inner forces that once bound each boy to his mother. At first, the mother-child relationship reappears in 'transference' in the relation between the guardian and his novice, as is clearly indicated in many of the conversations that take place between the two. But later the emotional aspect of the relationship between novice and guardian is also gradually 'neutralized', as the guardian must be treated by the novice with respect and not with dependent love. Only during seclusion is the boy to his guardian 'like a wife' (*ngwe mpwevo yendi*), i.e. dependent on him. Later relations with the guardian are expected to become increasingly formal.

Everything that emphasises action within the group of novices, and this includes dance movement, promotes the development of strong group attachment and reduces mother attachment. In the early stage of *mukanda* the novices do not feel like talking much to each other. Later they form a happy crowd with all sorts of relations between them.

The novices' dances are part of a culturally tested repertory of techniques designed to achieve a lasting change in an individual's patterns of behaviour. The co-ordination of movement patterns that are performed by more than one person is both an expression of and a tool for the creation of a feeling of group identity, group attachment and solidarity. In the *mukanda* this is further emphasized by the fact that the novices eat the same food, walk about together and share the same festive dance-dress and body paint when they finally return to the village. This does not mean that age-group experience in a male individual's life starts with *mukanda*: children also play in peer groups, but in *mukanda* the age-group assumes a more exclusive role.

The second way in which relief from tensions comes about and in which movement patterns exert a therapeutic effect in the *mukanda* seems to depend on the structure of the movement patterns themselves. It is well known in psychology that in a state of severe and serious inner experience, structures of strong inner coherence have a compensatory effect (Jung 1974). A structure of great inner order, such as Jung's *mandala* figures, may be produced spontaneously by individuals in a state of psychic danger.

If such states are provoked by ritual, the ritual itself must contain patterns and symbols which trigger off the necessary process of reconstitution and reintegration. Mbwela people acknowledge that there is a danger of *kuzaluka* (to run mad) if certain interdictions are not observed by those involved. Taboos are barriers designed to avert the danger of psychic and social disintegration. Some of the movement patterns performed in the *mukanda* function as compensation for emotional disorder, characterized by fear and anxiety during the first stage. These movement patterns are structures of harmonious order and beauty.

The culture of the Mbwela thus provides for external structures of a compelling logic, which must be absorbed physically by the novices and in this way internalized. The movement patterns to be learned by the novices are structurally regular. None of the asymmetric 'additive' rhythm patterns which are common in the entertainment music of the village, occur in the *mukanda* school. Most combinations in Mbwela boys' initiation music are made up of interpolations between two and three, built up to form compound units defined by the numbers 6 and 12. Though I have not, so far, come across any explicit statements by Mbwela people to point out that the novices or the other members of *mukanda* actually conceptualize these units, their presence is overt (compare transcriptions). Apart from the inner structure of the patterns, the formal layout of dance figures is also determined by regular constellations emphasizing the numbers 3, 4, 6, 8, 9, 12, 16.

By playing stick patterns as a group, performing *kuhunga* and *kundekula* and other dance movements, the *tundanda* are able to 'get better'. But this works only if the individual identifies himself with the coercive inner order of the patterns *through body participation*, through his personal incorporation of the pattern.

Patterns of body movement in musical performance

One of the first musical activities of the young boys in the *mukanda* is to learn concussion stick patterns. They are urged to do this during the first stage after circumcision, when they are 'still at the leaves'. At sunrise and sunset the novices assemble by the long fire-place (*lwowa*), which is one of the secrets of the *mukanda* never to be revealed to the women. No member of *mukanda* may jump over it under penalty of later impotence. Each novice holds a stick of about 20-25 cms length in each hand. Striking the sticks together, they sing symbolic songs of welcome to the rising sun and songs of farewell to the setting sun.

The most important communal activity during *mukanda* is the performance of songs inside the lodge at night. Men and youths from the village form a circle looking towards the fire, and perform the 'songs of the *mukanda*' (*myaso yakumukanda*) in multi-part voice combinations (Kubik 1968). The novices are required to participate in these sessions, which are referred to as *kukuwa mingongi*. Each performer holds two concussion sticks (*mingongi*) which are secret instruments of the *mukanda*. Women and uncircumcised boys in the village must not see them. They are told that the novices have their legs cut by magic and beat the calf-bones together, or alternatively that they are bones from dead corpses in the cemetery. One of the promises extracted from the boys at the end of the seclusion period during the oath of *mwongwa-luvanda* is never to reveal that the *mingongi* are just wooden sticks.

The *mingongi* are cut by the guardians, and on the closing day of *mukanda* they are burnt together with all the other secret things inside the lodge, the 'tree of the dead', some masks, and so on.

Besides *mingongi*, other secret instruments are the *tutanga* (sing. *katanga*). These are wooden slats vertically suspended from a string that runs through a tiny hole near the upper edge of the slat. The player holds his *katanga* in the left hand. The beating stick, which is usually one of the *mingongi*, is held in the right hand, and strikes the slats in the middle (Figs 1 and 2).

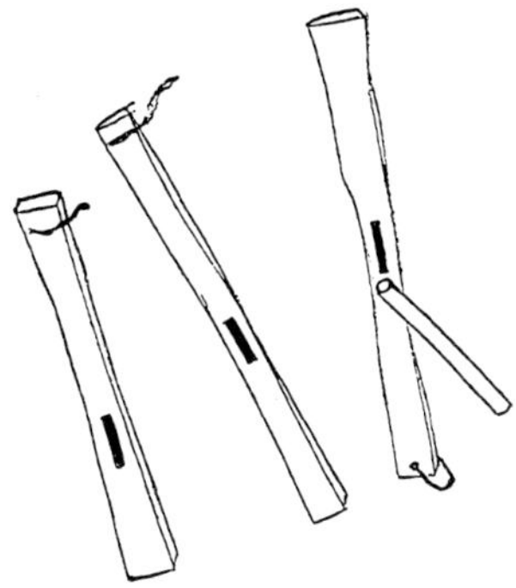

Fig. 1 *Tutanga* wooden slats.

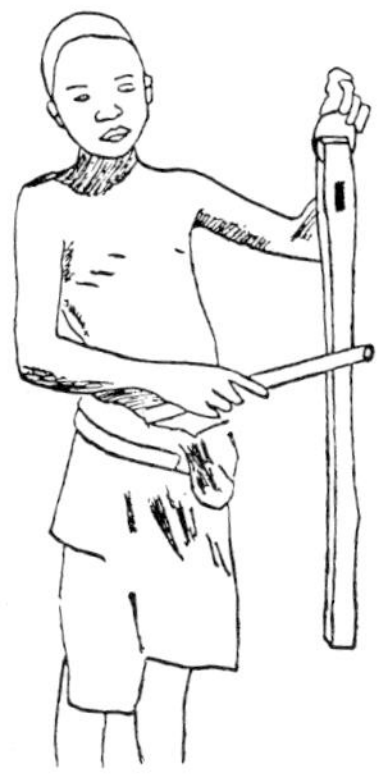

Fig. 2 Position of striking a *katanga*.

The name *tutanga* was not used in the four *mikanda* I attended in south-east Angola. Sometimes the slats were just called *mingongi*, and also less elaborately shaped.

Mingongi and *tutanga* are interesting examples of instruments that would never be seen by ethnomusicologists unless they stayed in the lodge; nor would they appear in the archaeological record.

During performance the *mingongi* are not struck together in the same way by all players. They organize themselves into groups striking different patterns, which they combine while singing the songs of *mukanda* (See Figs 3 and 4).

The following is a description of the music in terms of body movement rather than sounds. In Mbwela there is no single term corresponding to the Western term 'music': there is a generic term for dancing (*kukina*), and there are many specific

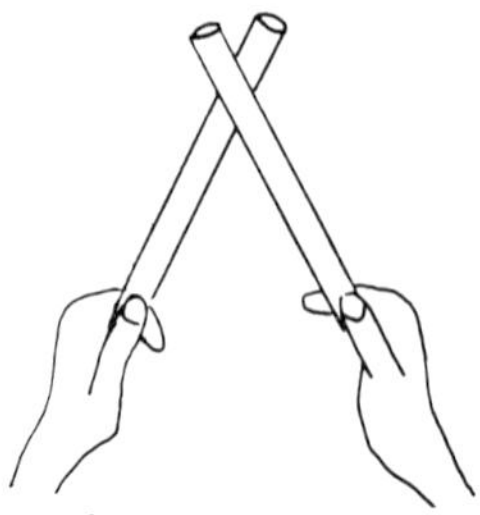

Right stick hits left stick
from above
(transcription symbol: A)

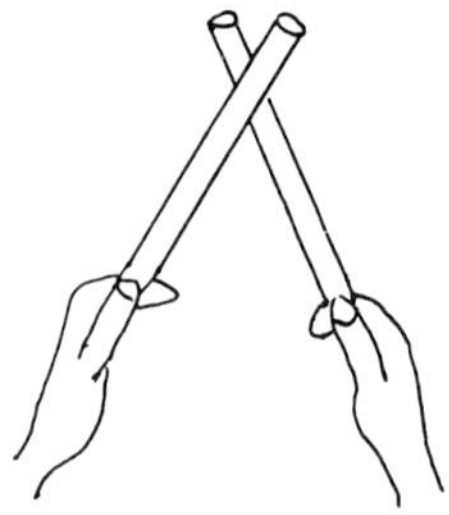

Right stick hits left stick
from below
(transcription symbol: B)

Fig. 3 *Concussion stick patterns and their combinations (viewed from above)*

Group I (tundanda):	12	A . . B . . A . . B . .
Group II (men and youths):		A . B . A . B . A . B .
Group III (men and youths):		B . A . A . B . A . A .
Individual men playing *tutanga:*	(I)	x x . x x . x x . x x .
	(II)	. . x . . x . . x . . x

Note: x is a stroke. The points are 'empty' pulses along the time line. They are filled in by portions of the song melody.

Fig. 4 *Transcription of* kukuwa mingongi.

terms describing various dance movements; there are the words meaning 'song' (*mwaso*, pl. *myaso*), 'to sing' (*kwimba*), to play an instrument (*kusika*), and a rich vocabulary referring to specific 'musical' actions.

As is often the case in African music there is a difference between a listener's auditory impression and a performer's motor concept of sound patterns. Fig. 4 above would be appreciated by listeners as a pattern covering six elementary pulses (cycle 6), while for the players the total pattern repeats every 12 pulses (cycle 12).

The basic stick pattern is allocated to the novices (Group I in the transcription). Though others may play the novices' pattern, I have never seen novices play the stick patterns of the other performers. This may be because the basic stick pattern is the easiest to play, or because of the novices' status.

No one made any explicit statements about this.

The sticks are struck vertically, up and down, in slow, duple motion (see Fig. 3): first a down-stroke of the right stick, hitting the rising left stick from the top, and then a down-stroke of the left stick hitting the right stick from the top; the two sticks meet about half-way. The metre of the songs is related to the novices' stick patterns.

While the novices play the basic duple patterns, another group of men and older boys produce the same movement pattern a little faster, so that the strokes of the two groups combine in a 2 against 3 relationship (see transcription, Fig. 4). After six strokes of group II and four strokes of group I, the two movement patterns meet. The third group performs a pattern which *sounds* identical with that of group II, but is *produced* differently. While group I and II play down-up-down-up with each stick in contrary motion, group III strikes up-down-down with the right stick and consequently, down-up-up with the left stick, performing a pattern in threes.

The left-hand stick is hit by the right-hand stick on the first stroke from below and on the second and third from above: one could also say, the right-hand stick is hit by the left on the first stroke from above, on the two subsequent ones from below. In transcription we use two symbols only: A (right stick hits left stick from above), B (right stick hits left stick from below). Both sticks move, however, completely evenly.

This combination is then further expanded by the addition of patterns struck on the vertically suspended wooden slats (*tutanga*). The *tutanga* patterns are fast strokes which interlock with the other patterns. *Tutanga* are never played by novices.

This combination of stick patterns gives a 'bi-metric' result. The novices learn in those sessions to play their basic patterns *against* the other patterns and also to keep time throughout the song.

Patterns of body movement in dance performances

After the symbolic ceremony of *kulonda ntsimba* (to climb the genet), which takes place when the wound is healed and during which the boys are allowed for the first time in weeks to take a bath in the river, a new phase of instruction begins. From then on teaching is strictly formal. I define as formal instruction a situation in which a clear distinction can be made between a teaching body (in this case the guardians and dance instructors) and learners or pupils, and where the two meet regularly.

One kind of formal instruction in *mukanda* is to teach recitations (*kutangesa* or *kutanga*). These are arranged in the form of an alternation between drum beats and the reciting voices of the boys. During an evening teaching session the novices usually practice *kutanga* before dance instruction. One low-pitched long drum (*chipwali*) is used by the instructor. The texts of the recitations are symbolic and also rather long, in order to train the novices' memory. (An account of *kutanga* is given in Kubik 1974, with text transcriptions and analysis.)

For dance instruction, three single-skinned long drums are needed, the *vipwali* (sing. *chipwali*). These drums are played by the instructors and are not secret instruments, but are brought over from the village, where they are also used in other dances not connected with this institution. They are used in sets of three and are played with the hands. They are tuned with tuning paste. The individual drum names are: for the instrument with a 'big voice' (*lizi lyakama*), also used for the recitations; *chimpululu* for the middle-voiced drum, and *mundengu* for the one with a 'small voice' (*lizi lzyalindende*).

The boys put on dance skirts made of the reddish *ndzombo* fibres of the inner bark of *Syrygium guineense*. These skirts are made for them in the *mukanda* by elder brothers before the feast of *kulonda ntsimba*, and the novices wear them for the first time after the ritual bath in the river. People in the *mukanda* say: 'Now they will begin to learn *kuhunga*, a dance movement.

Until the end of the period of initiation, the novices learn and practise various movements and dances that they perform in public during the final feast of the *mukanda* when they return to the village. The most common and most typical movements for this period are called *kuhunga* and *kundekula*, verbs that refer to the form of the movements. These are linked with two types of songs of the novices which are called *myaso yakuhungisa* (songs to make *kuhunga*) and *myaso yakawali* (songs of Kawali).

The movement pattern kuhunga. *Kuhunga* means 'to shake'. It conveys the idea of a sideways shaking motion. The word is also used, outside *mukanda*, to refer to the women's work of winnowing a basket full of bullrush millet (*masangu*). A woman holds a small oval basket in her hand and performs a swinging movement, forming circles in contrary motion with each hand.

People say that the novices' dance pattern alludes to the women's work of winnowing, though the movement patterns are *not* completely identical. In dancing, *kuhunga* means a horizontal shaking movement of the hips and pelvis in a right-left alternation that causes the dance skirt (*ndzombo*) to swing out.

(Fig. 5). The arms are lifted and the elbows are bent at right angles and held away from the body. Arms, head and shoulders are kept still in this position. The dancer holds his rattle-stick in his right hand and shakes his hips sideways. His legs and knees are kept elastic and perform a complementary movement together with the shake. When the left knee is slightly raised, the right hip is thrust forward and the dance skirt swings counter-clockwise. Next, the left leg is stretched and at the same time the right knee slightly raised. The left hip now pushes forward, the pelvis appears twisted towards the left in relation to the torso, and the dance skirt swings back clockwise.

This can be clearly seen in cinefilms taken when the novices put on ornamental patterns of white and red body-paint during the last rehearsal, a few days before the return to the village, and on the great day of return. These patterns of paint are structured in a way that high-lights and emphasizes the dance movement to be performed. They make the movement centres of the body more conspicuous, so that body motion is transformed into a vivid display of contracting and pulsating ornamental motion.

While doing *kuhunga* the boys may move forward with tiny steps. But usually they stand on one spot. An analysis of the leg-and-foot work from film shows a 'pedalling' pattern, in which the dancer lifts his heels alternately. Lifting of the heel goes with raising the knee. This quick heel-lifting on the spot, together with simultaneous displacement of body weight from one foot to another, is necessary to allow the sideways shaking of the pelvis to develop (see Fig. 5 and 6).

Shoulders are not moved in this dance pattern, though the upper torso vibrates a little as a result of the pelvis shaking. Torso and arms remain rather immobile. This comparative immobility emphasizes the propeller-like swinging-out of the dance skirt.

Kuhunga is a duple motion: left-right, left-right. But it is superimposed on a triple pattern of the drums. When the dance skirt swings out, one notices a soft organized sound produced by the shaking of the fibres, at least with experienced dancers. People say the dance skirt sounds something like: *ka-cha-ka-cha-ka-cha* etc.

In the *mukanda*, sound patterns such as *ka-cha* etc. play an important role as teaching devices. They may be syllabic or verbal. In dance instruction they are used by the teacher to characterize movement patterns and to correct the novices' mistakes. These mnemonics are a kind of oral notation comparable to what is also found in other parts of Africa and in India. There are also verbal patterns employed in the *mukanda*

Dancer's position:

Transcription from 8-mm-film, black-and-white, X 4/D/1 (short extract) 16 frames/sec.

Symbols: (corner points)

- < pelvis shake to the left, left hip thrust forward
- > pelvis shake to the right right hip thrust forward
- ▲ left knee and left heel raised, leg bent
- ■ left leg stretched, body weight on left foot
- △ right knee and right heel raised, leg bent
- □ right leg stretched, body weight on right foot

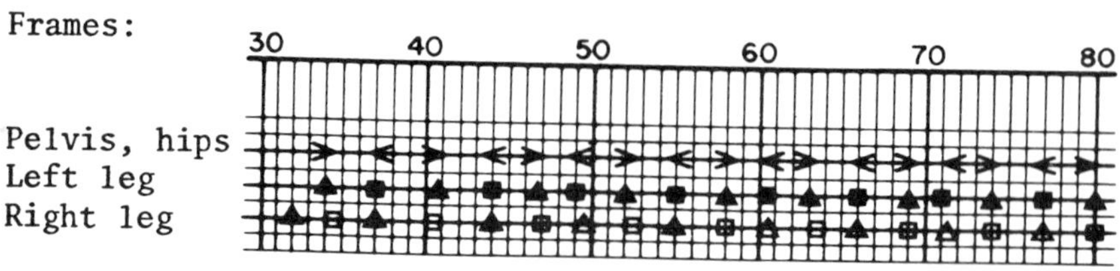

Learners' syllables: ka-cha ka-cha (etc.)

Fig. 5 *Illustration and transcription of* kuhunga

which go with patterns of body movement or drumming. For example: *kuvamba kuli masika*, a drum pattern for the accompaniment of novices' dances, means: 'in the lodge there is coldness'. The timbre and movement structure of this phrase is directly related to the drum pattern. The drum 'speaks' these words. These are verbal associations that occur among speakers of tonal languages when they hear pitch patterns. One of several possible verbal associations fits the drum pattern so well, that it becomes a kind of label for this pattern. In this case, the verbal content is an appropriate comment, because temperatures in July and August may well reach zero degrees in exceptionally cold years in south-east Angola, and the stay in the lodge may become very uncomfortable for the novices and their guardians.

Kuhunga is considered to be the most important dance-pattern of the novices, always mentioned first in conversation, with *kundekula* second. The dance skirt is designed for *kuhunga*, but is not used in *kundekula*. The guardians, as well as the dance master, take great care to improve the performance standards of their pupils. Conversation, criticism and discussion, reflect their values and attitudes: for example, in a Zambian *mukanda* one initiate was known as *kaliselwa* (little cloud), because he was so excellent in performing *kuhunga*. A good dancer, say the people, should make the dance skirt swing out continuously in perpetual motion so that it seems to stand still like a 'little cloud'.

The movement pattern kundekula. This is the second pattern characteristic of boys' initiation music. It goes together with a well-known song:

Kawali'ee! Nana yow'e èéè! (2x)	*Kawali ee!* My mother is over there, *èéè!*
Kawali Mwanandonga!	*Kawali Mwanandonga!*
Njichukatale vanana kwimbo.	I should like to go and see my mother in the village
Kawali'ee! Nana yow'e èéè!	*Kawali ee!* My mother is over there, *èéè!*

The elements of the *kundekula* pattern are rolling the shoulders with a corresponding movement of the shoulder-blades, and foot work. The novices form a circle, facing the centre, bodies very relaxed, knees bent. Then the leg-and-foot complex is moved in the following way:

1. Left leg (and foot) slightly lifted off the ground.
2. Foot turned a little outward.
3. Leg stretched.
4. Foot put forward on the ground *without* displacement of body weight.

Symbols (corner points)

● impact of left foot (put on the ground)
○ impact of right foot (put on the ground)

Impact — foot rests of the ground — lifting off

\ foot turned outward to the left
/ foot turned outward to the right
| foot put into straight position

Frames:

Left foot put forward:
Initial position:
Right foot put forward:
Initial position

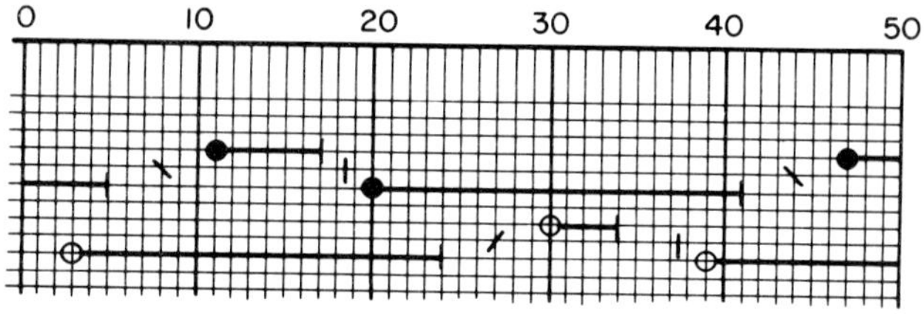

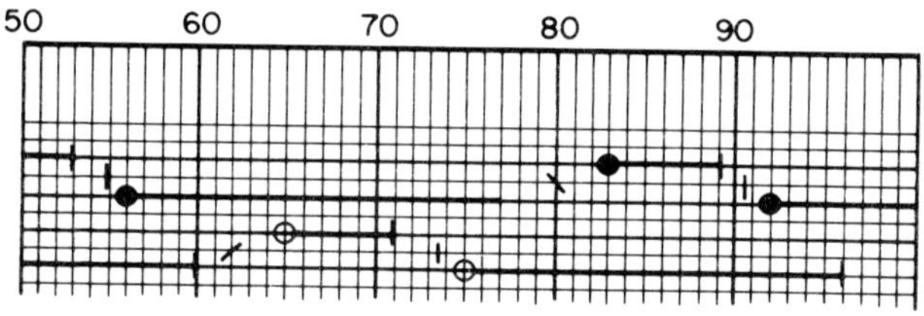

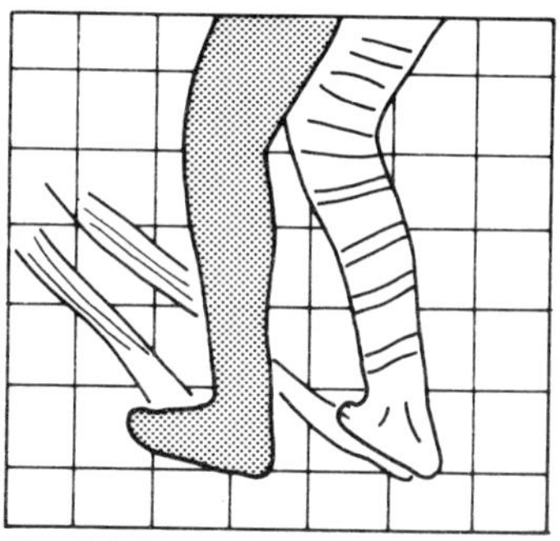

Kundekula step: initial position.
(left foot)

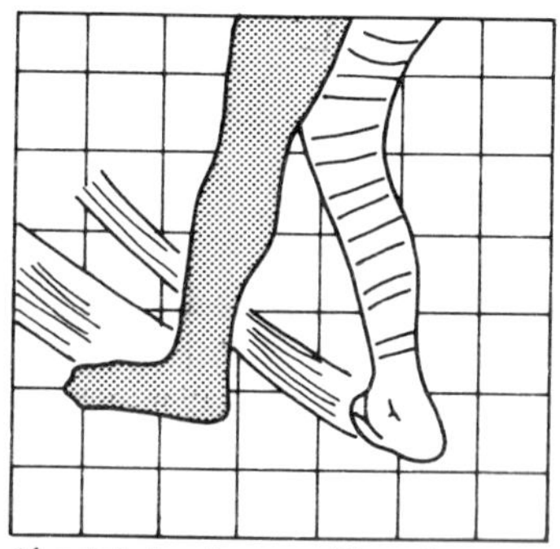

Kundekula step: position 4.
(left foot)

Fig. 6 *Illustration and transcription of* kundekula.
a) Analysis of leg and foot work, 8-mm-film, black-and-white, X 4/B/11, frame-by-frame transcription, 16 frames/sec.

Additional symbols:

△▲ vertex in circular shoulder movement (the highest point). Right and left symbol.

▽▼ bottom position in circular shoulder movement (the lowest point). Right and left symbol.

(Both shoulders move backwards describing circles).

C contraction of shoulder blades, stomach put slightly forward, pelvis thrust backward

Ↄ release of shoulder blades, stomach contracted, pelvis forward, back appears broad and stooped.

● placement of body weight into left foot

○ placement of body weight into right foot

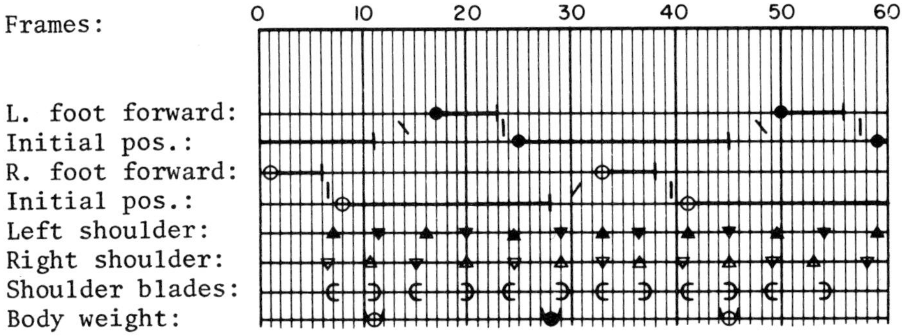

Fig. 6 *Illustration and transcription of* kundekula:
b) Combination of movement patterns of various body parts in *kundekula:* leg-and-foot complex, shoulders and arms, shoulder blades and back. Transcription (extract) from 8-mm film, black-and-white, X 4/B/7

---

5. Leg (and foot) lifted again.
6. Foot placed back into initial position (compare illustrations).
7. Displacement of body weight.

The same is done with the right leg. In this way, left and right foot are put forward alternately. Simultaneously, the dancers perform the rolling shoulder motion in such a way that the shoulders always take opposite positions (one up - the other down, one behind - the other in front), like the pedals of a bicycle as if they were being turned backwards. It is a constant, circular motion. This is combined with a periodical retraction and release of the shoulder-blades.

When one looks at the dancing boys from behind, one could believe that shoulders, shoulder-blades, and back were made of rubber, especially with the body paint on, so elastic appears this movement when it is really well performed. Interaction between legs, feet, shoulders and shoulder-blades can be shown in a transcription.

From time to time the novices stop this pattern, tap the ground with their rattle-sticks several times and move on a little further, counter-clockwise, stepping sideways. While doing so, they continue facing one another, looking towards the centre of the circle.

Another pattern called *chikoka* is occasionally performed in *mukanda*. This pattern is based on the *kundekula* movement, which is stopped from time to time and interspersed with brief episodes of skipping forward, legs and feet closed. Then there is *vutotola*, a movement style named after the mask *Katotola*, whose movement patterns the novices imitate. *Vutotola* is danced by the novices at the end of one particular kind of *kutanga* recitation. The basic dance patterns *kuhunga* and *kundekula* may be displayed in various formations in which the novices are seen in kneeling position, front support position, and also forming other figures.

TRANSCRIPTION AND ANALYSIS

In dance transcription I follow two principles. First, the total motion can be divided for the purpose of analysis into several movement areas, i.e. shoulders, shoulder-blades, pelvis etc. Secondly, movement can be defined and analysed in terms of *corner-points* or *points of inflexion* (Kubik 1972:35). Corner-points are those extreme positions in the evolution of a motor pattern at which a phase of movement is aimed. Corner-points are something like marks at the end of a section of movement. They are visible on cinefilm: a change in or complete reversal of the direction of movement follows after a short stand-still or running-down phase.

This is a concept of movement analysis based on the idea of use of body energy. What I call corner-points, corner-positions, turning points, or points of inflexion, are those moments in a movement pattern where kinetic energy reaches zero intensity and where new energy has to be injected by the dancer to keep the movement going. The employment of new energy often results in a change of direction, hence the name *corner*-points. As the direction changes, these points are clearly visible in a film or in a live-performance. Exceptions are some kinds of circular movements (pelvis circle etc.) which require a constant flow of energy. The body can be divided up into parts that have constant energy flow and those that are segmented.

Movement aims at reaching certain positions in space, for example a hand is slowly lifted until it reaches the vertex, then new energy has to be employed to move it in another direction, such as sideways. The path of movement in a pattern shows several 'cracks', angles, corner-points. These positions appear to be determinants and the really formative elements in a movement pattern. There is a direct relation between corner-points and the inner experience of a pattern by the dancer.

Most of the novices' movement patterns can be transcribed according to this system. In the largely three-dimensional movements of the dancers and in complex work patterns, several corner points can usually be discerned. Before transcription, it is necessary to discover all corner-points in a movement pattern from watching a cinefilm in slow motion, and draw up an inventory. Then each corner-point or corner-position must be defined verbally and in some cases represented with an illustration, either with a characteristic photograph, or still better, with the appropriate frames reproduced out of the film. Later, frame-by-frame transcription may be used in order to represent the structure of the movement pattern in a graph.

For example, in *kuhunga*, the two corner-points of the movement centre called 'left knee' are 1. knee raised, 2. leg stretched. Through frame-by-frame analysis we now find out in which time-sequence these positions occur. We see, for instance, that the rising knee reaches its corner-point approximately every sixth frame of the film. The stretching of the leg also, but the movements of the two legs are related in opposition to each other (see transcription).

In the graph transcription (Fig. 6) the corner-points of all motion centres are marked systematically at crossing points of the time-lines, i.e. the vertical lines of the graph paper which stand for the frames, and the horizontal lines which stand for motion centres or body areas. The entries use a limited number of symbols such as points, dots, rings, crosses etc. according to what should be expressed. The key to each symbol is the illustration of the respective corner-point position which goes with the sheet of graph notation, or a verbal explanation of the symbols used. In this way the temporal organization of movement patterns can be represented very accurately in a graph transcription, and thus make clear the inner structure of a movement pattern.

CONCLUSION

On the final day of *mukanda*, the novices return to the village and the lodge is burnt, together with the secret musical instruments, some of the masks, the 'tree of the dead', and the plates containing medicine which are attached to it. Surrounded by their mothers and all the villagers, they form various dance figures (lines, circles, squares etc.) and begin to display the patterns *kuhunga, kundekula* and others, while singing the didactic songs of the *mukanda* which they have learned during seclusion. This is accompanied by the three lon drums, played by men. The women become very excited during the performance: some respond by imitating the dance pattern of the boys for a few moments; some throw ground-nuts at the dancing novices; and others bring gifts in large baskets for the novices and their guardians.

Instruction and learning does not come to an end with the closure of a *mukanda* school and the return of the boys to their village. A year or two later another *mukanda* is established nearby and the former novices, who are now graduates, return to it and assume other roles. When they are sufficiently grown-up to wear mask costumes, new dance patterns and pantomimes are learned. Each mask has its own distinct repertory of dance movements and theatrical behaviour; for example *mutenya* is a rotating pelvis motion with sexual meaning, and *kâtonda* the name for jerky hip thrusts. These patterns are part of the movement repertory of the comic mask called *Chileya* (literally: the fool), a kind of jester or court fool which is very popular especially with women. *Chileya* uses the 'female' pelvis motion to parody the women, both to their own delight and that of onlookers.

When a boy ultimately starts to learn masked dancing at the age of sixteen or seventeen, he already has a good movement repertory, because of earlier instruction received in the *mukanda* and subsequent observation. For example, *kuhunga* also occurs in the dance style and theatrical pantomime of certain masks. The Chokwe-originated *Chikunza* which represents the ancient Chokwe king Kanyika, and *Kalelwa* his assistant, dance *kuhunga*, though the leg part of the pattern is different from the novices' version. There is also a mask called *kandanda*, which represents a novice who died in the camp. Usually a boy, though he must be a graduate of the *mukanda*, wears this mask costume. The *kandanda* mask behaves exactly like an initiate of the lodge and performs the patterns *kuhunga* and *kundekula*.

ACKNOWLEDGEMENTS

My field work in Angola was done with a research grant by the Junta de Investigacões do Ultramar, Lisbon, on kind recommendation by the late Professor António Jorge Dias and his wife Margot Dias. My follow-up studies were financed by the Foundation for the Advancement of Scientific Research in Austria (Fonds zur Förderung der Wissenschaftlichen Forschung in Österreich), Vienna. While in Zambia I worked as an affiliate of the Institute for African Studies, University of Zambia, Lusaka, and in cooperation with the Department of Cultural Services, Lusaka. I am most grateful to all who assisted me.

REFERENCES

Baumann, H. (1932). *Die Mannbarkeitsfeiern bei den Tshokwe und ihren Nachbarn*, Berlin.

Baumann, H. (1930). *Tänze der Frischbeschnittenen. N.O. - Angola*, Cinefilm, 4 min. C 122, Göttingen, Institute for Scientific Cinematography.

Jung, C.G. (1974). *Man and his symbols*, London: Jupiter Books.

Koolwijk, M. van (1963) Entre os Ganguelas. I. Circuncisão dos Rapazes, II. Festa da inciacão das raparigas. *Portugal em Africa, XX*, No. 117 and 119.

Kubik, G. (1968). *Mehrstimmigkeit und Tonsysteme in Zentral und Ostafrika*. Vienna: Austrian Academy of Sciences.

Kubik, G. (1969). Masks of the Mbwela. *Geographica* (Revista da Sociedade de Geografia de Lisboa), *5*, No. 20.

Kubik, G. (1971). *Die Institution mukanda und assoziierte Einrichtungen bei den Vambwela/Vankangela und verwandten Ethnien in Südostangola*. Dissertation, Institut für Völkerkunde, University of Vienna.

Kubik, G. (1972). Transcription of African music from silent film: Theory and methods. *African music, 5*, No. 2.

Kubik, G. (1973). Music and dance education in *mukanda* schools of Mbwela and Nkangela communities. *Proceedings of the Lusaka Music Conference June 15-22, 1971* Lusaka; reprinted in *RE - Review of Ethnology, 4*, No. 7-9, 1974.

Kubik, G. (1974). Sozialisierungsprozess und Gesänge der Initianden in *mukanda*-Schulen (Südost-Angola). In *In Memoriam António Jorge Dias*, Lisbon.

Murdock, G.P. (1959). *Africa: its peoples and their culture history*. New York: McGraw Hill.

Mwondela, W.R. (1972). *Mukanda and Makishi: Traditional education in northwestern Zambia*. Lusaka: NECZAM.

Turner, V. (1962). Three symbols of passage in Ndembu circumcision ritual: an interpretation. In Max Gluckman (ed.) *Essays on the ritual of social relations*. Manchester University Press.

Turner, V. (1967). *The Forest of Symbols*. Aspects of Ndembu Ritual. New York, Ithaca: Cornell University Press.

White, C.M.N. (1961). *Elements in Luvale Beliefs and Rituals*. The Rhodes-Livingstone Papers, No. 32, Lusaka.

MOVEMENT PATTERNS IN PLAYING THE HERATI *DUTĀR*

John Baily

In this paper I shall describe the performance techniques used to play three types of *dutār*, used in and near the city of Herat, Afghanistan. The *dutār* is a member of the long-necked lute family of instruments, and the three types of *dutār* considered here are the 2, 3 and 14 stringed *dutār*-s. These represent three successive stages in a process of change in the morphology of the *dutār* that began in about 1950. In this process, the 14 stringed *dutār* has become a technologically more sophisticated instrument, and is used to play a different kind of music to its 2 stringed predecessor. The performance techniques of the three kinds of *dutār* will be compared in terms of how well the movement patterns they employ 'fit' the innate anatomical, physiological, and psychological organization of the human motor system. It will be shown that *dutār* playing has, in absolute terms, become a considerably more complex and difficult perceptuo-motor skill as a consequence of the changes in morphology.

The way the human body is organised to move is, in certain respects, a crucial element in the structure of music. A musical instrument transduces patterns of body movement into patterns of sound. The morphology of an instrument imposes certain constraints on the way the instrument is played, favouring certain movement patterns that are, for ergonomic reasons, easily organised on the instrument's spatial layout. Thus, the interaction between the human body and the morphology of the instrument may shape the structure of the music, channelling human creativity in predictable directions. Musical styles vary widely with respect to the degree to which the motor patterns of the performance technique embody the constraints imposed by the morphology, and utilize movements and sequences of movements that are intrinsically easy for the human sensorimotor system to organise. The greater the *compatibility* between movement and morphology, the nearer the music comes to being a transformation of the human body.

An outline of the plan of this paper may help to orientate the reader.

Section 1 describes the morphology of the three types of *dutār* and outlines the way in which the changes in the *dutār*'s form came about.

Section 2 describes the movements used in the performance techniques of the three instruments.

Section 3 deals with the interaction between the human body and the morphology of the *dutār*, in an attempt to isolate some of the constraints on performance which arise from the spatial layout.

Section 4 extends the arguments developed in the previous section through an examination of two types of *dutār* music: traditional Herati music formerly played on the 2 stringed *dutār*, and Afghan classical music played on the 14 stringed *dutār*.

In Section 5 the major findings are summarised and discussed in terms of the social implications of the recent developments to the *dutār*, and several conjectures are made about the significance of the motor structures that underlie the performance of music.

1. CHANGES IN THE MORPHOLOGY OF THE *DUTĀR*

Herat is a 'traditional Islamic city' (English 1973) located in the fertile valley of the Hari Rud, a large oasis in Western Afghanistan. The *dutār* (two strings) has been associated with this region for centuries and is depicted in miniature paintings of court life from 15th Century Herat. Little, if anything, remains of this courtly music tradition; the Herati *dutār* in the early part of the 20th Century was an instrument of rural 'folk' music. The changes in the *dutār*, the events which led to these changes, and more general aspects of musical change in Herat since the 1920s, have been discussed in detail by Baily (1976). While other transitional stages are found, three main stages in the development of the *dutār* may be distinguished.

(a) *The two stringed* dutār

The type of *dutār* which was played in Herat area until about 1950 is shown in Fig. 1. The body (*kāse*) is carved from a single block of mulberry wood (*tut*), and the belly or lid (*rutāl*) is made from a thin sheet of the same material. The traditional shape of the body is long, narrow and deep, often described as spoonshaped (*qāshoqi*), with a ridge (*tez*) running

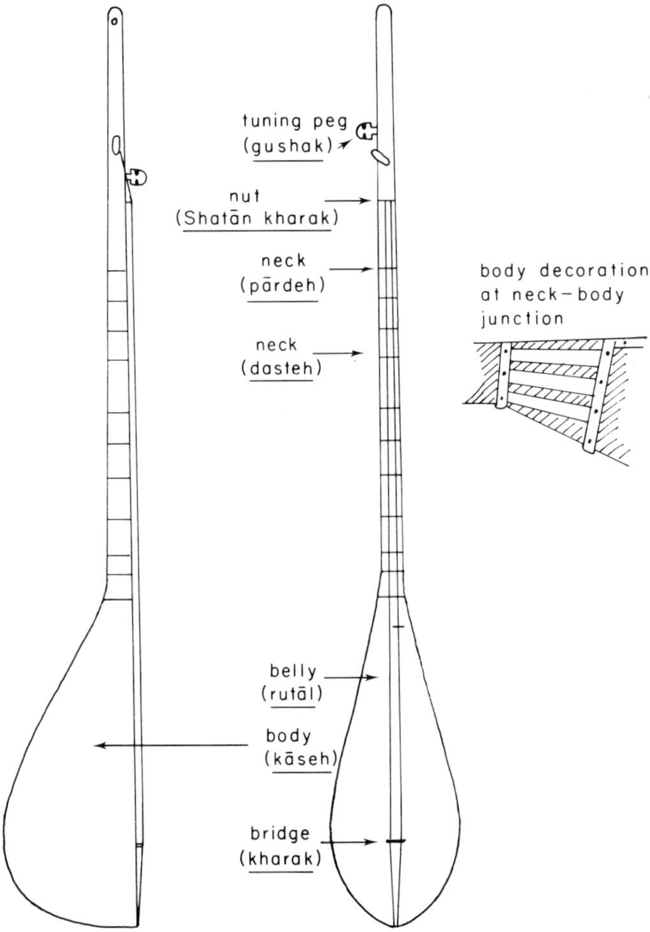

Fig. 1 *The 2 stringed* dutār.

along the back. The neck (*dasteh* ≐ handle) is usually of mulberry, apricot (*zerdalu*), or walnut (*jouz*) wood. Small pieces of bone or plastic inlay (*sadafkāri* = sea-shell work) are used to decorate the fretboard and the sides and back of the body at the junction with the neck. The string length is in the region of 65-76 cm., and in former times the two strings (*tār*) were made of gut (*rodeh*), as were the frets (*pardeh* = curtain), which were tied round the neck, much as certain Iranian lutes are fretted today. The strings are tied to a bone or horn string-holder (*tārgir*) attached to the bottom of the body, and are attached to wooden pegs (*gushak* = little ear) at the top of the neck. The bridge (*kharak* = support, lit. little donkey), which stands on the belly, is made of wood, while the strings are supported at the other end of the neck by a second bridge, or nut (*kharak Shetānu* = the Satan support). A Herati folk-tale connects the *dutār* with Satan.

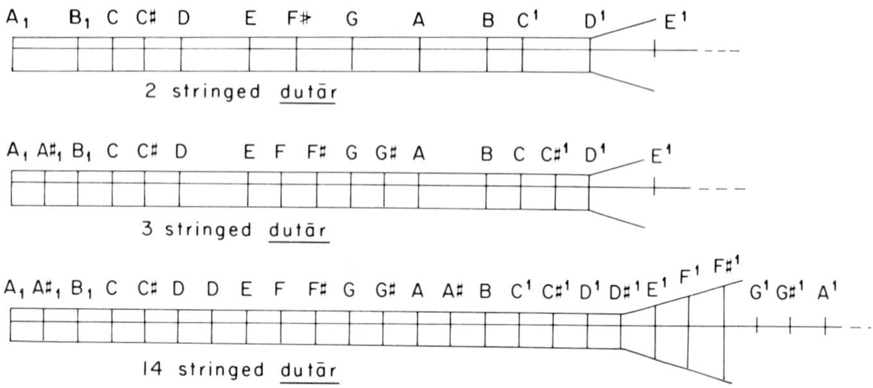

Fig. 2a *Arrangements of frets on the three types of* dutār.

Precise details of the arrangement of the frets and the tuning of the two strings used in former times are not known, but a system of fretting which is said by Heratis to resemble the old system is still sometimes encountered today. This system, as fretted for me on a number of occasions by a *dutār*-maker in Herat, is shown in Figure 2a. It has 12 frets;

the highest is not tied round the neck but consists of a wire staple inserted into the belly (the same is done for the 3 stringed *dutār*). The system that I found probably has more frets than had the old 2 stringed *dutār*, for informants speak of 2 stringed *dutār*-s with eight or ten frets. The addition of the extra two or four frets may have been influenced by the greater range of the 3 stringed *dutār*.

These complications notwithstanding, the spatial layout of the 2 stringed *dutār* has the following important features. (i) Only the 1st string is fingered with the left hand: the 2nd string, tuned a 4th below the 1st, functions as a drone. (ii) The note a 5th above the open note of the 1st string usually functions as the finalis (i.e. the tone centre, the chief tone around which melodies are ordered and on which they normally come to rest). This position is marked *f* in Fig. 2. The finalis note is reinforced by the open 2nd string, which gives the note an octave below the finalis. (iii) From the finalis the system of fretting gives a single mode which is approximately the same as Shur, the principal mode of Iranian music:-

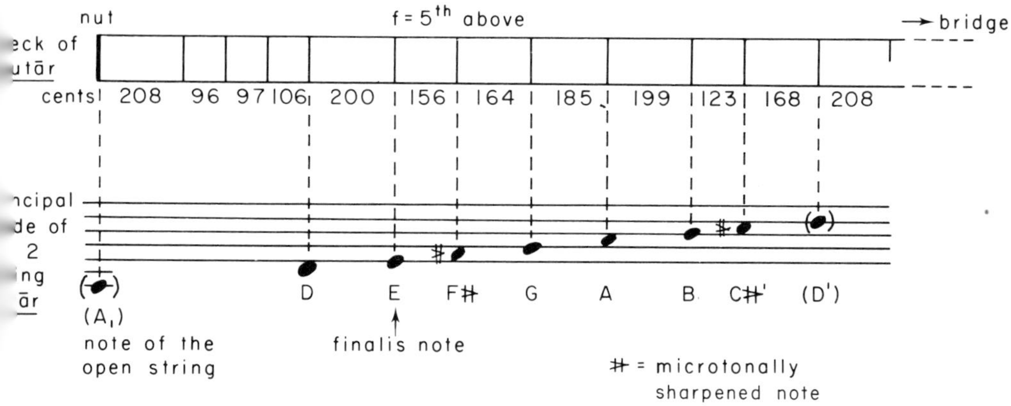

Fig. 2b *Tuning of notes masked by frets on the 2 stringed* dutārs. *1 cent = 1/100 of a tempered semitone.*

The size of the intervals in cents was established by Stroboconn measures of seven *dutār*-s fretted by Painde Mohammad, a *dutār*-maker in Herat (in the cents system of Ellis the octave is divided into 1200 equal intervals, so that the size of a tempered semitone is 100 cents). A feature of the mode shown in Fig. 2b is that not all the intervals are tones or semitones; some are neutral 2nds, which are intermediate in size between a minor 2nd (100 cents) and a major 2nd (200 cents). For example, the interval between E (finalis) and F# is 156 cents and between F# and G 164 cents. Neutral 2nds are characteristic of Iranian modes (Farhat 1965).

Another feature of this system of fretting is that the lower frets (i.e. those below D) are hardly used by *dutār* players in the Herat area. Their presence may be attributed to the fact that the 2 stringed *dutār* is an instrument which has enjoyed an increase in popularity over the last fifty years which started across the border in Iran. *Dutār* music played in Iran does make use of these lower frets, it also employs a second tuning in which the strings are tuned a 5th apart, and uses other positions for the finalis, such as a 2nd and a 4th above the note of the open string. It appears that although the instrument diffused across the border to Herat, some ot its repertoire was not transmitted. Although further research is required to clarify this point, it is the case that virtually all the old tunes associated with the *dutār* are in the *dutār* mode described above, and this is the most common mode of Herati music.

During my period of fieldwork (1973-4) I never encountered a 2 stringed *dutār* with gut strings, though I found several *dutār* players who played this instrument with metal strings. (The 3 and 14 stringed *dutār*-s use metal strings). In its 2 stringed form the *dutār* was predominantly, if not exclusively, an instrument of rural 'folk' music. As far as can be ascertained, it was used both to accompany male song, and as a solo instrument to play a repertoire based on the melodies of songs. The structure of these melodies is considered in more detail in Section 4. In the rural context, the *dutār* was played by amateur musicians for their own enjoyment and to entertain small gatherings of friends.

(b) *The three stringed* dutār

This form of *dutār*, shown in Fig. 3a, developed in about 1950. It is generally somewhat larger than the 2 stringed *dutār*, with a string length of 75-85 cm, though often 2 stringed *dutār*-s have simply been adapted. The instrument has three metal strings and is played with a metal plectrum which fits tightly over the 1st finger like a thimble. The frets were formerly made of metal but now nylon is used, tied round the neck as already described for gut frets. A number of frets have been added to the 3 stringed *dutār* which gives it an approximately chromatic system (i.e. composed of semitone intervals approximately 100 cents in size), with gaps between the 4th and 5th and between the 8th and 9th above the open note of the 1st string. These larger intervals are approximately

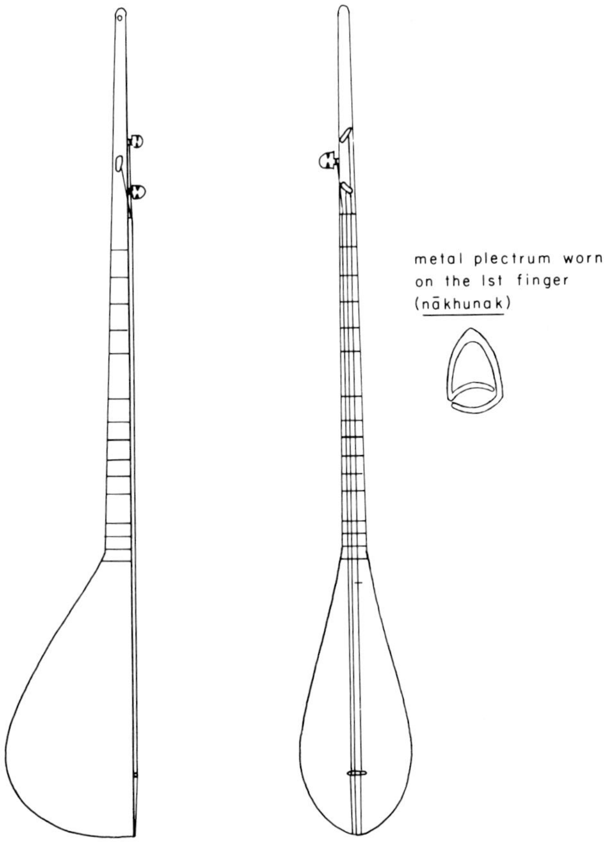

Fig. 3a *The 3 stringed* dutār.

a tone (200 cents) in size. The total number of frets is usually 16. The change towards a system of 12 approximately equal semitones to the octave is explained by the fact that music in Herat was falling under the influence of Kabuli music, dominated by the keyboard of the equal temperament harmonium, originally brought from Europe and introduced into Afghanistan from North India in the late nineteenth Century.

The change in the fretting eliminated the neutral 2nds that characterise the fretting of the 2 stringed *dutār*. The added frets allowed the 3 stringed *dutār* to be played in many modes, but only four of these are at all commonly played on the 3 stringed *dutār*. The most important mode is Bairami, which is like the mode of the 2 stringed *dutār*, but with the neutral

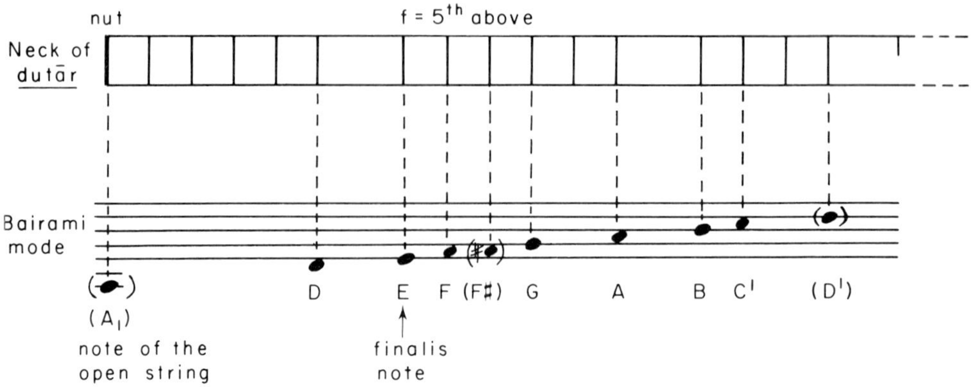

Fig. 3b *Tones, masked by frets, used on the 3 stringed* dutār.

2nds 'ironed out' and with an additional F#, sometimes used instead of F in ascending melodic movement. If we exclude the 'accidental' F# this mode is like the E-mode on the piano.

The change to the 3 stringed *dutār* occurred at a time when both popular radio music broadcast from Kabul and Indian films (with a strong element of Indian popular music) were making a strong impact on musical preferences and attitudes in Herat. It was at that time that the *dutār* became an important instrument of amateur music-making in Herat city. The change in fretting allowed the instrument to play the more important modes of radio songs, while the change in the material and number of strings may be ascribed to the influence of the metallic and drony sound of the Afghan *tanbur*, a large multi-stringed plucked lute that was often played over the radio. The metal plectrum used in playing the 3 stringed *dutār* was certainly derived from the *tanbur*. The 3 stringed *dutār* is particularly associated with music in tea-houses, of which many became famed for their music sessions in the 1950s. Like the rural players of the 2 stringed *dutār*, the urban players of the 3 stringed *dutār* were recruited from non-musician backgrounds and played for their own enjoyment, not for financial gain. They described themselves as *shauqi*, a word that means 'enthusiast' and has a number of secondary connotations, such as that one is an amateur, as distinct from being a professional musician.

(c) *The fourteen stringed* dutār

This form of *dutār*, shown in Figure 4, is again larger than the 3 stringed *dutār*, with a string length of 85-95 cm. The shape of the body has changed, becoming wider, rounder and

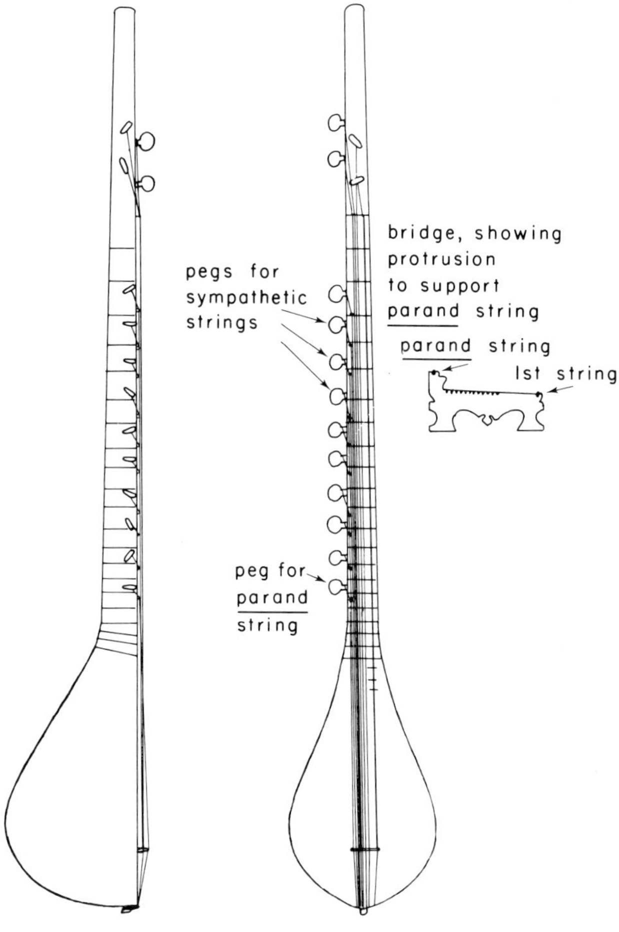

Fig. 4a *The 14 stringed* dutār

shorter in proportion to the total length of the instrument. Not only is the neck longer, but its width and depth have been increased to withstand the tension of the additional strings. More frets have been added, giving a chromatic range of two octaves (see Figure 2). Nylon is used to tie the frets; the elasticity of this material allows small adjustments to be made.

The highest frets are usually made by sticking thin strips of wood to the belly. Still only the 1st string is fingered with the left hand, and the added strings function mainly as sympathetic strings. The instrument is often highly decorated, with inlays made from mother-of-pearl or plastic. A special feature is the raising of the shortest sympathetic string (called *parand or bajgi*) by a protrusion on the bridge, to allow it to be struck in isolation with a special right hand technique.

The tuning of the 14 stringed *dutār* is complicated because of the many modes used for the instrument. In addition to the modes commonly played on the 3 stringed *dutār*, there is a wide range of other scale types, in almost every case derived from the Hindustani music system. Thus we find scale types such as Yeman, Piloo, Bihāg, Bhairav, Talang, Madhubanti and many others (see Kaufmann 1968). In general, the 4th above the open note of the 1st string is taken as finalis, but often the modes are transposed to other positions. For playing in the Bairami mode (the Herati mode described above) the finalis is often taken at the 5th above, as on the 3 stringed *dutār*. The sympathetic strings are tuned to the notes of the mode being played. Two tunings for the 14 stringed *dutār* may be illustrated. Firstly, the Herati Bairami mode played from the 5th above. Secondly, the Yeman mode played from the 4th above.

The 14 stringed *dutār* was invented in about 1960 by Mohammad Karim Herawi, a Herati *dutār* player who worked in the National Orchestra of Radio Afghanistan in Kabul from 1957-1970. This change in the *dutār* was motivated by both the need to increase the volume of the instrument so that it could be heard within the context of a radio orchestra and a desire to make the *dutār* into a technologically more sophisticated instrument. The invention of the 14 stringed *dutār* was strongly influenced by the use of another plucked lute, the short-necked Afghan *rebāb*, shown in Fig. 5. Its body consists of two inter-connecting sound chambers; the belly of the lower chamber is made of skin. There are 3 main strings of gut or nylon; 2 or 3 metal drone strings; and 13-15 metal sympathetic strings. It is played with a wooden plectrum. The following structural features of the *rebāb* have been adopted on the 14 stringed *dutār*: the sympathetic strings (these are also present on other plucked and bowed lutes used in Kabul), the raising of the shortest sympathetic (*parand*) strings by the bridge, and certain motifs used in the decoration of the instrument, see Fig. 6. The connection between the *rebāb* and the *dutār* is not limited to morphological features. Part of Karim's motivation in developing the 14 stringed *dutār* was to play music from the repertoire of the *rebāb* on the *dutār*, which will be discussed in Section 4. Similarities in performance technique are discussed in Section 2.

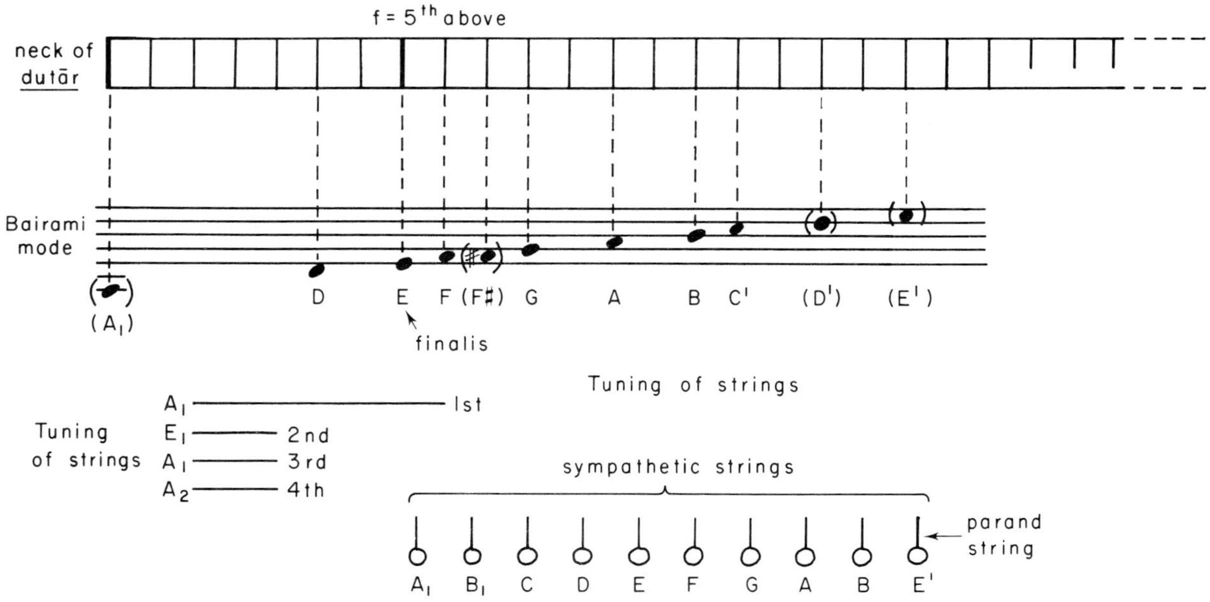

Fig. 4b *One method of tuning the fourteen string* dutār

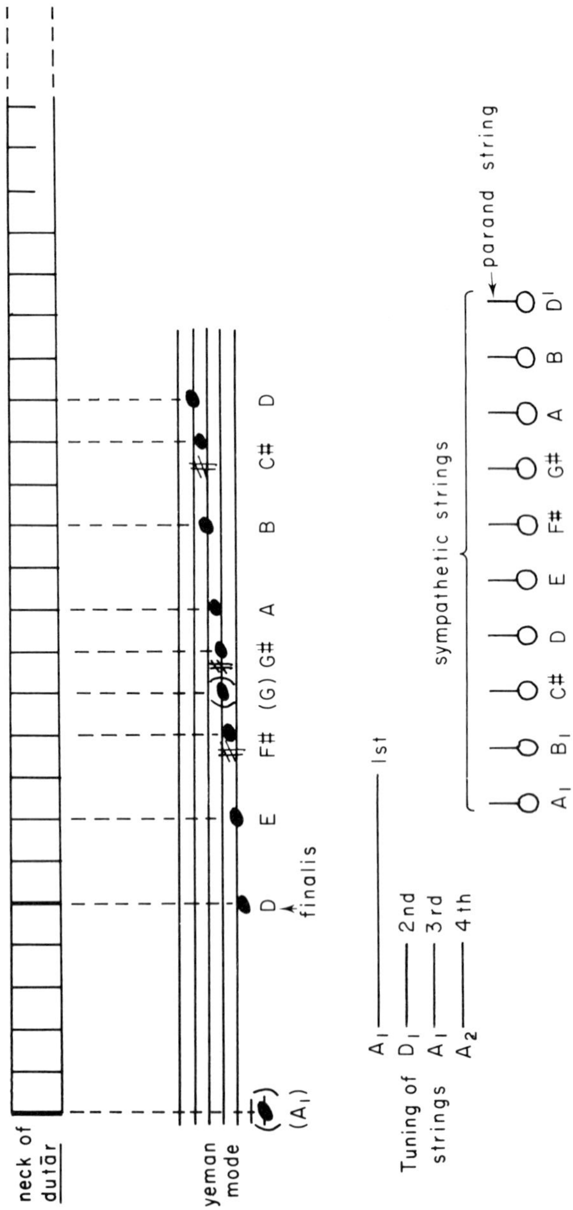

Fig. 4c *A second method of tuning the fourteen string* dutār.

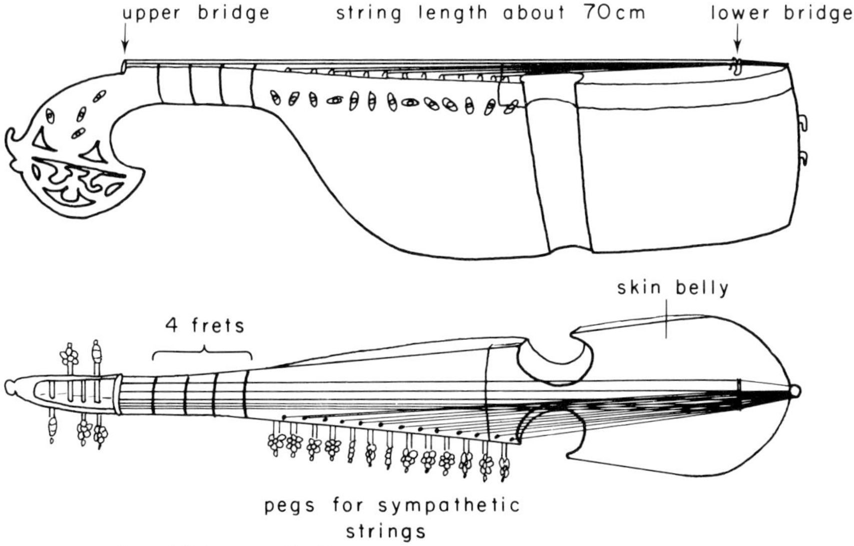

Fig. 5 *The Afghan* rebāb.

In its 14 stringed form, the *dutār* has been absorbed into the mainstream of music played in Herat for public entertainment by bands of hereditary professional musicians. Most of this music comes to Herat from Kabul and is played in a distinctly Kabuli style on instruments that have been derived mainly from Kabul. Although hereditary professional musicians have not taken up the *dutār* themselves, regarding it with scorn as an 'amateur's instrument', they employ *dutār* players in their bands. This situation has created a new type of professional *dutār* player who, through contact with hereditary professionals, has become a highly competent musician with an extensive knowledge of the musical repertoire, concepts and terminology of Kabul.

2. MOVEMENT PATTERNS IN THE PERFORMANCE TECHNIQUES FOR THE *DUTĀR*

(a) *Body posture while playing*

As with other types of plucked lute, the right-handed *dutār* player strikes the strings with his right hand and grasps the neck and stops the strings with his left hand. In a sample of forty *dutār* players only three played left-handed.

Players of the 2 and 3 stringed *dutār*-s usually sit on the ground in a cross-legged position. The body of the *dutār* rests on the player's right thigh. The left hand supports the neck

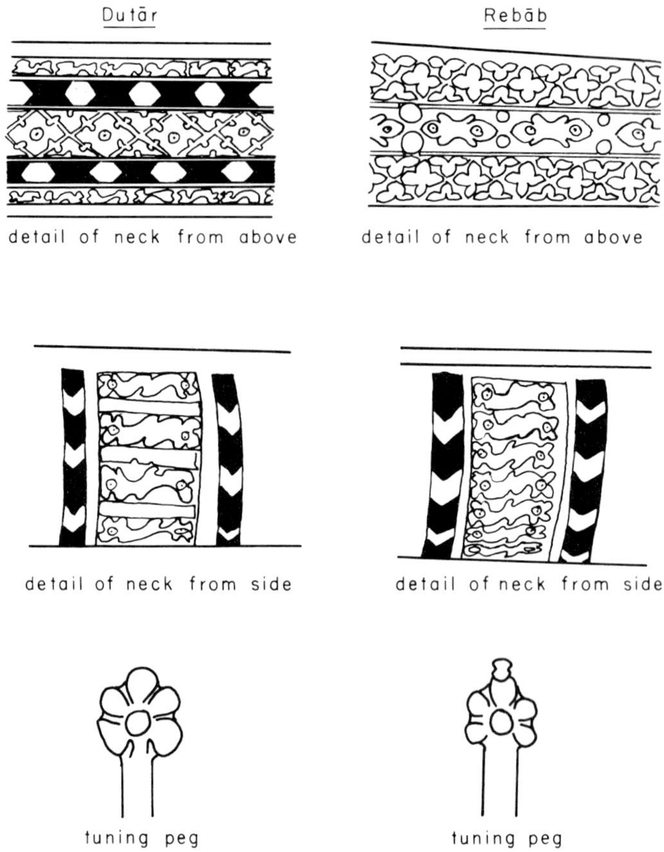

Fig. 6 *Decorative features of* **rebāb** *and 14 string* dutār

at an angle of between 10-60° from the horizontal (usually around 30-40°) and about 25-45° relative to the frontal plane of the player's body, so that the head of the instrument is positioned forward of his body. In this posture the left hand takes the weight of the neck. This posture is shown in Fig. 7. Sitting cross-legged on the ground is a common posture, as chairs are a modern innovation in Herat. Because of the increased size of the 14 stringed *dutār*, this posture is unsatisfactory. The heavy neck imposes undue strain on the left arm, which makes it a considerably more difficult instrument to hold than its 2 and 3 stringed predecessors. Many 14 stringed *dutār* players have adopted a new posture, see Fig. 8, which is a mixture of two common resting positions: the cross-legged and squatting positions. The right knee, still flexed, is raised and pushed across the midline of the player's body. The body

Fig. 7 *Body posture for playing the 3 stringed* dutār.

Fig. 8 *Body posture for playing the 14 stringed* dutār.

of the *dutār* is clamped between the right thigh, the side of the abdomen and the right arm, and left arm support is not required to hold up the neck of the instrument. This is a more dynamic and alert position than the cross-legged posture. Musicians, including *dutār* players, often play for several hours with only one short rest and no breaks between different items. Considerable stamina is required.

(b) *Right hand technique*

Right hand techniques have been more affected by the changes to the *dutār* than left hand techniques, and it is necessary to consider the three types of instrument separately.

(i) *The 2 stringed* dutār. The 2 stringed *dutār* is played without a plectrum and the strings are struck in an 'open handed' manner, with the thumb and fingers extended. There are four important strokes, illustrated in Fig. 9.

Downstroke with the 1st, 2nd, 3rd and 4th fingers (V with F1234)
Downstroke with the 1st finger (V with F1)
Upstroke with the 1st finger (Λ with F1)
Upstroke with the thumb (Λ with Th)

In these strokes both strings are struck together, the 2nd string acting as a drone. The forearm is supported by resting it on the upper part of the body of the *dutār*. The fingers and thumb are extended. Movement consists mainly of a rotation of the wrist, but alterations in finger posture are also important because they determine which part of the hand strikes the strings when the hand as a whole is moved down and up. This

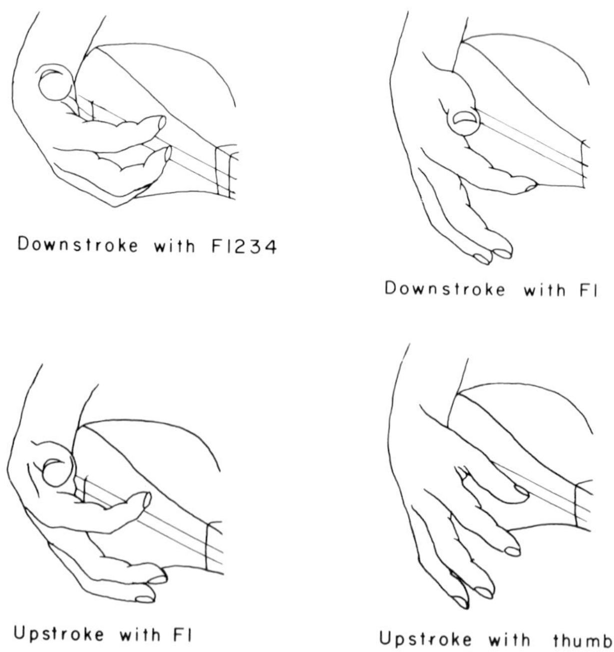

Fig. 9 *Right hand positions for playing the 2 stringed* dutār.

performance technique takes advantage of the morphology of the hand in the organization of rhythmic patterns. A rather similar technique is used in the *flamenco* style of guitar playing, where in the *rasquerdo* the fingers are strummed in rapid succession down across the strings, in the sequence 4th 3rd 2nd 1st. In this movement the fingers move in relation to the palm of the hand, each finger in turn extended; but in the downstroke with four fingers on the 2 stringed *dutār* the fingers do not move relative to the palm, and are brushed across the strings.

As in almost all *dutār* playing, the downstroke is stronger and rhythmically more important than the upstroke. On the 2 stringed *dutār* the downstroke with the 1st, 2nd, 3rd and 4th fingers is the strongest stroke and is used to delineate the main beat in a rhythmic pattern, though it may be used for other beats too. The usual strategy for playing a tune is to adopt a particular motor pattern and to maintain it throughout the piece. This also applies to playing the 3 stringed *dutār*. The right hand is used in a percussive manner, almost as though it were playing a repetitive rhythmic figure on a drum, A variety of right hand patterns for the 2 stringed *dutār* have been observed, for example:

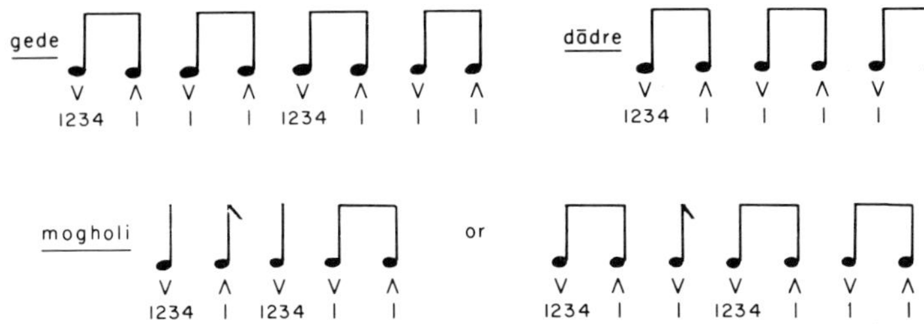

These show the three main Herati rhythms: *gede* (4 or 8 beats), *dādre* (3 or 6 beats) and *mogholi* (7 beats). (Only a few *dutār* players are familiar with these terms, and they have usually had close contact with hereditary professional musicians). So far the patterns have not involved the thumb, whose use appears to be something of a refinement. It is used to provide a triplet pattern, as in the following examples:

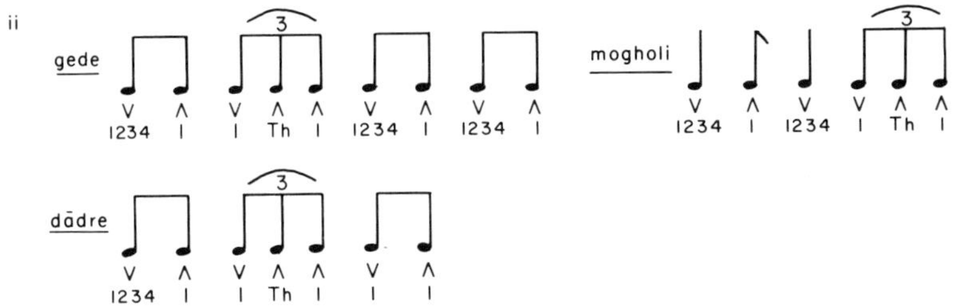

In an important genre of the traditional repertoire of the *dutār*, called *chaharbeiti*, the time structure is 'unmetred', lacking a regular pulse. A distinctive feature of such pieces is the use of a fast sustained tremolo, called *riz* (acceleration). This is produced on the 2 stringed *dutār* in two ways, either by a fast down-up strum with the 1st finger, or in a triplet pattern in which the upstroke with the thumb is also used:-

iii

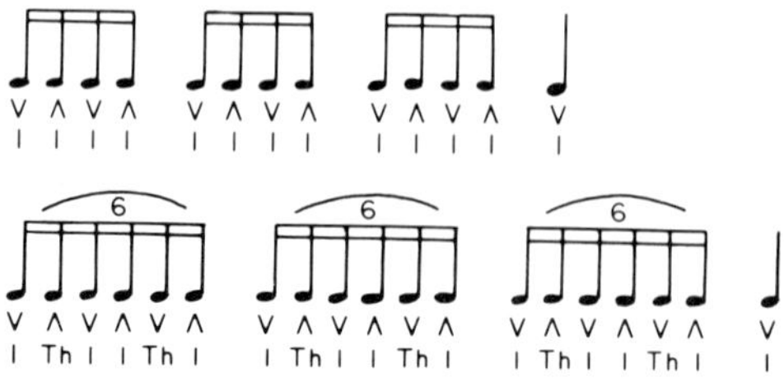

The tremolo is a feature of Iranian lute styles and its use on the *dutār* is one of many links between the traditional music of Herat and Iran. It is employed on all three varieties of *dutār*

(ii) The three stringed dutār. The 3 stringed *dutār* is played with a metal plectrum that fits tightly over the distal segment of the 1st finger and requires no support from the thumb or other fingers. This is significant, because the development of the correct hand posture to hold the plectrum is a major problem in the early stages of learning many other plucked lutes. The basic right hand pattern consists of a repeated down-up strum across all 3 strings. This is achieved in several ways. In one technique the hand is supported by the thumb, which rests on the edge of the belly, and movement consists of alternate extension and flexion of all the fingers, only the plectrum actually striking the strings, see Fig. 10.

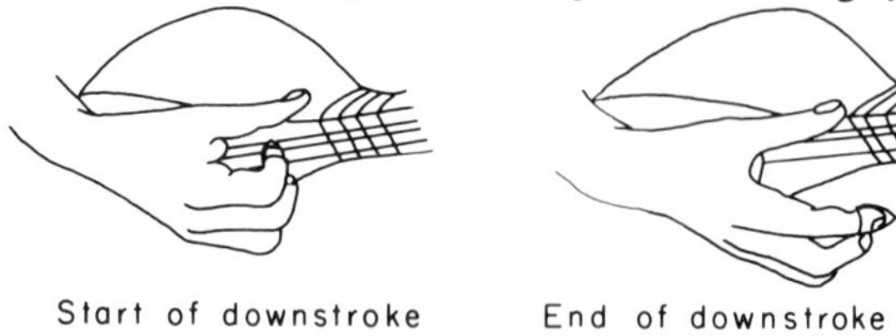

Fig. 10 *Strum on the 3 stringed* dutār *achieved by extension and flexion of the fingers.*

In a rather different technique, movement consists of a rotation of the wrist. Here the hand is held in a rigid posture, with the 1st finger flexed and the other fingers and thumb extended, see Fig. 11. This technique is often favoured for tremolo (*riz*) patterns. Other techniques lie in between these two and involve a combination of wrist and finger movement.

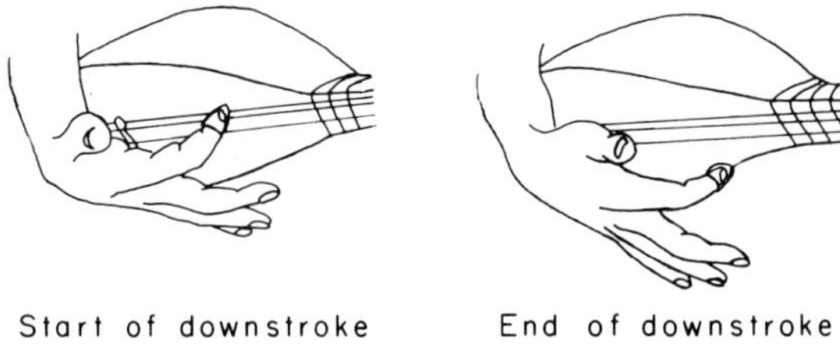

Fig. 11 *Strum on the 3 stringed* dutār *achieved by rotation of the wrist.*

A prominent feature of the right hand technique is a scratching of the plectrum across the belly of the instrument, which adds a percussive component to the sound. This is an almost inevitable result of the right hand technique and use of the plectrum. When three of the best 14 stringed *dutār* players in Herat were asked to play the 3 stringed *dutār*, it was noticeable that they minimised this percussive element by playing very near the neck, for the loudness of this element increases as the strings are struck near the bridge. Although these experts did not like the percussive element, other players, especially those who mainly play the 3 stringed *dutar*, make much use of it and vary it to add dynamic range to their playing. The triplet pattern of the 2 stringed *dutār*, using the upstroke with the thumb, is not usually used in 3 stringed *dutār* technique.

Like the 2 stringed *dutār*, the 3 stringed *dutār* is usually played using a constant right hand pattern throughout the piece. The instrument gives a sharp and clear sound and is often played extremely fast. Some examples of right hand patterns are:-

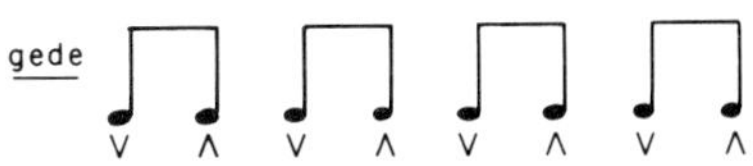

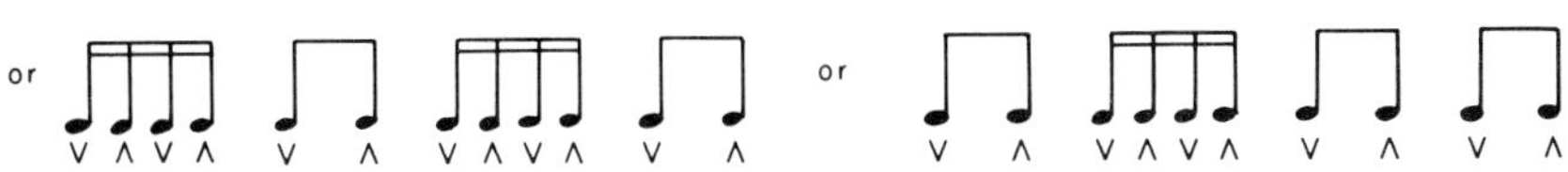

In the *gede* rhythm there is often a tendency to delay the

upstrokes so that the rhythm becomes with a percussive stroke on the 3rd beat.

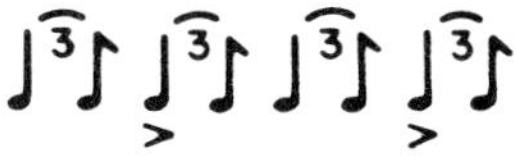

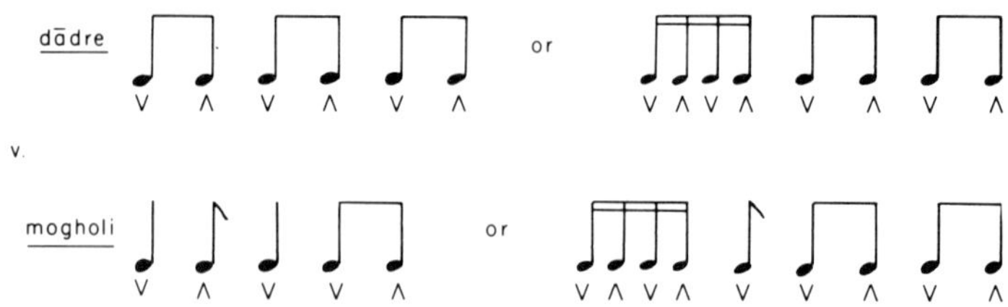

(iii) The fourteen stringed dutār. Two separate right hand techniques are employed for the 14 stringed *dutār*: in one, the upstroke (towards the body) is the strong stroke (*rāst*); in the other, the downstroke is the *rāst* stroke. The term *chap* is used for the weak stroke; opposite in direction to the *rāst* stroke. In both techniques the plectrum of the 3 stringed *dutār* is worn on the 1st finger.

The upstroke technique is employed principally by the inventor of the 14 stringed *dutār*, Mohammad Karim Herawi. In his technique, the right hand is supported over the strings by the 3rd finger and thumb resting on the belly of the *dutār*, and the strokes are delivered by the 1st finger. Movement of the finger occurs at the 2nd and 3rd phalangeal joints. On one occasion Karim told me he had borrowed this technique from the North Indian *sitar* (which is played by a few individuals in Kabul), an instrument he greatly respects. However, in the right hand technique for the *sitar* the hand is supported only by the thumb, the whole hand moving about this pivot, and not just the 1st finger. In fact, it seems likely that Karim's technique was influenced by the technique for the Afghan *tanbur*, where a very similar hand position is employed in which the upstroke is the strong stroke. All other 14 stringed *dutār* players in Herat use the downstroke technique, though a few individuals occasionally use Karim's technique as a special effect. It is worth noting that the 14 stringed *dutār* is not generally played in the way intended by the man who invented it. We shall see later that Mohammad Karim is also exceptional with respect to his left-hand technique.

In the downstroke technique the 14 stringed *dutār* is normally played in a *punteado* or 'single string' manner, in which only the 1st string is struck (except in *parand* patterns, see below). Some players may sometimes also strike the longest sympathetic strings along with the 1st string - this is a feature of Karim's upstroke technique - but the emphasis is always on the 1st string. Occasionally a player will strum across all 14 strings as though he were playing a 3 stringed *dutār*, but this is only used as a special effect. It is clear that this right-hand technique for the 14 stringed *dutār* has been borrowed from the *rebāb*, which is also played *punteado*. The right-hand positions for the two instruments are very similar, as shown in Fig. 12. The *rebāb* player holds a flat wooden plectrum between the 1st finger and the thumb, while the *dutār*-player retains the metal plectrum used for the 3 stringed *dutar*, but holds his hand in a new position, in which he pinches the plectrum between the thumb and 1st finger. For both instruments the other fingers are flexed and tucked into the palm. The wrist is flexed at an angle of 60-90° relative to the forearm. The forearm rests near the bottom of the body of the instrument to provide support for the hand. Movement of the plectrum is brought about by a rotation of the wrist. The main stroke is not only downward, but also inward towards the belly of the instrument. On the *rebāb* the hand tends to hit the skin belly on the downstroke, which acts to check the movement; on the *dutār* the plectrum often clicks against the wooden belly. With a large rotary wrist movement a forceful downstroke can be achieved. The upstroke, in contrast, is much weaker.

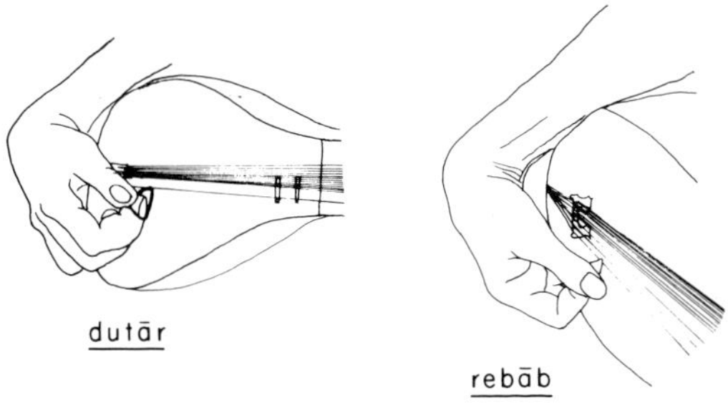

Fig. 12 *Right hand positions for playing the 14 string* dutār *and* rebāb.

Right-hand patterns in playing the 14 stringed *dutār* are more complex than for the 2 and 3 stringed *dutār*-s. The steady strum is not used on the 14 stringed *dutār*, except by those who are learners or otherwise unskilled players. Instead, the players tend to articulate the metric structure of the song they are playing, as in the following examples.

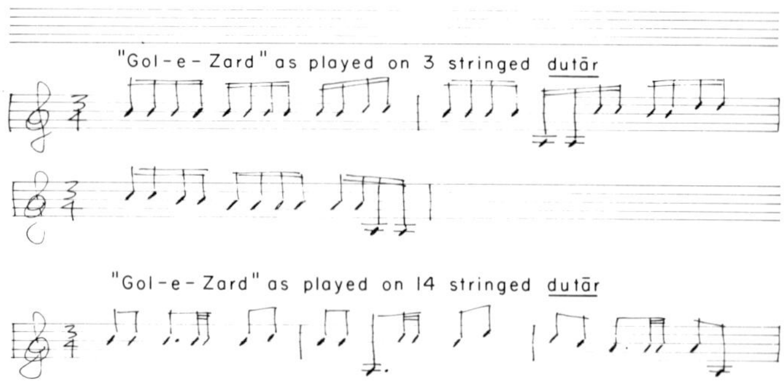

Rebāb playing is characterised by a rich variety of right hand patterns and to a considerable extent these have been adopted by 14 stringed *dutār* players.

For the *gede* rhythms we find combinations such as:

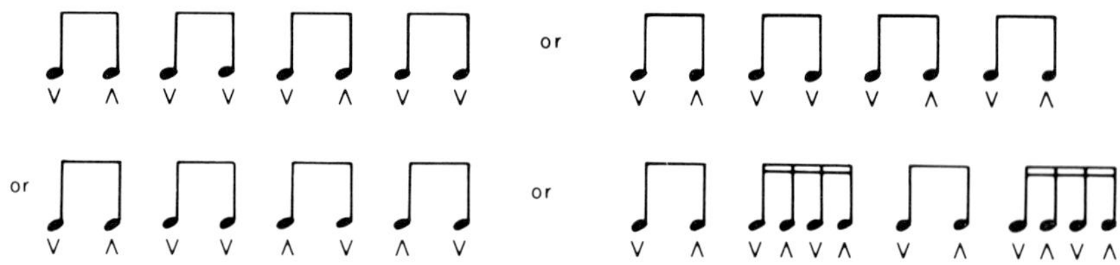

For *dādre*

For *mogholi*

In the examples above the transcriptions of the rhythms are not quite accurate, partly because the downstroke and upstroke tend to produce slightly different sounds and also because there is a tendency to delay the upstroke following a downstroke, so that  tends towards  and might be better written as  (Sherbourne, private communication). As a result, there are slight time differences between patterns which use different arrangements of down and up strokes, such as:

x

The *rebāb* and *dutār* exploit this difference (which follows from the difference in the strength of the down and upstrokes) to generate a variety of patterns whose possibilities multiply as the rhythmic units increase in length.

It is in the matter of *parand* patterns that the technique of the *dutār* owes most to that of the *rebāb*. In these the shortest sympathetic string (*parand*), tuned an octave above the finalis, is struck in alternation with the 1st string in complex patterns of high and low notes, the former being given by the *parand* string and the latter by the melody fingered on the 1st string. This technique is difficult to execute and is perhaps the crucial element of 14 stringed *dutār* playing where individual virtuosity can be developed.

A *parand* pattern such as:

xi

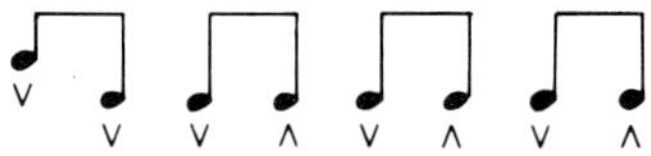

or

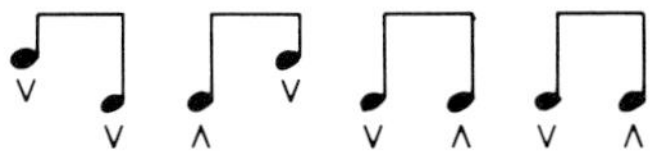

may be played throughout a melodic section. In a typical use of *parand* borrowed from the technique of the *rebāb*, the frequency of the *parand* strokes is progressively increased. The following is in the six metre known as *dādre*. The player starts with a simple *dādre* pattern on the main string, using a down-up-down, down-up-down movement:

xii

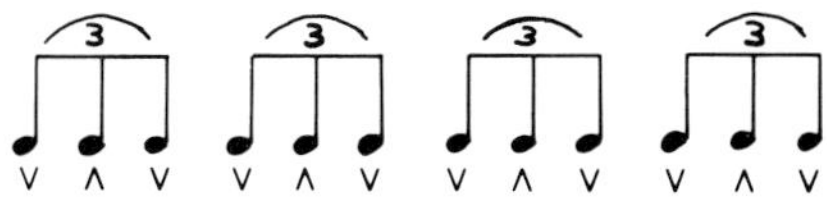

then adds single *parand*-s

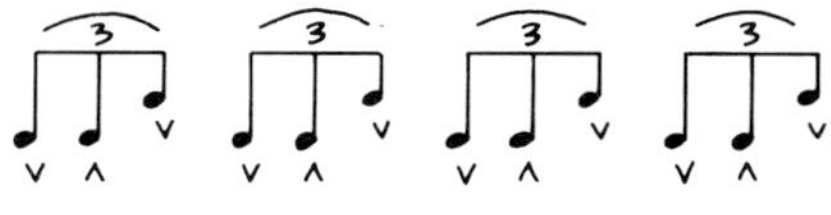

then makes it more complex

and then again more complex

Such a technique often terminates on a strongly emphasised beat (*gor*), followed by a brief rest (*istād*). This common device of *parand* - *gor* - *istād* is used to build up 'tension' and then to release it. *Parand* techniques are clearly similar to the *jhala* techniques of North Indian plucked lutes such as the *sitar* and *sarod*, although on these instruments several drone (*chikari*) strings are struck together to provide what corresponds to the *parand* string component.

(c) *Left hand techniques*

Left hand techniques for the *dutār* have not changed as much as right hand techniques, and it is not necessary to consider the three types of *dutār* separately. The increased girth of the *dutār*'s neck means that the thumb is extended further away from the fingers. The increased length of the *dutār*'s neck, and the imposition of extra frets, means that the fingers have to stretch greater distances relative to each other. There is, however, no evidence to show that alterations in hand posture have had a significant effect of patterns of fingering.

Dutār playing relies heavily on left hand ornamentation. In most *dutār* performances it is possible to distinguish (through analysis) the notes that form the essential melody from those that are added to it as decorations. This distinction is found not only in the music but in the motor patterns of the left hand, where two kinds of movement may be isolated: the primary positioning movements, and the secondary movements for ornamentation.

In primary positioning movements, the fingers 'take' the frets to produce melodically important notes. A fret is 'taken' by pressing the string against the fingerboard close by a fret. Many *dutār* players use only the 1st and 2nd fingers to do this. The typical posture of the left hand in relation to the neck is shown in Fig. 13.

The back of the *dutār*'s neck rests in the fleshy web between the thumb and the 1st finger and does not touch the heel of the palm. The thumb, which is extended, projects round the neck and is usually placed opposite the position of the tip of the 1st finger. The 1st finger is moved up and down the neck by hand movements, while the 2nd finger is used additively above the 1st finger. By 'used additively' I mean that when the 2nd finger is used the 1st finger is kept in position on the fret-

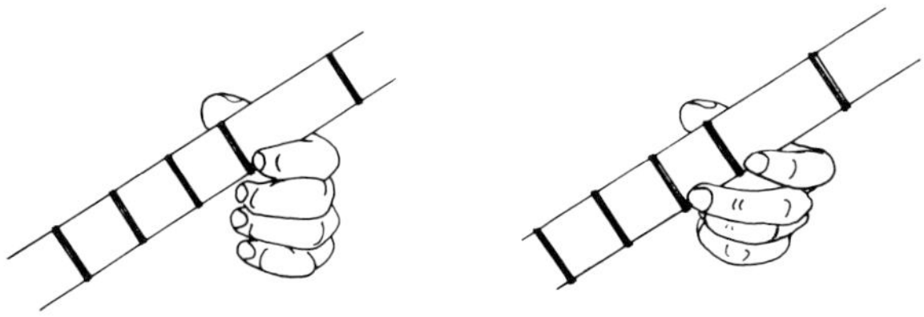

Fig. 13 *Posture of the left hand.*

board and the 2nd finger is placed on the fretboard above it, as shown in Fig. 13.

The additive character of finger movement is an important feature of *dutār* playing. Some players use the 3rd finger in the same way, i.e. when the 3rd finger is employed, the 1st and 2nd fingers are kept in position on the neck. However, three-finger additive patterns are not very frequent in *dutār* playing. The question of finger use in primary movements will be examined in detail with reference to several versions of the same melody, but secondary movements for ornamentation must first be considered.

Most ornamentation falls into the category of legato, a term which refers here to a significant vibration of the string set up by the left hand. Any change in position of the fingers on the string generates some audible effect, but this is usually masked by the right hand striking the strings. Legato is used deliberately to create extra sounds. The terminology of legato used here is derived from modern Western guitar tutors, and does not correspond to Herati terminology, which is considered below.

The most important secondary movements used on the *dutār* are the hammer and the pull. In the hammer, the 1st finger is used to hold one fret, while the 2nd finger strikes a rapid blow on a fret above, producing a clear tone.

xiv

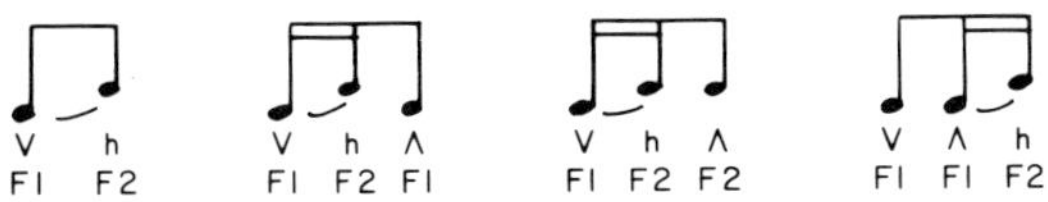

In a special variant of the hammer - the damped hammer - the movement is essentially similar but the 2nd finger does not strike the string sufficiently forcefully to vibrate the string against the next fret and the string is damped, giving a muffled upward lift to the preceding note. Pull is the opposite process to hammer. Here the 2nd finger is quickly removed from the string with a slight lateral pluck, leaving the 1st finger taking a fret below.

xv

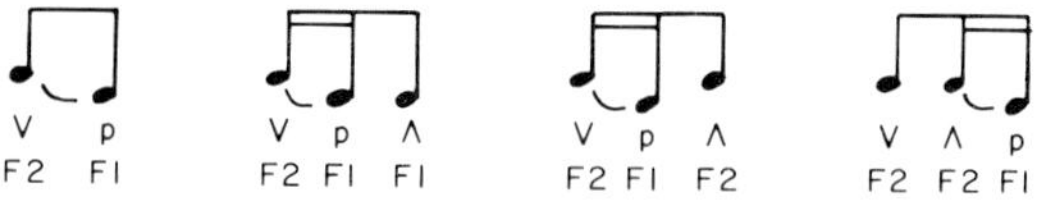

Hammer and pull are sometimes used in combination, and sometimes strings of hammer-pull movements are run together.

xvi

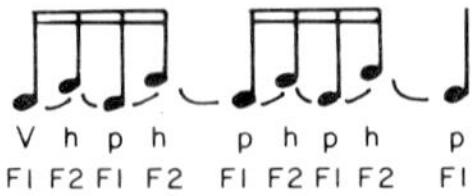

An ornament that is rather rarely used on the *dutār* is the slide, in which the finger, usually the 1st finger, is pushed from one fret to another by a hand movement, while pressing down on the string.

xvii

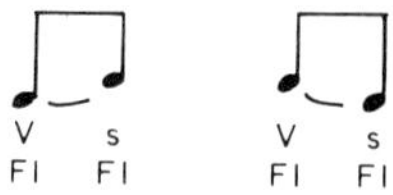

Another device that is occasionally used is the squeeze, in which the string is deflected slightly to one side by the finger pushing it across the fretboard, raising the pitch by a microtone.

The hammer and pull legato effects described above are termed by Heratis *larz-e-panje* ('fluttering fingers') or *pardeh kashi* ('pulling the frets') but specific terms for what I have termed the hammer and the pull were not found. The North Indian terms *gamak* and *meend*, which describe particular types of ornamentation in India, are also occasionally used with reference to the *dutār* in Herat, where they usually refer to either the slide or the squeeze.

An ornamental device that is used extensively but does not fall into the category of legato, is to release the 1st string altogether and to strum it open to provide a pedal effect.

xviii

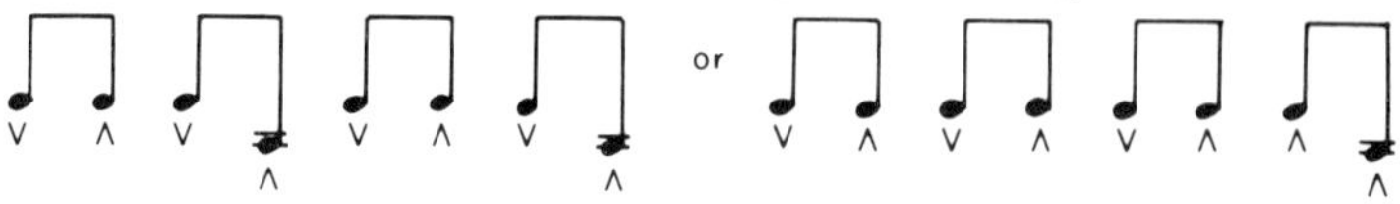

This is commonly used as a 'punctuation mark' at the end of a melodic line. In a special use of this technique the open string is sounded by a pull (p).

xix

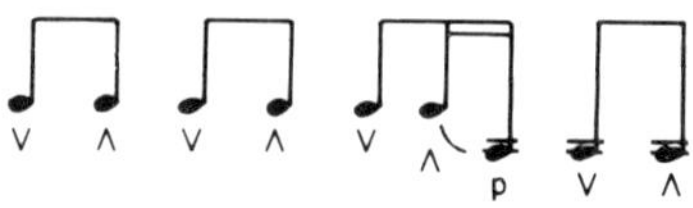

In order to examine the question of finger use in more detail, I shall consider four instrumental versions played on the 3 stringed *dutār* of the 'verse' section of 'Mullah Mohammad

Jan' a very well-known traditional Herati song. The four musicians are: 1) Safar, a weaver and amateur *dutār* player who plays only the 3 stringed *dutār*; 2) Aziz, a professional player of the 14 stringed *dutār* who usually plays in small groups with a singer and *zirbaghali* (clay pot drum) player; 3) Gada Mohammad, a highly rated professional 14 stringed *dutār* player who has worked with bands run by hereditary professional musicians for 8 years; and 4) Mohammad Karim, the inventor of the 14 stringed *dutār*. The data were collected in test film/sound recording sessions.

The patterns of finger movement revealed by this analysis of 3 stringed *dutār* playing are in general the same as those used on the 14 stringed *dutār*. That is, a performer such as Aziz, who uses two fingers of the left hand for playing the 3 stringed *dutār*, uses the same two fingers and uses them in similar ways in playing the 14 stringed *dutār*.

Examples 1 and 2, by Safar and Aziz, show typical use of the 1st and 2nd fingers in what may be termed a 2-finger 2-component technique (explained in more detail below). Their versions are very similar, differing only in small details of fingering and ornamentation. In Safar's version (Example 1), note the following: (a) frequent use of the open pedal effect, (b) hammer patterns in bars 3, 7 and 11, (c) pull patterns in bars 8 and 11, (d) use of the steady down-up right-hand strum. Many of these comments apply to Aziz's performance (Example 2): frequent use of the open pedal, hammer patterns in bars 3, 5 and 9, a pull pattern in bar 11 and a steady right-hand strum.

Both versions illustrate the partial development of a technique that I call *playing by position*, and which I observed in the playing of a number of *dutār* players. Playing by position means the tendency to take certain frets with a particular finger in order to allow for ornamentation with other fingers. The simplest example of this is the tendency to take the note F with the 2nd finger, allowing the 1st finger to be used for legato patterns involving the note E, which is a semi-tone below.

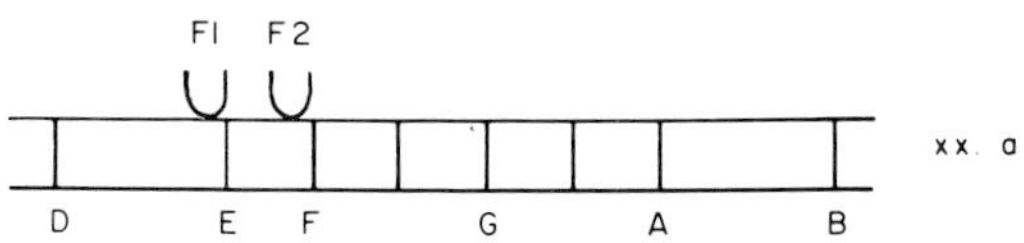

Playing by position for F

In bar 8 Safar takes this position to pull onto E; Aziz does the same thing in bar 11. The significance of playing by position is that often the position is taken without being exploited. In bar 10 Safar takes the F with the 2nd finger without decoration, Aziz does the same thing in bars 8 and 10. It

Example 1 by Safar

Example 2 by Aziz

Example 3 by Gada Mohammad

Example 4 by Karim Dutari

seems clear that this kind of 'position habit' is established because it is sometimes useful. It shows the beginning of some kind of motor schemata underlying the movement patterns. Aziz shows some development of playing by position for the note G: in bar 11 he takes G with the 2nd finger in order to pull onto F#.

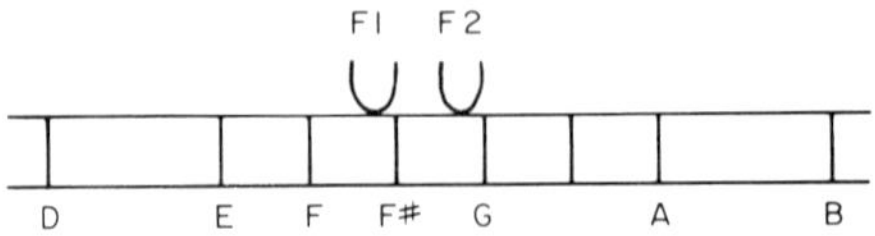

xx. b

Playing by position for G

In Karim's playing, this principle of playing by position is applied much more extensively and consistently.

Gada's version in Example 3 illustrates what may be termed a 3-finger 2-component technique, in which the 2nd finger is used against the 1st finger when the note to be taken lies immediately above, and the 3rd finger is used against the 1st finger when the next note is non-adjacent. However, the three fingers are not used independently because the 3rd finger is not used against the 2nd finger, (i.e. the three fingers only operate as two independent motor components). This version shows a considerable increase in complexity over Examples 1 and 2, with short ornamental sections added in bars 4, 8 and 12 where there is otherwise no melodic movement. In much of the piece Gada uses the rhythmic pattern:

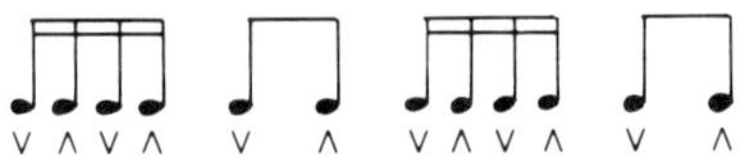

xxi. a

He uses hammer and pull patterns and more elaborate compound legato patterns, hammer-pull (bars 1 and 7) and pull-hammer (bar 7). In bar 2 there is a double hammer pattern in which the 4th finger is used for a damped hammer movement. He frequently changes note in the middle of a short *riz* (bars 3, 4, 5, 8 and 11). In bars 8 and 9 he shows a position habit when he takes the note F with the 2nd finger. In bar 7 we find a curious situation where he changes finger while taking the same note:

xxi. b

He takes the A first with the 1st finger, providing an opportunity for upward legato to the B, then he changes to take the same note (A) with the 3rd finger, leaving the 1st finger free for a possible legato pattern down to G (a pattern which is not in fact used). This is probably another aspect of a position schemata.

In Karim's version in Example 4 we observe the emergence of a 3-finger 3-component technique in which the 3rd finger is used against the 2nd finger in the same way that the 2nd finger is used against the 1st finger in the 2-finger 2-component technique. When the 3rd finger is used in this way, the 1st and 2nd fingers are normally kept in position on the string. We find the use of open pedal and various hammer and pull patterns, but the version is less complex than Gada's. The most striking aspect of Karim's playing is the extension of the playing by position principle, made possible by the full use of the 3rd finger. He uses two basic positions for the left hand:

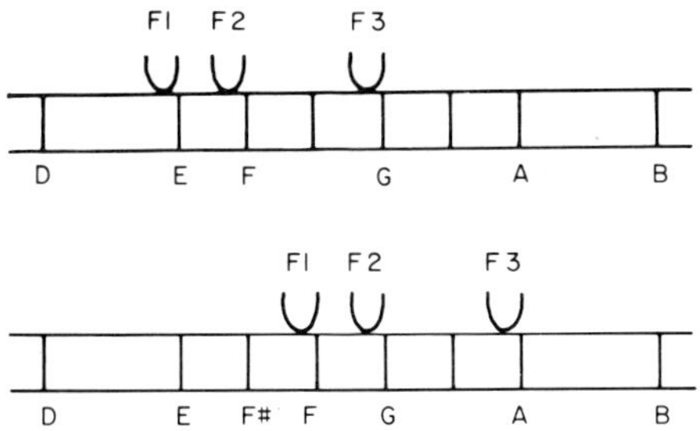

a

In most cases when Karim plays the note G he takes it with the 2nd finger, with the 1st finger placed on F#, even in cases where the F# is not sounded. The use of three components allows for various legato patterns which are not possible with two components, such as

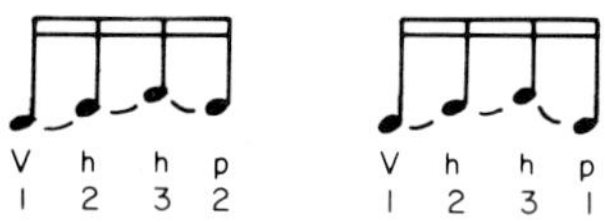

b

In fact, in Example 4 there are no patterns of this type. In bars 7 and 11 we find the following patterns, which are typical of 3-component technique.

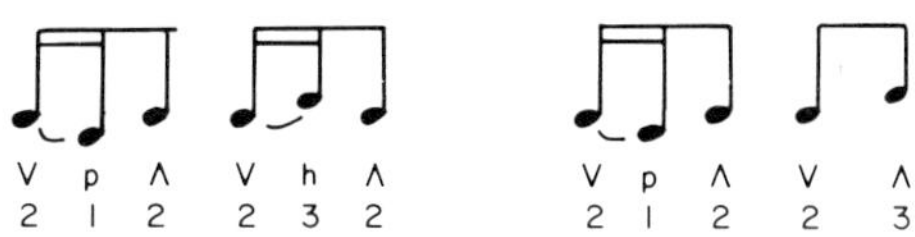

c

Only Karim has really developed the use of three components to the point where this method is integral to the style of playing. He uses this technique on both the 3 and 14 stringed *dutār*-s (which shows that the length of the neck of the 14 stringed *dutār* does not present insurmountable problems to the use of three fingers as three components). Other *dutār* players sometimes bring out a 3-component pattern, but it is not the basis of their techniques. Some *dutār* players use only the 1st and 2nd fingers, others the 1st, 2nd and 3rd, and even the 4th, but mainly as two components. It is concluded, therefore, that *dutār playing is based on a 2-component technique for the left hand.*

3. THE INTERACTION BETWEEN THE BODY AND THE *DUTĀR*

In this section I am concerned with the interaction between the body of the player and the morphology of the *dutār*. Many features of the *dutār*'s morphology could be taken into account here. Factors such as the shape and size of the instrument, the material of the strings, the tension to which they are stretched, the height of the frets, and many other seemingly small details, are significant when we consider how the body interacts with them. For example, it is important that the string tension of the *dutār* is low, for this facilitates the use of legato techniques such as hammer and pull, so that a light touch produces a clearly audible effect. For the purposes of the ensuing discussion I shall concentrate on just one aspect of the morphology of the *dutār*, the spatial layout of the 'keyboard' with which the left hand interacts.

As a preliminary step, I shall try to define the nature of the *dutār* player's task in spatial terms. The frets of the *dutār* may be considered as a set of targets to which left hand responses are directed in patterns that are spatially and temporally isomorphic with the structure of the music produced. The basic left hand response consists of 'taking a fret', that is, placing a finger on the string behind a fret so as to prepare that note to be sounded, while at the same time the right hand makes a coordinated movement and strikes the strings (secondary movements for ornamentation may be ignored for the present discussion). The player of a musical instrument can be described as an operator in a discontinuous tracking task. In the case of solo *dutār* playing the task is self-paced, and the sequences of targets that give rise to the relevant melodic movement are organised from information in a memory store. This mode of operation in playing would seem to be different from that involved in 'sight-reading' from written notation (as in some Western music), and different again from improvisation from a set of generative rules (as in some North Indian music).

It seems reasonable to suppose that, in the early stages of learning an instrument such as the *dutār*, the left hand is controlled principally by visual information. The learner understands and conceives of the music he is playing in terms of visual patterns: a melodic sequence is represented at a cognitive level as a sequence of visual targets. At a later stage in the acquisition of the skill, motor sequences become grouped and routinised, and kinaesthetic control assumes increasing importance. This state of increased automatisation of performance is shown by the ability to play without error with less attention to the actual mechanics of playing. It is perhaps at this stage that auditory control of performance becomes more important (Szende and Nemesuri, 1971). The relationship between audition and hand and finger movement is not at present clearly understood, but it appears as though some kind of 'auro-motor coordination' is established in learning a musical instrument, in which direct links between auditory inputs and motor outputs are built up. This kind of learning allows the musician to reproduce immediately musical patterns that he either hears or experiences as auditory images. Auro-motor coordination is also presumably the basis for transfer between auditory and kinesthetic perceptions sometimes reported by Western musicians. Hearing music on an instrument he plays, a musician may experience (illusory) perceptions of the hand and finger movements appropriate for playing that piece. Conversely, simply fingering an instrument without making it sound may give rise to an (illusory) auditory perception of the music that would be produced.

The complexity of the total pattern of information processing involved in skilled performance on musical instruments has received insufficient recognition. What is so remarkable about perceptuo-motor skill in musical performance is the variety of types of information that can be used to control behaviour, and the way that these are integrated by the performer. It would be fruitless to try to show that musical performance is controlled exclusively by one sensory modality, such as vision, audition, or kinaesthesis, or within one kind of conceptual field. A range of possibilities are open which may be stressed to different degrees in different musical cultures, by different individuals in the same musical culture, and by a single individual at different moments in a single performance. Nevertheless, despite this multiplicity of representations and the possibility of switching between them, it is quite clear that spatial relationships are very important at the cognitive level, being directly linked with the physical operations to be performed. Thus, one important conceptual model that the musician has at his disposal is actually a spatial model in which movements are planned and experienced in visual, kinesthetic, and tactile terms.

An important characteristic of the *dutār* is that, since only one string is fingered with the left hand, all the notes used to give the melodic movement are arranged in a straight line, which we may term a *linear array*. This makes the *dutār* of special interest, for most lute types have a *tiered array*, consisting of two or more linear arrays and often with multi-representation of notes.

A comparison between linear and tiered arrays reveals some important ergonomic principles. The *rebāb* (described above), is an excellent example of an instrument with a tiered array, with which to compare the *dutār*. The *rebāb* has 3 strings, tuned in 4ths, giving the instrument a fretted range of a chromatically divided octave plus two semitones. In most *rebāb* music the note given by the 1st fret on the 3rd string is taken as the finalis (chief tone), and this note is usually tuned to D (with reference to the harmonium). Thus the fretted range of the *rebāb* might be represented in this way:

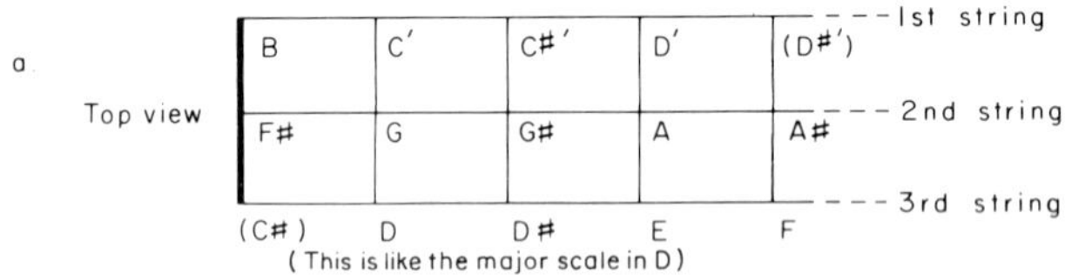

Each tone is represented only once. By entering the more difficult unfretted area of the neck on the 1st string, a skilled player can extend the range of the *rebāb* to 1½ or even 2 octaves. Here, however, we are concerned only with the instrument's fretted range. The same range (and more) is fretted on the linear array of the 14 stringed *dutār* (in which only the 1st string is fingered with the left hand).

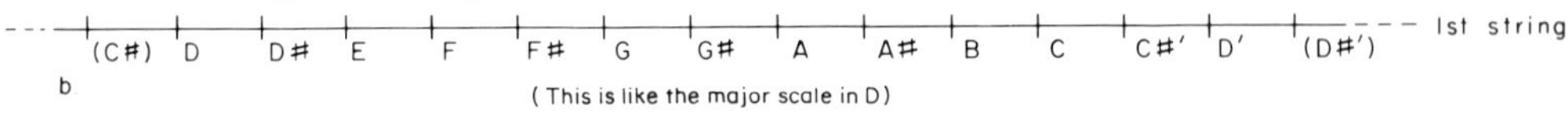

The linear array is of great significance from the ergonomic point of view, and we may isolate two types of factor that are important in this respect: the relative frequency of hand and finger movements and spatial confusion.

(a) *Relative frequency of finger and hand movements.* All the notes in the fretted range of the *rebāb* can be covered by finger movements from two hand positions which are separated by 5-10 cm., depending on the size of the player's hand and the size of the instrument he is playing. On the 14 stringed *dutār* the same range, C# to D#', stretches over approximately 40 cm. On the *dutār*, only two or three frets are available to the fingers from a given hand position (except high up the neck). *To reach other frets a hand movement is required.* Thus, *on the dutār* hand *movements are required to exploit the same musical range that can be reached with* finger *movements on the rebāb.*

This difference has two important implications. Firstly, finger movements are generally faster and more accurate than hand movements. If a motor pattern can be produced by either hand movements or finger movements, the use of the fingers is usually a superior method. The difficult in producing an accurate hand movement increases with the size of the movement (the terms easy and difficult are used here to imply the amount of motor learning required to improve performance to a given criterion).

A second difference between hand and finger movements lies in the type of sensory information used in their control. When finger movements are involved, the hand can operate within a tactuo-kinesthetic framework, using active touch. The human hand is richly endowed with joint and cutaneous receptors, which form the sensory basis for this kind of control. The kinesthetic system serving the elbow and shoulder, on the other hand, has a significantly less acute position sense, and does not provide a good basis for accurate hand movements. The use of hand movements imposes a much greater demand for visual information to locate the target frets and to move to them accurately.

When I learned to play both the *dutār* and the *rebāb*, I found that I quickly advanced beyond the stage of needing to watch my left hand while playing the *rebāb*, but that I now still need to watch my left hand while playing the *dutār*. I do not usually need to watch my right hand when playing either instrument. Observation of Afghan *rebāb* and *dutār* players tended to confirm this point. Films of test pieces played on the *dutār* showed that some players watched their left hands nearly all the time. Others tried to face the camera in what they considered to be a 'formal' photographic pose, but turned their heads to look at the left hand every time a large hand movement had to be made. This reliance on visual information probably means that it takes longer for the *dutār* player to reach the stage of programmed automatic control with minimal attention load which is typical of skilled performance. Using visual control leaves the *dutār* player with less attention capacity for dealing with other aspects of his performance, such as listening to his own playing or to other musicians with whom he might be playing. This is important because too much reliance on visual control may delay the acquisition of auro-motor coordination.

(b) *Spatial confusion.* Another important feature of the linear array is that it is a confusing spatial framework in which to operate. On the tiered array of the *rebāb* the spatial framework is coded in two dimensions; on the *dutār*, in one dimension. In *dutār* playing the performer responds to a sequence of target fret positions which he in some way 'sets up' within the spatial framework with which he interacts. There are several kinds of confusion at work here.

Firstly, the musician has to identify, out of the total set that confronts him, the frets that make up the mode he is playing. He has to recognise those frets to which he should respond, and those he should avoid, and he must use this facility for recognition while playing at speed. The pattern of fret positions which makes up a mode is more easily dealt with when expressed in two dimensions than in one dimension. This point can be appreciated by comparing the fret positions for the Bairami mode on the *rebāb* and on the 14-stringed *dutār*. In these diagrams the circles represent the fret positions on the tiered and linear arrays, and a filled circle indicates that the fret position is fingered and is part of the mode.

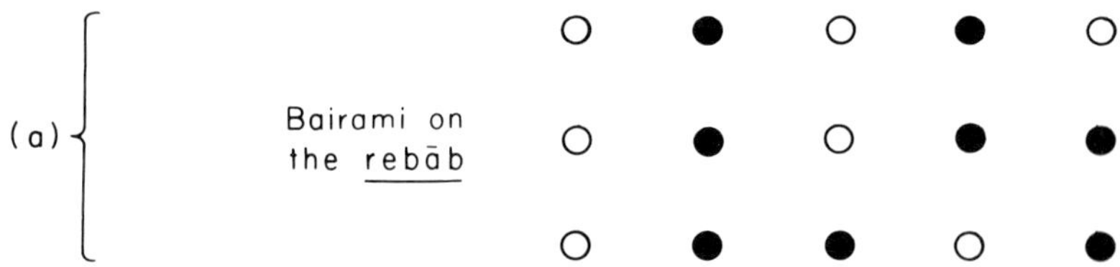

xxiv

Secondly, while in the act of playing, the setting up of each target in the sequence involves decisions in which the musician selects the appropriate target. It is clear that this decision process is quicker and more accurate on the tiered array than on the linear array, for the same reason that the modes themselves are more quickly identified on the linear array.

Thirdly, errors in identification due to confusion between fret positions on the linear array are likely to increase in frequency with the size of the displacement between the two fret positions. In the following example more errors will - occur with setting up Z as a target than with setting up Y as a target, with the finger already on fret X.

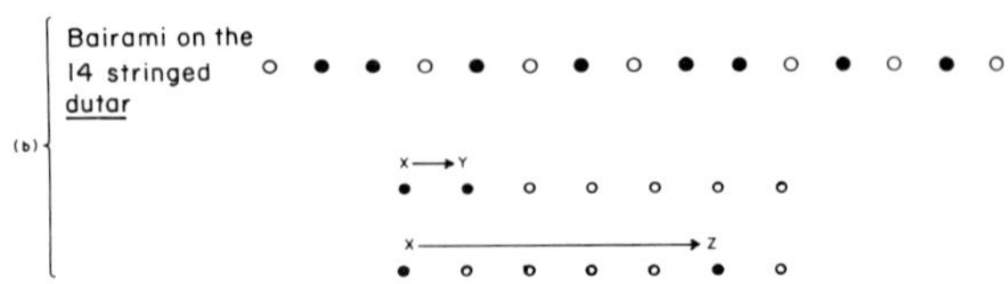

Fourthly, errors in setting up a sequence of fret positions

will be less when the sequence forms a coherent or complete pattern. It is well established that the human nervous system is organised to deal with complete, well structured patterns, and that there is a strong tendency to modify perception towards a good gestalt (pattern). Thus on the *dutār*, series of fret positions such as 12345 or 123234345456 form better gestalts and are more easily organised than a sequence such as 164352.

The three types of *dutār* do not present the same amount of spatial confusion. On the 2 stringed *dutār*, with its single mode, this factor is reduced because all of the frets in the range D-C^1 (see Figure 2 page 4) are part of the mode; there are none actually to be avoided, which makes recognition easier. The addition of frets to give a semi-chromatic system on the 3 stringed *dutār* introduces frets to be avoided and so increases the difficulty of recognition. However, the gaps in the fretting between the 4th and 5th and between the 8th and 9th above the note of the open string serve as anchor points which help to reduce confusion. On the 14 stringed *dutār* these gaps are filled in; almost invariably these added frets are distinguished by colour. This fact by itself would suggest the importance of the visual framework to the *dutār* player.

On two occasions where (for some reason) this cue was absent, I observed that errors in fingering were made. These errors were attributed by the players themselves to the fact that the colour cue was missing.

The ergonomic factors outlined above operate as constraints on performance with the linear array of the *dutār*. It is apparent that the melodic movement of a type of music well adapted to the morphology of the *dutār* would have a low frequency of hand movements in relation to finger movements. When hand movements occur they should be small, and sequences of notes should form good gestalts. It will be shown in the next section that the music traditionally played on the *dutār* is well structured to take these constraints into account.

So far, the discussion has been devoted to the left hand of the *dutār* player. There is less to be said about the right hand, not because this is in any way the less 'important' hand but simply because the special features of the linear array are less obvious in the way they affect the right hand. Very little is understood about the roles of the two hands in playing a lute type of instrument, where the two hands perform quite different tasks. The fact that the right hand normally vibrates the strings and the left hand stops them, points to important differences between the two hands as far as musical behaviour is concerned. It might be suggested that the right hand is the more rhythmic hand; the psycho-physiological advantages that the right hand has over the left hand may include a more accurate timing of movements.

The linear array has one important advantage over the tiered array in that it avoids spatial differentiation of right hand responses. On the tiered array of the *rebāb* the strokes of the plectrum have to be applied to three strings, and difficulties arise in moving from one string to another. Although these movements to make the necessary adjustments are small (and may require rather little in the way of visual control even in the early stages of learning), they have to be accurate. The linear array of the *dutār* avoids this difficulty because only one string is struck (except in *parand* patterns). The difficulty of moving the right hand from one string to another on an instrument with a tiered array such as the *rebāb* is greater when the jump occurs between a downstroke and the succeeding upstroke, than when the jump occurs between an upstroke and the succeeding downstroke. This difficulty is connected with the fact that the downstroke is the main stroke, the basic right hand unit being down-up.

This point may be illustrated as follows

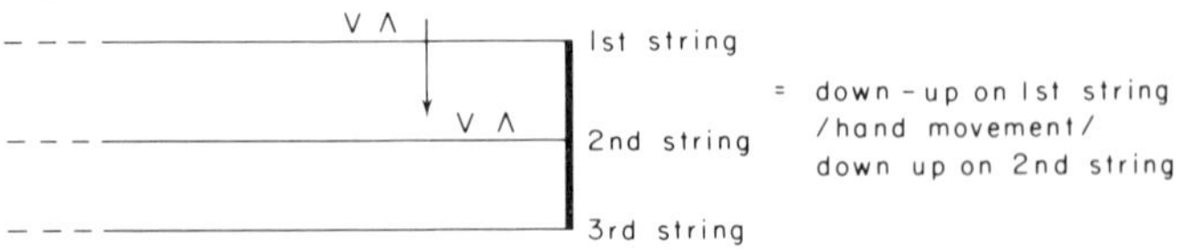

This constitutes an easier right-hand pattern than

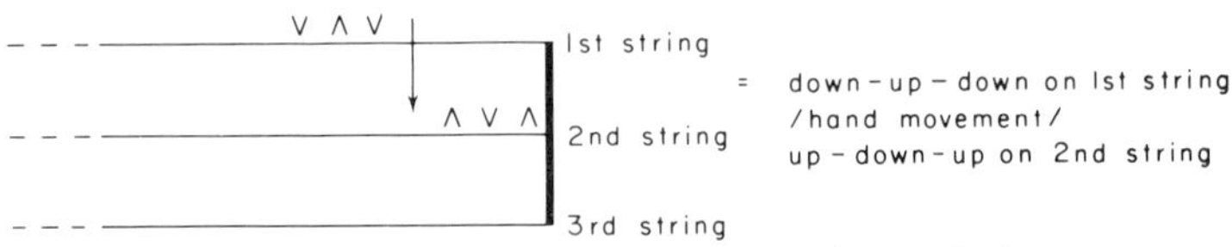

In the second case the change in hand position occurs between the downstroke and the succeeding upstroke.

Comparing the left and right hands on the linear array of the *dutār* and the tiered array of the *rebāb* we may conclude that, while for the left hand the tiered array has certain advantages over the linear array, for the right hand the opposite is the case.

4. THE ERGONOMICS OF *DUTĀR* MUSIC

Having discussed some of the constraints on performance imposed by the spatial layout of the *dutār*, we now proceed to examine how these constraints operate in the structure of two very different kinds of music played on the *dutār*: traditional music from Herat (*musiqi mahali*) and classical music from Kabul (*musiqi kelasik*). They will be compared in terms of the melodic movement they exhibit. These two types have been selected to provide as great a contrast as possible in styles of *dutār* playing. Whereas traditional Herati tunes associated with the *dutār* are known to most *dutār* players in Herat and are played on all three types of *dutār*, classical music is played only by some players of the 14 stringed *dutār*. *A priori*, traditional *dutār* music might be expected to be well adapted to the spatial layout of the *dutār* because it has developed in relation to it, while classical music might be expected to be less well adapted since it has been largely borrowed from the repertoire of the *rebāb*.

(a) *Traditional Herati music*

The instrumental pieces that may be regarded as belonging to the 'traditional' repertoire of the *dutār* are, with only one or two exceptions, the melodies of well known Herati songs. Herati song texts are based on a quatrain form known as *chahar-beiti*, which usually has the rhyme scheme *AABA*. Textually, Herati songs often consist of two sections, which operate as 'verse' and 'chorus', though this distinction in poetic terminology is not made by Heratis. The two sections usually have closely related melodic structures, and the section of the melody which corresponds to the 'verse' part of the song normally occupies a higher pitch range than the part which corresponds to the 'chorus'. More knowledgeable *dutār* players who are in contact with the musical terminology of Kabul call these two sections the *āstāi* and *antere* (from the terms *asthāyi* and *antara* used in North Indian music), the *āstāi* being the part in the lower, and the *antara* in the higher, pitch range. Although traditional *dutār* tunes are the melodies of songs, when played as *dutār* instrumentals they become an entity of their own: they are often played at a fast tempo, the rhythmic side of the music is stressed and the two parts of the tune are repeated many times.

The number of traditional *dutār* tunes that I have been able to isolate is small. They are all in the Bairami mode (the E mode), which corresponds approximately to the Shur-type mode in which the 2 stringed *dutar* is fretted. An important aspect of traditional *dutār* tunes is that melodic movement mostly proceeds in a stepwise fashion, moving from one tone to the adjacent tone. A simple count of intervals of different sizes in 13 traditional *dutār* tunes shows that approximately 85% of the intervals involve a movement from one tone to the adjacent semitone or tone above or below. The frequency of the various intervals is shown in the following table. Repetitions of the same note are excluded.

TABLE 1

Size of interval in semitones	1	2	3	4	5	6	7	Total
Ascending	34	128	16	21	14	-	8	221
Descending	68	199	4	16	-	1	-	288
								509

From a total of 509 melodic movements only 80 are larger than a tone. The majority of these are ascending, 59 as opposed to 21. The only common large descending interval is of 4 semitones. All the large upward leaps of 7 semitones occur at the start of an *antare* or *āstāi* section, moving from the finalis (E) to the 5th above (B).

Melodic movement of these tunes is largely confined within the range of a minor 6th (E-C^1), though in some cases this is extended to an octave by including D and D^1. From the melodic contours of these tunes it is possible to construct a generalised contour for melodic movement:

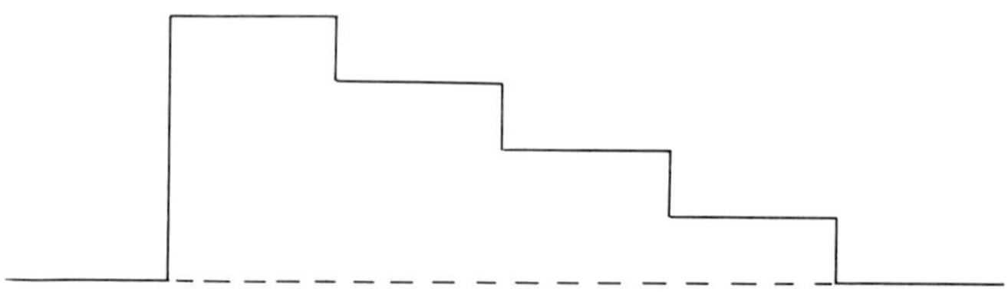

The *āstāi* or *antere* section starts with an upward leap of a 4th or a 5th above the finalis, followed by a gradual convoluted descending movement back to the finalis, as shown in the following example:

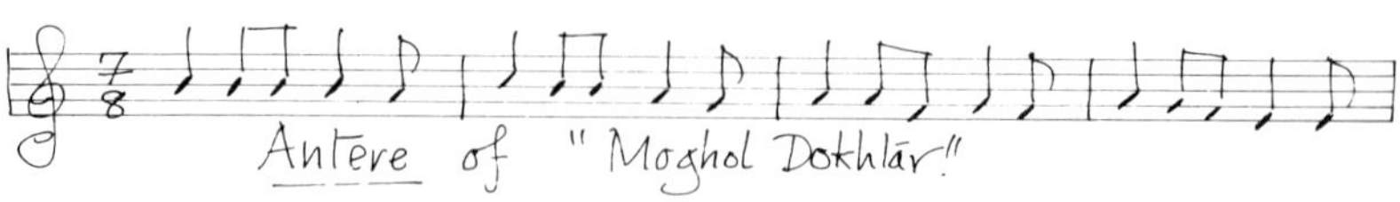

In some tunes, particularly in *āstāi* sections, the ascending movement at the beginning is not made in one single leap, but in an ascending sequence which often starts from the 2nd below the finalis.

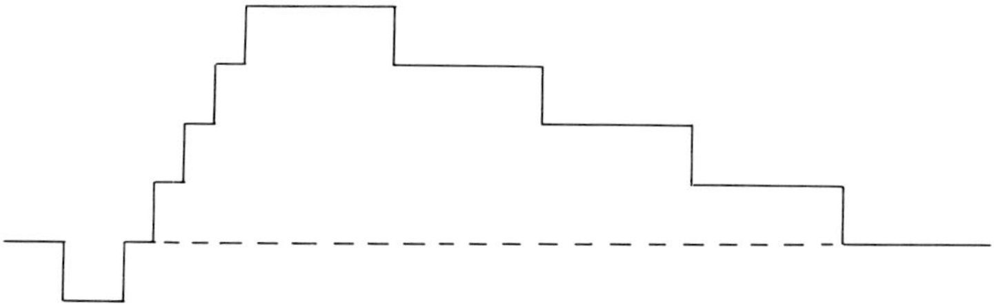

This type of movement is shown in the following example:

a.

Alternatively, the upward leap may start with an interval of 3 semitones (from E to G) followed by an ascending sequence, as in this example:

b.

The above example shows another feature of some tunes, which is that in the convoluted descent the melodic movement proceeds down to F and then leaps up again before starting another convoluted descent to E (the finalis). Another way of looking at this structure is that the same descending pattern is played twice, being a step lower the second time. This structure is shown very well in 'Mullah Mohammad Jan', the tune shown in Examples 1-4, where the same phrase is played three times in three descending positions.

It seems clear enough that the melodic movement of the tunes traditionally associated with the *dutār* is well adapted to the spatial layout of the instrument: there is a good fit between the constraints on performance expected from the spatial layout and the motor patterns actually used. To start with, these tunes only exploit the upper part of the tonal range available on the 1st string, which reduces the ratio of hand movements to finger movements. In this music very frequent hand movements do not occur. Most of the melodic movement proceeds in a stepwise fashion, many of the steps are carried by finger movements, and in those that are carried by hand movements the size of the movements is usually small. The motor patterns of this music tend to form good spatial gestalts, in the sense that the sequences of frets fall into units that perceptually are easily apprehended and easy to organise as movement. When a sequence of several hand movements does occur, typically at the start of an *āstāi* section, the movements are in a linear series. There is also a tendency to repeat the same patterns in two or more positions on the neck in a descending sequence, another situation where the patterns make good gestalts.

Having noted that a high degree of compatibility exists between the structure of traditional *dutār* music and the spatial layout of the instrument, the question arises whether the *dutār* has shaped this music or whether the *dutār* has been chosen by Heratis because it fits their music. It is not possible to evaluate which of these possibilities is the most likely, and both processes may have occurred. The following hypothesis has some plausibility. There is good evidence to suggest that the 2 stringed *dutār* enjoyed a new popularity from about the 1920s and that this began across the border in Iranian Khorasan. In spreading into the Herat valley the *dutār* lost part of its Iranian repertoire (including tunes played in the lower region of the neck, see page 5) and acquired a new Herati repertoire. This may have been derived from the repertoire of women's 'domestic' music (singing with frame drum accompaniment) which to a large extent overlaps with the male repertoire with which the *dutār* is associated. In the women's versions, the melodies of the songs are compressed into a smaller tonal range. These melodies, mostly in the same mode as the *dutār*, fit well onto the *dutār*, and could have been taken over and expanded by the men, using the *dutār* as an accompanying instrument. These melodic innovations would then have been shaped, in part at least, by the morphology of the *dutār*.

The motor structure that emerges from an examination of traditional *dutār* tunes reveals a good ergonomic solution to the problems posed by the linear array. A similar phenomenon is even more pronounced in two other types of long-necked lute, *tār* and *sehtār*, found in Iran. While these closely related instruments have a tiered array consisting of 3 linear arrays, there is a strong emphasis on playing only the 1st string, and they are to this degree comparable with the *dutār*. These Iranian lutes rely heavily on the use of three fingers as three components. The mode of operation in playing them is to move to a position on the neck, to perform a series of operations with finger movements at that position, then to move to a nearby position, to perform a series of finger operations there, and so on. Very often the sequence of hand positions forms a descending progression:-

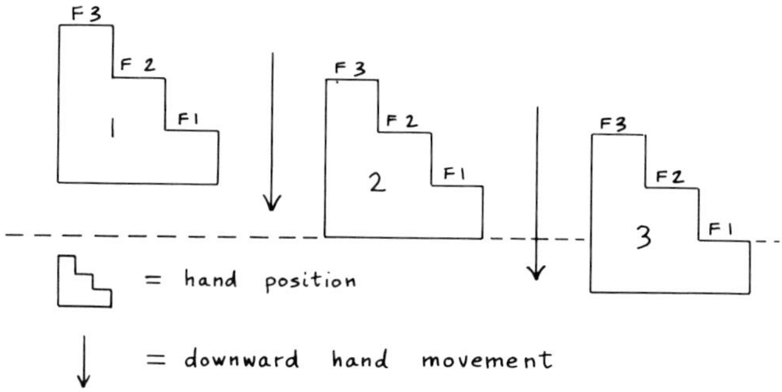

There is only one point on which it might appear that the motor structure of the *dutār* is not well adapted to the spatial layout; the general use of a 2-component technique for the left hand. From the ergonomic point of view it is clear that 3-finger 3-component action would be more suitable for the *dutār* because it would require less hand movement and more finger movement; movement patterns that we have concluded are more economical in terms of information processing, as well as providing for more ornamentation. The most likely explanation for the left hand 2-component use would seem to be in terms of finger dexterity. It appears that anatomically and physiologically the hand is arranged for the 1st finger to operate against the 2nd finger or the 3rd finger - the 2nd and 3rd fingers are anatomically and physiologically linked. 3-component playing with the 1st, 2nd and 3rd fingers demands the development of a high degree of dexterity in the use of the 3rd finger against the 2nd, and it may be this that limits the use of three fingers on the *dutār*. It is easier for the beginner to employ small hand movements to avoid this difficult finger movement, even though hand movement itself is a factor which is constrained. Once this habit is acquired in the early stages of learning, it remains as a basic feature of the motor structure, even though it later acts as a limitation on the fullest development of the technique.

(b) *Classical music*

Classical music in Afghanistan is derived from the Hindustani music system of North India. Indian instruments, scale types, rhythmic cycles and the *rāga* form are used by musicians in Kabul and, to a lesser extent, in Herat. Although classical music in Afghanistan is predominantly vocal, instrumental performance is also practised. In this respect the *rebāb* is one of the leading instruments of classical music. In comparison with instruments such as the North Indian *sarod* and *sitar* the *rebāb* is technically limited, for its effective range is small and its fretting gives little possibility for playing the microtonal alterations and ornaments considered necessary for the correct performance of classical music in North India.

Despite these limitations, and possibly because of them, there is an important classical repertoire for the *rebāb*, and the classical music played on the *dutār* is largely borrowed from this repertoire. The term *rāg* is used in Herat to describe three types of piece: (i) *khyāl*, which corresponds to the usual *rāg* form used in North India, (ii) certain standardised instrumental pieces usually played by a group of musicians at the start of a musical performance, which bear the names of *rāg*-s such as Kumaj and Piloo, (iii) light instrumental pieces that are loosely based on the scales of *rāg*-s. All three types of *rāg* are included in the term *musiqi kelāsik*. We may note that Mohammad Karim is the only *dutār* player who has full control of the *khyāl* form and the ability to improvise within the framework of a rhythmic cycle.

Classical music on the *dutār* is structured in terms of ascending and descending scale movements, usually within the range of an octave, and is in a variety of modes. These characteristics are evident in the following examples, which are the fixed compositions of two *rāg*-s played on the 14 stringed *dutār*:

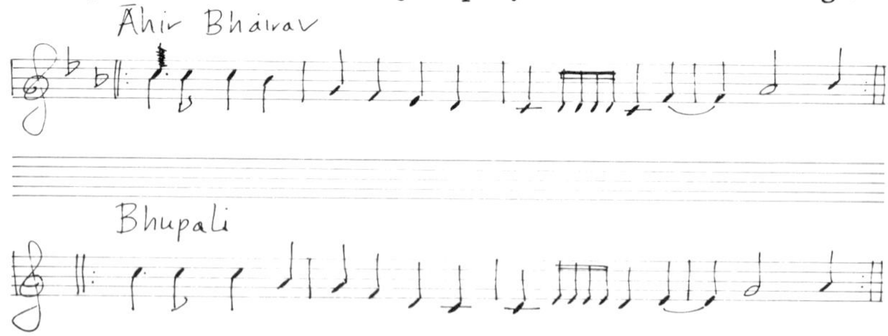

(The use of sharps and flats in the notation of these Kabuli scale types requires some explanation. The convention has been adopted here of regarding the key of C as the basic scale type from which the other scale types are derived. Thus all pieces in this section are written in in the key of C rather than D, although the instruments are generally tuned so that D is the tonic. Following Indian convention (see, for example, Kaufmann 1968), the transformations of the major scale consist of flattening the 2nd, 3rd, 6th and 7th, while the 4th is changed to a sharpened form. It is important to note that the Bairami, or Bhairavi, scale type is notated in two ways in this paper, firstly, as though it were the E mode (key of C without accidentals) and secondly, in the key of C with four flats, Db, Eb, Ab and Bb.) Here is an example of the 1st of several standardised sections of a group instrumental piece played on the *dutār*:

a.

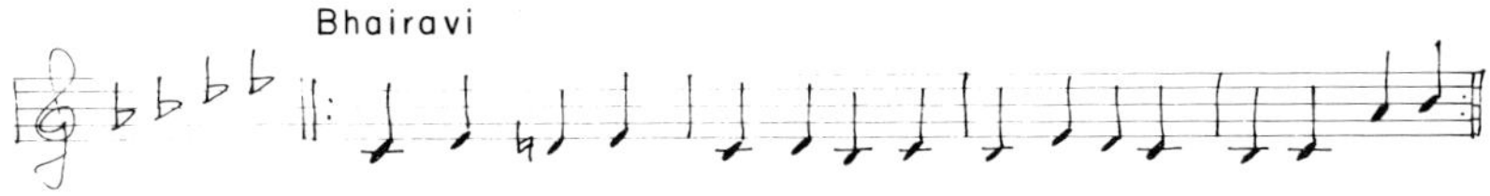

Here is an example of a light instrumental piece based on the scale of Rāg Madhubanti:

b.

The first criterion on which to compare traditional *dutār* music with classical music is on the basis of hand and finger movements. In both types melodic movement tends to be stepwise but, while Herati music tends to be contained within the range of the minor 6th, the melodic movement of classical music covers the range of an octave. Moreover, as in some of the examples above, often the movements up and down are larger than 2 semitones. In *dutār* music the frequency of hand movements is relatively low, but in classical music it is high because of the continually ascending and descending scale configurations. It is evident that the melodic movement of classical music is easily produced on the *dutār* because it can be obtained with finger movements, but on the 14 stringed *dutār* many hand movements are required, and as a result this music is difficult to play.

A second criterion for looking at the compatibility between classical music and the sensori-motor system is in terms of perceptual confusion. The sequences of classical music form good spatial gestalts in terms of orderly ascending and descending sequences, but confusions arise with regard to selecting the appropriate frets for the mode being played. The problem is acute because of the wide variety of modes used for playing classical music. On the *rebāb* these modes form patterns or configurations which are easily apprehended, because they are expressed in two dimensions, while on the *dutār* they are much more difficult to apprehend, as the following examples show:

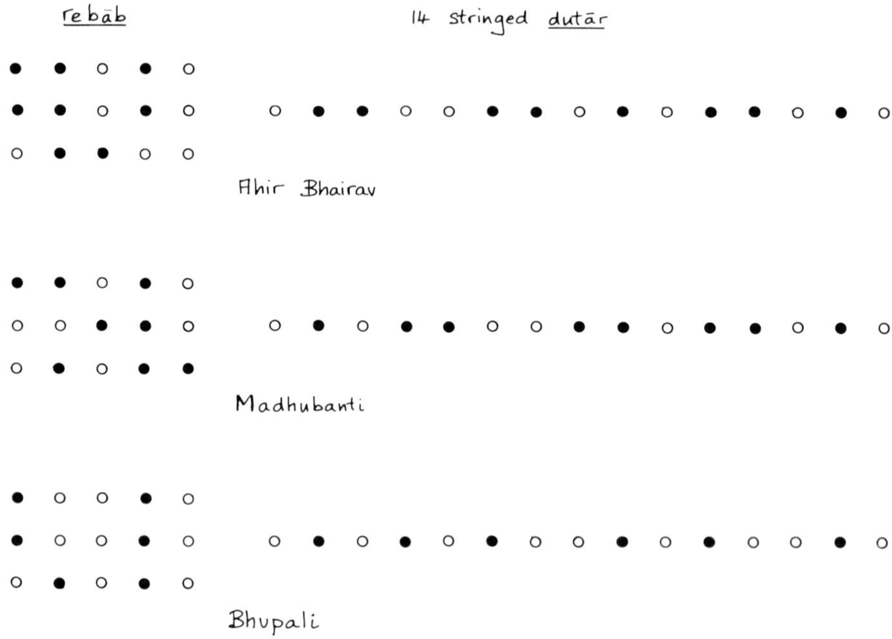

It should be clear from the foregoing dicussion that classical music is much less compatible with the layout of the *dutār* than Herati music. Classical music is not structured to take into account the constraints imposed by the linear array in the way that Herati music is. The example of the North Indian *sitar* illustrates this point well. The *sitar* is a long-necked lute which, although it has a tiered array with 4 strings, stresses use of the 1st string for a music based on scalar melodic movement and which necessitates frequent hand movements. Some schools of *sitar* playing advocate the use of the 1st finger to take nearly all the frets. The motor structure of the *sitar* contrasts strongly with that of the Iranian *tār* and *sehtār* considered above. It is perhaps significant that the *sitar* players spend literally years practising exercises based on scales, presumably to overcome precisely those difficulties posed by the linear array that we have already discussed. Traditional musicians in Iran do not practise formal exercises of this, or any other kind. This suggests that music which has a low compatibility with the linear array *can* be played satisfactorily on a long-necked lute, but that special training is required.

(c) *Traditional Herati music played on the* rebāb

Further insight into these problems is provided when we consider the transfer of Herati music onto the *rebāb*. The *rebāb* player tends to take the open note of the 2nd string as the finalis of the Bairami mode, when playing Herati music so that the spatial layout of the notes of the mode is as follows:

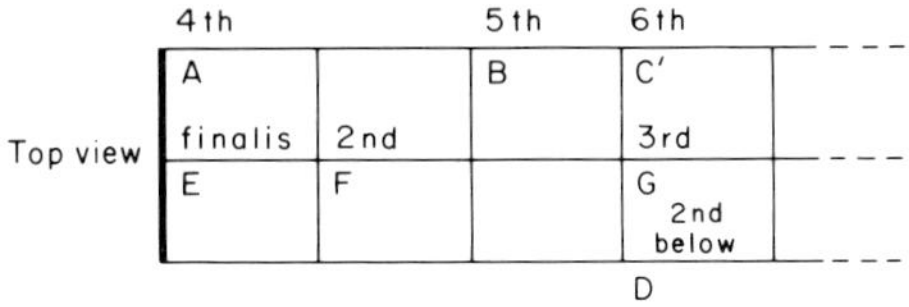

There are several problems with this transfer. In the first place, the weight of the thick gut or nylon strings on the *rebāb* does not lend itself to the legato ornamentation effects that are so much a part of *dutār* playing, and the break in the continuity of the notes as the pitch range passes from the 2nd to the 1st string, means that the 3rd above the tonic (labelled as G in the diagram) has to be ornamented using the unfretted part of the neck. Secondly, the *rebāb* requires spatial differentiation of right hand responses, for the plectrum has to be moved from one string to another. In most of the music which is adapted to the spatial layout of the *rebāb*, we find that a change from one string to another usually follows an upstroke and proceeds the succeeding downstroke but, in playing *dutār* tunes on the *rebāb*, the situation often arises that the melody jumps from one string to another after a downstroke and before the succeeding upstroke, a spatial differentiation which is difficult to achieve on the *rebāb*. The open pedal effect, which is used extensively in traditional *dutār* playing and which is easily achieved simply by removing the fingers from the 1st string, can be rendered on the drone strings of the *rebāb* if they are retuned. Again, this effect is less easily achieved on the *rebāb* because of the movement required to bring the plectrum from the main strings into proximity with the drone strings. It is perhaps significant that very few *rebāb* players have specialised in playing Herati music.

The difficulties of playing *dutār* music on the *rebāb* further illustrates the point that, while classical music may be highly compatible with the spatial layout of the *rebāb*, traditional Herati music is not. This shows how closely these two kinds of music are adapted to the instruments on which they are habitually played and suggests that in both instances the instrument has to some degree shaped the music.

5. CONCLUSIONS

(a) *General remarks*

The change in the morphology of the *dutār* is one element in a far more widely generalised pattern of change in the structure of Herati music and in the place of music in Herati society. These two facets of what may be broadly termed 'socio-musical change' will be discussed briefly.

Firstly, there is the change in music structure. In the course of this change the music of Herat has become less like that played in Iran and very close to that played in Kabul, a city that lies within the zone of Hindustani music. As recently as the 1920s, Iranian instruments such as the *tār* and *santur* were the most important instruments played in Herat city. They have been replaced by instruments such as the harmonium and *rebāb* which are typical of Kabuli music. The Herati *dutār* has become less like the kind of *dutār* played in Iran and more like a Kabuli instrument. The changes in the morphology of the *dutār* have been accompanied by changes in performance techniques, which follow the same general trend.

Secondly, music has to a considerable extent been released from negative social controls which were backed by the condemnation of music on religious grounds. Heratis distinguish unaccompanied song (*khāndan*), which can be used for religious purposes, from music played with instruments (*musiqi*). It was the latter which was formerly categorised as a somewhat sinful activity. Music is now played in a wider range of social contexts than before, and the frequency of performance has increased. In the course of this process of release the *dutār* has been accompanied by a change in the professional status of *dutār* players. The *dutār* has changed from being the instrument of the amateur to that of the professional musician, but professional *dutār* players remain socially distinct from members of hereditary professional musician families, even though they may play together in the same groups.

The changes in the *dutār* may be related to a growing awareness of national identity in Afghanistan, an awareness that finds expression in a type of music originating from Kabul which is now common to all urban centres in Afghanistan. The *dutār* has been adapted in order to equip it to play this music. It has been suggested by Baily (1976) that the development of the 14-stringed *dutār* in the context of the National Orchestra at Radio Afghanistan served to increase the contribution of Herat in a group that represented an integration of the diverse cultural elements in Afghanistan.

Along with the changes in morphology, the prestige of the

dutār as a musical instrument has increased. The *dutār* has been accepted within an ensemble consisting of sophisticated instruments with correspondingly sophisticated performance techniques. The greater prestige of the *dutār* is shown in the lavish use of mother-of-pearl inlay work to decorate it. This practice is probably derived from the *rebāb*. Some *rebāb* players spend considerable sums of money on having their instruments adorned in this way, sometimes to the extent of having gemstones inlaid. It would seem to be no coincidence that the actual designs used to decorate the *dutār* are skeuomorphs of those used for the *rebāb*, specifically extending the high prestige of the *rebāb* to the *dutār*.

(b) *Absolute changes in the skill of* dutār *playing.*

In this paper I have been concerned in some detail with a description of individual movements and sequences of movements employed in playing three types of *dutār*, and with comparing these motor patterns with those used for the *rebāb*. Qualitative differences between the performance techniques of these instruments have been noted. The question arises whether *quantitative* comparisons are also possible.

I claim that they are possible. It may be assumed that the physiological processes underlying skilled movement are universal human attributes. Likewise, it can be argued that the psychological processes that underlie human skill, such as perception, recognition, memory, attention, decision making, motor programming etc. are also universal. While movement in different cultures may have variable form, while different skills are practised and their goals and other performance criteria vary, the underlying processes of motor control and skill acquisition are, I believe, the same. If so, the principles of skilled performance adduced by experimental psychologists and other researchers working mainly in the West can be validly applied in the study of skills in other cultures. Undoubtedly, cultural and social factors are likely to have effects on the performance of a skill, and we have to contend with the complication of *relative* skill, as shown in the phenomenon of habituation, whereby the motor problems of one generation become the motor routines of the next. But for the analyst even to conceive of socio-cultural factors affecting the *apparent* difficulty of a motor performance to its practitioners implies that an absolute assessment of the skill can be made.

In theory, it should be possible to devise absolute measures of skill that have cross-cultural validity, but these analytical tools are not available at present and we cannot yet apply a numerical scale to variations in skill. It is, however, possible to apply certain tests to the kind of data described in this paper, which allows us to compare the skill of two performance techniques. Four such tests can be applied (and there may be others). They are:

I) *Intrinsic ease of movement*. Some movements are easier than others because they fall within the repertoire of innate movements or because they are part of the cultural repertoire of utilitarian, communicative or other movements.

II) *Degree of precision*. Motor learning is, in part, a refinement of muscle action and, in general, the more refinement required by the task the more skilled the movement.

III) *Number of stimulus and response categories*. It is well established that the time taken to process information in organising a response to a stimulus is proportional to the number of stimulus and response choices available.

IV) *Extent to which the sensory and motor patterns form good gestalts*, configurations which are quickly handled by the central nervous system.

In many situations these tests do not allow us to make an overall evaluation of two skills because trends measured in different tests are in opposite directions. For example, we cannot decide on this basis whether the *dutār* or the *rebāb* requires more skill. For playing classical music the left hand of the *dutār* is more difficult than the left hand of the *rebāb*, while the opposite holds for the right hand. We cannot measure whether what the right hand does is more or less difficult than what the left hand does. Only by measuring acquisition curves in a rigidly controlled experimental situation could we find out whether, overall, the *dutār* is easier or more difficult than the *rebāb*.

This kind of difficulty poses less of a problem when we look at the changes in skill associated with the change in the *dutār*, and this is because the trend on almost every criterion is in the same direction. The data show that *the process of change from a 2 to a 14 stringed dutār has been accompanied by an increase in the skill required to play the instrument*. The only exception relates to one aspect of right hand movement (see below).

The evidence supporting this statement may be reviewed as follows; the appropriate test is indicated in each case:

(i) The size and weight of the instrument have increased, and it is more difficult to hold in a stable position that does not impose undue stress on the left arm (Test I).

(ii) The increase in the length of the neck, and the imposition of extra frets, have both operated to increase the separation of the frets, and this means that the fingers have to stretch greater distances. Stretches that cannot be covered have to be obtained with hand movements. Thus the frets do not fall 'conveniently' under the fingers on the 14 stringed *dutār* as they do on the 2 stringed *dutār* (Test I).

(iii) Right hand techniques have been affected in several ways. The right hand technique for the 2 stringed *dutār* relies on movements that are more or less within the normal repertoire of body movement. The technique for the 14 stringed *dutār*, on the other hand, relies on an 'unnatural' hand position with the wrist sharply flexed (Test I). The strokes on the 2 stringed *dutār* need not be very exact, while the strokes on the 14 stringed *dutār* must be very precise (Test II). In playing the 2 stringed *dutār*, behaviour is not spatially differentiated, while on the 14 stringed *dutār* there is differentiation between hitting the 1st string and the parand string (Test III). The rhythmic patterns used in 14 stringed *dutār* playing are more diverse and produce more complex sequences than those used on the 2 stringed *dutār* (Test IV).

(iv) The 14 stringed *dutār* has more frets, providing a greater number of stimulus and response categories (Test III) and making the linear array a more confusing spatial framework in which to operate (Test IV).

(v) The 14 stringed *dutār* is played in many different modes, each of which utilises a particular combination of frets out of the total set of those available, whereas the 2 stringed *dutār* is played in only one mode (Test III).

(vi) The melodic movement of the music traditionally associated with the *dutār* is much more compatible with the linear array of the *dutār* than is the melodic movement of the classical music associated with the 14 stringed *dutār* (Test IV).

The one exception to the general trend concerns the fact that the 2 stringed *dutār* employs several different strokes with the fingers and thumb of the right hand, whereas the 14 stringed *dutār* uses only one kind of downstroke and one kind of upstroke. In this respect the 2 stringed *dutār* therefore has more response categories than the 14 stringed *dutār* (Test III).

The developments in the *dutār* have resulted in an extension in the use of the inherent potential of the human body to organise musical behaviour. Playing the 14 stringed *dutār* requires more of the sensori-motor resources of the player to be utilised than playing the 2 stringed *dutār*. Musical instruments are sometimes described (in the West) in terms of the challenge they present. Part of the reinforcement received by the musician is undoubtedly the sheer pleasure of indulging in a complex sensori-motor activity. Practitioners within a music culture often compete with one another, not only because the best musician is likely to get more work, but also because of social rivalry. This type of competitiveness is characteristic of *dutār* players, and is shown, for example, by a strong pride in being self-taught, in not being the pupil of another player. Through extending the body the 14 stringed *dutār* provides more scope for the development of individual virtuosity than its 2 stringed predecessor.

(c) *Social changes pertaining to* dutār *playing*

The increases in the skill of *dutār* playing can be correlated with other sociomusical trends that are involved with the changes in the *dutār*. Until recently, the distinction between professional and amateur musicians in Herat was very important. Professionalism was not simply a matter of making a living from music but of being a public, as distinct from a private, performer. Professionals were recruited from hereditary musician families, who kept their musical knowledge to themselves, as though it were a trade secret. For those who played in the Kabuli style this 'folk' theory of music included (among other things) a knowledge of the modes and rhythmic cycles, and of performance techniques for instruments who few but they knew how to play. Professional musicians trained their sons in music from an early age, exposing them purposefully to live music from infancy, and starting them playing instruments from the age of about nine.

In contrast, amateur musicians received no such training and were largely ignorant of the theory and terminology of the professionals. In the past, *dutār* players learned to play by watching others and by practising quasi-secretly. *Dutār* players usually started to play at the age of about fifteen. The changes in the *dutār* have been accompanied by the increasing tendency for *dutār* players to perform in public, and with the coming of the 14 stringed *dutār* a number of *dutār* players from amateur backgrounds have joined bands otherwise recruited from hereditary musician families. Through this contact, *dutār* players have become conversant with the musical culture of the professionals. It is not surprising that in this process playing the *dutār* has become a more skilful activity, for the information base that supports the skill has become much richer. Other significant factors are that the hereditary professional musicians set new standards of musicianship for the *dutār* players, and that, through becoming professional musicians, *dutār* players probably spend far more time playing than they would as amateurs, and thus have sufficient practice to develop their skills further.

(d) *Implications for a motor theory of music structure*

In this paper it has been demonstrated that a style of instrumental music is characterised by particular patterns of movement that are specific to that style. These motor patterns, which in a sense 'lie behind' the music, may be said to constitute the *motor structure* of that style. The motor structure consists of certain overlearned patterns of movement which are established by learning structurally related pieces of music that embody these patterns. It has been shown that there is a basic difference in the motor structures of two important types of long-necked lute. On the one hand, the Iranian *tār* and *sehtār* rely on the repetition of 3-finger 3-component patterns at ascending or descending sequences of positions; on the other hand, the North Indian *sitar* relies on continual ascending and descending movements, with one or two fingers. The motor structure for the *dutār* lies somewhere between these two extremes, and with changes in the morphology of the *dutār* its motor structure has become less like that of the *tār* and *sehtār* and more like that of the *sitar*.

Different motor structures show varying degrees of adaptation to the instruments for which they are employed. This analytical dimension is referred to as *compatibility*. An examination of two cases (Herati music played on the *dutār* and classical music played on the *rebāb*) shows how close this adaptation can be; which suggests that the music in each case has been shaped by the instrument. What we are dealing with here are body factors in the structure of music. It has been shown that traditional Herati music played on the 2 stringed *dutār* has a high degree of compatibility with the human body and to this extent may be regarded as a transformation or isomorph of body structure.

Once the player's motor structure has been established and a set of skilled movements acquired, it can function in a generative manner. It appears that in some way the motor system 'learns the grammar' of the motor structure and can generate 'grammatically' correct novel sequences with a minimum of conscious planning by the player. This capacity is exploited in certain styles of music that deliberately cultivate improvisation, such as North Indian classical music, but is probably operative in any instrumental musical skill. According to this model, musical creativity often involves using the 'motor grammar' to generate novel melodic sequences, some of which are then selected by the creating musician to form his new 'compositions'.

These results from the *dutār* indicate the need for studies of other musical instruments in terms of the relationships between the morphology of the instrument, the motor structure of the performance technique and the structure of the music played. Through this approach we may come nearer to understanding how elements of music structure are rooted in the human body. Music can be viewed as a product of body movement transduced into sound. It is subject to the same psychophysiological laws that determine the organization of body movement. Only with modern techniques of generating synthetic sound has music become uncoupled from the organizing structure of the human motor system.

ACKNOWLEDGEMENT

The research described in this paper was carried out in Herat in 1973/4 and was supported by a Social Science Research Council research grant to John Blacking and the author. I wish to thank John Blacking for encouraging my interest in the uncharted areas that lie between experimental psychology, anthropology and musicology, for generating many stimulating ideas, and for valuable comments on several drafts of this paper. Thanks also to my wife Veronica for assistance with preparing the manuscript. I am indebted to the Government of the Republic of Afghanistan for permission to carry out this research, and to the musicians of Herat.

REFERENCES

Baily, J.S. (1976). Recent changes in the dutar of Herat. *Asian Music*, Vol. VIII, No. 1 1976, pp. 29-64.

English, P. (1973). The traditional City of Herat, Afghanistan. In L. Carl Brown (ed.), *From Medina to Metropolis*. Princetown, New Jersey: The Darwin Press.

Farhat, H. (1965). *The Dastgah concept in Persian Music*. Doctoral dissertation, UCLA. Ann Arbor: University Microfilms.

Kaufmann, W. (1968). *The Ragas of North India*. Calcutta, New Delhi and Bombay: Oxford and IBH Publishing Company.

Szende, O. and Nemessuri, M. (1971). *The Physiology of Violin Playing*. London and Wellingborough: Collet's.

THE BENESH MOVEMENT NOTATION

Rudolf and Joan Benesh

Editorial Note: I had the privilege of meeting the late Rudolf Benesh in 1974, and was able to discuss with him his ideas for the wider use of his movement notation, which was originally devised for ballet and is now used by dance companies throughout the world. The notation has been used for recording ethnic dance, and I had engaged Dora Frankel to transcribe Venda dances from films. I have for long been convinced that some kind of universal notation is also essential for studies of nonhuman primates and for nonverbal behaviour in general. I thought it would be useful to devote time at the conference to a comparison of uses in anthropology of Benesh notation and Labanotation, which is used in Lange's essay, and of various idiosyncratic notations that have been devised so that they can be read without special training, such as Kubik's notation in his essay on *mukanda* dance movements. In the end, we decided to concentrate on Benesh notation, and Rudolf agreed to introduce it, with his wife, Joan, assisting. In the meantime Rudolf Benesh fell ill and was unable to come to the conference, and so Joan Benesh introduced the notation, assisted by Fiona Mitchell. My short acquaintance with Rudolf Benesh was such that I felt a deep sadness and sense of personal loss when he died only a few weeks after the conference. What follows is a summary of his notes and of Joan Benesh's presentation. Brian Street has kindly contributed a short account of his reactions to his discovery of the notation.

The Benesh Movement Notation, completed in 1955, is only an alphabet from which has arisen many 'languages' and research techniques. Since 1955, use of the notation has extended outside the field of ballet. In 1959, for instance, it was applied to work-study in collaboration with the Centre of Technical Studies for the Clothing Industry, Paris. And between 1964 and 1966, it was used in medical research in collaboration with the

Centre di Educazione Motoria, Florence.

The assumptions and approaches of the notation system are linguistic, cybernetic and ergonomic. The first axiom is a linguistic notation principle based on the fact that different forms of communication or language are perceived through one or another of our senses, and the need for the appropriate notation in all cases to be geared to the working of that particular sense organ. It must also have a logical structure derived from the particular medium. For example, speech is perceived through the ear and the logical structure is phonetic. Consequently, the only effective speech notation must be phonetically based. Music is also perceived through the ears and is therefore an oral art and its logical structure is based on pitch, interval and duration. Movement, on the other hand, is perceived through the eyes, and dance must accordingly be one of the visual arts, since contact with the audience can only be through their eyes. Therefore, a movement notation must be visual and based on the logical structure of visual perception.

As we record what we see, we must decide upon a view-point. This can either be from in front of the subject or behind him. From the very start, recording from behind the subject's back has been established as a convention, since this enables the writer and reader to identify themselves with the subject, which in itself has advantages in medical and scientific records and studies.

Supposing you were to ask a person to press himself face inwards against a wall and take up any position in which every part of him, so far as possible, is in contact with the wall. By marking the exact spots where his four extremities contacted it and also of a bent knee or elbow, he will be able to go away and return to the posture at will. The tilt and turn of the head and body could also be added by single lines. If now we calibrate the wall by drawing horizontal lines so that the top one is level with the top of the head, the next is at shoulder height, the next at the waist and another at the knee, the floor being the bottom line, we have the five-line staff and no longer need the subject to return to show us the position but can accurately read it when written. If the subject were to hold a piece of chalk in one hand and move his arm about, we would have a record of the path of movement on the wall.

The second axiom therefore arises from the first. It is the principle of linear perspective and utilizes the manner in which the eye projects three-dimensional objects on to its two-dimensional retina. This has long since been reduced to mathematics, and the notation simplifies the rather complex subject by the application of parallel projection on to a 'picture plane' which is the staff of the notation (see Figs. 1 and 2). This 3-D representation gives rise to the need for three basic signs (signs, not symbols), which plot points in three dimensional space on the matrix or five-line staff (see Figs. 3a and b). This utilizes simple Cartesian geometry with its x, y, and z co-ordinates for plotting positions. Since the limbs are of fixed length, the z co-ordinate is not required. That is to say that the distance in front or behind the body is conditioned. By adding to this three-dimensional representation of space the dimension of time, it is possible to trace the path of movement of an object. These four-dimensional representations are called movement lines, which by manipulation can reduce enormous complexity of movement to simplicity (see Figs. 4 and 5).

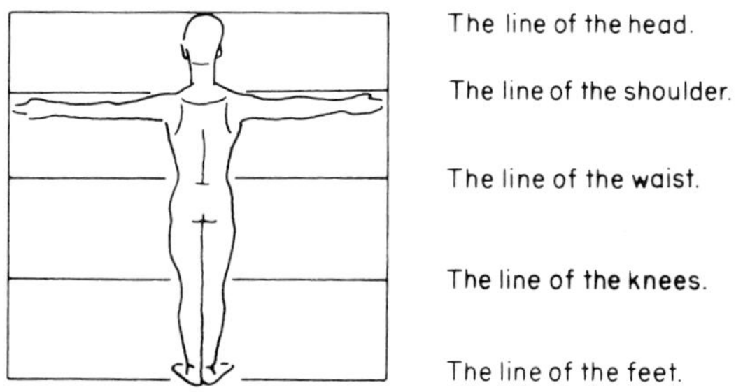

Fig. 1. *The five-line staff as a matrix for the human body.*

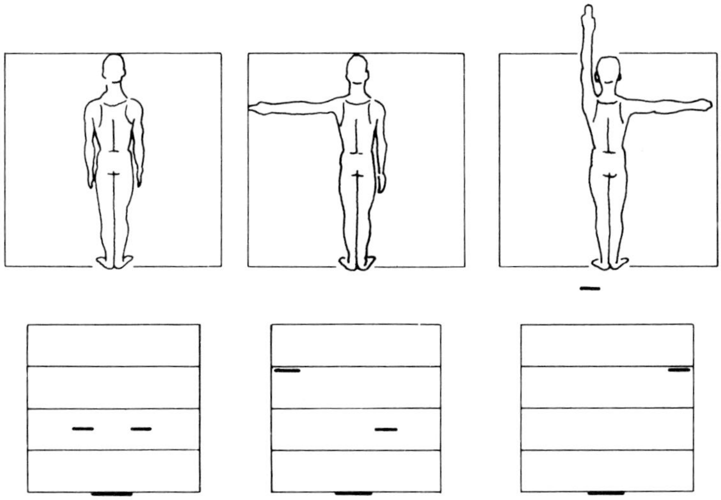

Fig. 2. *Positions of the feet and hands level with the body.*

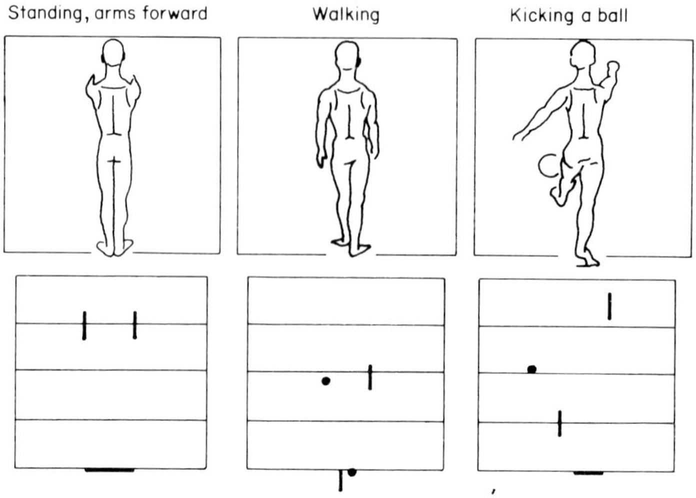

Fig. 3a. *The three basic signs. Horizontal strokes show positions of the hands and feet in the vertical plane of the body. Vertical strokes show the positions of hands and feet in front of the plane of the body, and dots, behind it.*

	Hands and feet	*Elbows and Knees*
Level (with the body)	—	—+—
Infront (of the body)	|	+
Behind (the body)	·	X

Fig. 3b. *If the limbs are bent, the positions of the elbows and knees are marked with crossed signs. This gives six signs.*

The premises are the three basic signs and the matrix or five-line staff. By the application of the concept that a cross equals a bend or tilt, and arising from this, that a tilt to the left means left and vice-versa, and the use of strict logic, consistency and ergonomics, the manipulation of signs gives rise to a vast vocabulary. One further concept introduces contact with objects and this gives rise to a vocabulary of literally many thousands of possibilities, with no addition of new signs.

However complex the situation, the resulting sign remains simple and necessarily tiny, since it still plots a position in space or on the body. The use of the five-line staff as used in music has many advantages. Writing and reading from left to right has its obvious ergonomic advantage and is also

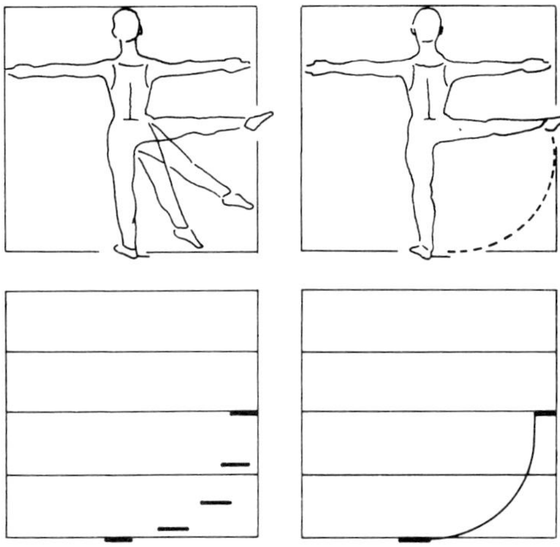

Fig. 4. *The movement line shows the leg rising to the side, summarizing a series of salient positions of the joint in the plane of the body.*

the generally accepted convention for the flow of time. This is of paramount importance where correlation with other notation, such as music, is required. Moreover, the five lines usually used in music give the half and quarters ideal both for visual purposes and for representing the human figure.

So far this deals with the notation or alphabet. The 'linguistic' structure of movement language is another fascinating subject with the correct 'spelling' to be determined for postures (words), steps (compound words), syntax (the order of signs, words, etc.) grammar, with its principles of redundancy

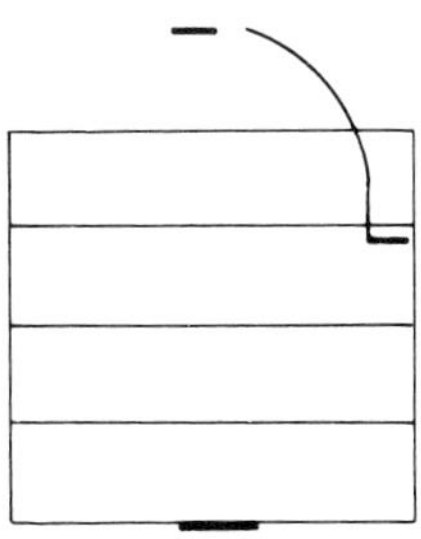

Fig. 5. *The arm moves from directly above the shoulder out to side, finishing just below shoulder height.*

and conditioning, and the resulting vocabulary which must be ergonomic - that is, easily written with a line in a cursive manner from left to right.

The signs, as so far described, account for two-dimensional paths of movement. It is a simple matter to incorporate the three dimensions of space. As we plot points in space, so these must also be plotted as points in time. The concept of rhythm in dance and movement generally is very different from that of music. Firstly, the scale is different. Furthermore, whereas in music one has a note or chord with duration in time followed by another, in movement this has no meaning whatsoever. In movement notation we have a time signature giving the metronomic pulse beat and the division into metric cycles, such as bars in music. Within this matrix the salient points in space are fixed as points in time, whether these be on the pulse beat or on a sub-beat. All this is basically simple, but it is because of the basic simplicity of the notation that the full complexity of movement can be dealt with simply (see Fig. 6).

Fig. 6. *Mazurka step showing a 'jump-line' indicating a jump in the air. The blocked in triangle above the staff indicates an audible sound.*

The notation is elastic. The normal ½ inch staff, giving a frame of 1/4 square inch, is as accurate as a trained recorder can cope with, and leaves a marginal reserve for all general purposes. For medical and neurological research and cerebral-palsy patients' clinical records, where details of body displacement as well as very precise differences of movements are important, a 1-inch staff is used. This increases the matrix area by four times and its volume by eight times without increasing the size of the signs used. This accordingly increases the accuracy by eight times. In work study, where details of finger, hand and arm movements are important, the use of a so-called waist or hip staff is used, according to whether the operator is sitting at a desk or standing at a bench. This makes use of the normal staff, but effectively increases the volume by eight times. These techniques more than cover the accuracy required, and no further modification has been called for, but it can be seen that by combining both these techniques an increase in volume, and therefore of definition or accuracy, could be by eight times eight, or sixty-four, times.

The notation also takes into account movement qualities as well as movement quantities. Qualities of movement are closely tied up with dynamics and interpretation. Much of the dynamic quality is intrinsic in the precise recording of the movement, its exact path, pace and change of pace, so that with a precise notation there is often very little to add. We found that during the course of developing a new movement language, we used a large number of dynamic and quality indications which with progress gradually fell away.

Even in noting the movement of spastic children, where one would expect a great deal, co-workers are always surprised by the lack of such markings. The reason also lies in the proper understanding of dynamics. The body is a very efficient machine, especially when used in dance and athletics. That is to say, it expends the minimum amount of energy for the maximum of movement. In a well executed movement, only that amount of energy is used to move the limb, body or object, no more and no less. If more is used in order to convey, say, emotional tension, when acting, or less in order to convey relaxation or floppiness, or the extent of spasticity or flaccidity, then the degree of these must be indicated. Floppiness in a movement, that is the relaxing of muscle tone, will result in a decrease of efficiency and a probable increase in energy used, like lifting a floppy rather than a rigid body.

As an objective movement notation, the system is universally applicable to all forms of movement. It can be used for recording details of facial expression, eye and finger movements, as well as movements of complex groups of people, stage patterns, teams, and work groups. It can record babies from birth, children, athletes, dancers, spastics, other primates, the multiple arms of Indian deities, and it can be adapted for animals.

We are often asked how much training is needed to become competent in the use of the system. This question cannot be directly answered because one is not taught notation as such but the 'languages' to which it has been applied. One does not teach a child the alphabet, one teaches him to read and write a particular language using the alphabet. The alphabet of the notation is very simple, the number of signs being basically three, from which all the others are derived. But this alone is of little use. A notator must be an expert in the 'language' he is recording and familiar with its style and conventions. Two basic movement styles in common use are ballet and modern dance (based on American modern dance), but others include jazz, folk, ethnic, historical, and Indian dance styles. Those concerned with noting movement in other contexts can soon learn special adaptations of the basic styles. (See Figs. 7 and 8).

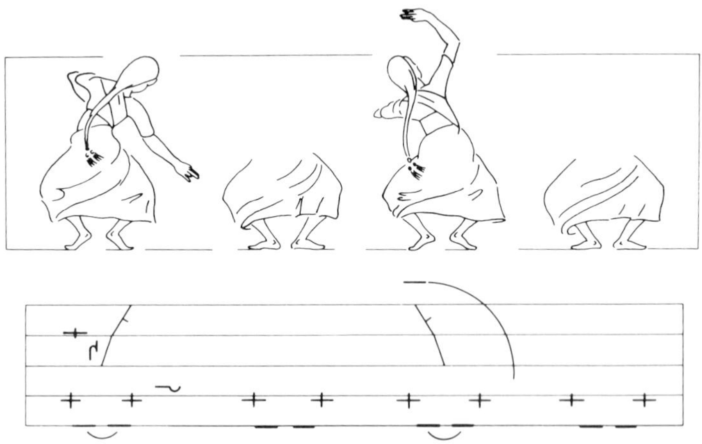

Fig. 7. *Part of an* adavu *from Bharata Natyam. From the opening position with right hand down by the right side, the dancer lowers her heels, then carries her right hand in a circular sweep to a position over her head. The small curve below the staff indicates a little jump in place. The slanting line in between the top line and the second line indicates an inclination of the head to the left, and that between the second and third lines an inclination of the top half of the trunk to the left. The curves added to the basic hand signs for the opening position indicate direction and flexion of the wrist.*

With the increasing availability of film, videotape, and other technical recording devices, one may also ask why a written notation system is necessary for recording and studying movement. In the fields of ballet and modern dance, there is no doubt about its uses, since the creation of dance takes place in two stages: firstly, the creation by the choreographer, and secondly, its interpretation by the dancers. No one would give a musician a disc or tape and say 'Perform this', since it would immediately restrict the creative act of interpretation by providing only one particular interpretation as a model.

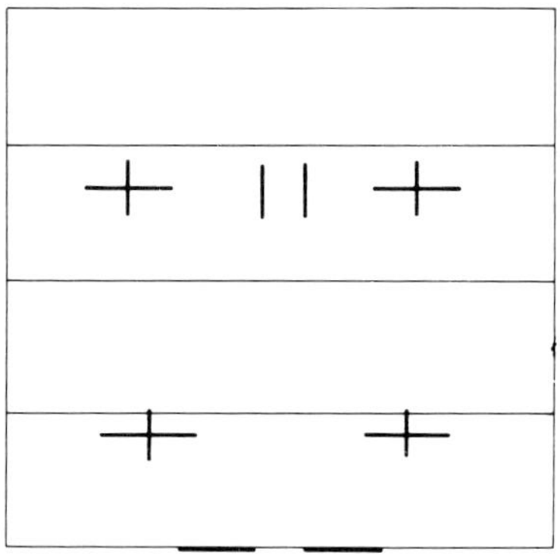

Fig. 8 *Bharata Natyam:* mudra *Sikhara.*

Film is indeed one means of storing visual phenomena, and in a sense it is no less subjective than movement notation, since its record depends on the film makers' perception of relevance and consequent use of angle, as much as notation depends on the notator's selection of movements to record. Film records should be taken together with movement notation, but the notation is particularly suitable for mental manipulation and analysis. Notation is the tool of the brain, and its efficacy in any research or study determines the progress of that subject. Developments in science, engineering and technology have only been made possible by their notation, mathematics, and the same is equally true in the case of music. Similarly, a computer can only work through a notation.

A completely objective redording is as impossible in notation as it is in film. The notator must decide what to put in and what to leave out. It is necessary to know the purpose of the recording, but initially, when noting a new style of movement, it is advisable to note as much movement as possible. For instance, a simple operation could be noted in terms of layout and general action, or it could be part of a psychological study in which the characteristics of the individual performer(s) are of prime interest. Thus, one may record on one staff details of movements of the head, direction of the feet, precise timing, and relationships with objects and spatial layout of the environment. On a second staff one records the more personal characteristics of the performer's action.

As a tool for the cross-cultural study of movement, the notation has unlimited possibilities of adaptation and use.

REFERENCES

Benesh, R. and Benesh, J. (1969) (C. 1955). *An Introduction to Benesh Movement Notation: Dance*. New York: Dance Horizons.

Causley, M. (1967). *An Introduction to Benesh Movement Notation*. Its general principles and its use in physical education. London: Max Parrish.

Hall, F. (1967). Benesh notation and ethnochoreology. *Ethnomusicology, XI* (2), 188-198.

Moyle, A.M. (1973). Sound Films for combined notation. The Groote Eylandt Field Project, 1969. *1972 Yearbook of the International Folk Music Council*, 104-118.

A COMMENT ON THE BENESH SYSTEM OF NOTATION

Brian Street

The lecture on the Benesh notation system appeared as though it was intended as a primarily technical matter and for this reason not many people attended. In practice, it was one of the most intellectually stimulating and exciting sessions of the conference. This stemmed from the fact that the system generated a whole series of questions about the nature of the body, its relation to the social context and the different presentations of it which more abstract and ambitious papers failed to achieve.

In order to be able to make the notations the observer is obliged to make certain critical distinctions between kinds of action and sources of movement; for example in a particular movement is the hip or the thigh the significant feature? In other words, the system is not just a neat way of 'describing' but a source of analysis that by its very concreteness and precision generates questions and problems that might be less clear were the movements being put into words immediately.

The system has many shorthand ways of presenting postures of the body and indicates quite clearly what it is possible for the body to do; the observer is assumed to be behind the actor so that if the actor has his arm bent at the elbow and his hand straight out in front, there is no need to record the hand - it can only be in one place; there is a certain path that can be traced for the hand according to whether the knees, trunk or arms are extended or bent, a concrete map of the physical possibilities of movement against which can be assessed those movements that (certain) actors most frequently select. The anthropologist using such a notation in the field is not only 'recording' activities in such a way that they can be reproduced and studied later: to be able to reproduce them correctly from the notation involves him in studying that activity more closely at the time, asking questions that he might otherwise miss, analysing aspects of body movement that might otherwise be taken for granted and so increasing his ability to relate them to wider categories of thought in that society. Such distinctions as that between 'upper' and 'lower' when applied to perceptions of the body might be more clear to the observer when reproduced by the notation system than in lengthy verbal description. One anthropologist at the conference had fifty pages of description of a dance which he considered this system could have dealt with more accurately in half a dozen pages, besides generating more questions about it.

The system has been mostly used by choreologists for recording dance movements that could be reproduced later, rather like musical scores. Since all of the choreologists are well acquianted with classical and modern ballet and dance conventions and thus 'know the language' in which dancers are expressing themselves, there is very little problem for them in setting down notations of these performances that others can read, recognise and reproduce accurately. When asked to record a dance movement in an anthropologist's film, however, the notators were at first producing different records and taking much longer than usual to record. As they got to know the society, under the instruction of the anthropologist, they found themselves asking questions about movement which were not part of their classical repertoire and discovering new principles to assess and record (an anthropological truism of course); in one instance the problem of differing records was resolved when it was realised that the hip and not the thigh was the essential source of the movement. The significance of this for the anthropologist, concerned with indigenous classification of the body, of movement and of 'social space', is obvious: it might provide him with a link with other categories that he learnt of in other ways, such as the distinction that Sutherland's paper notes between 'upper' and 'lower' parts of the body, or the relation emphasised in another between circular dance movements, mosque design and the central features of Islamic cosmology.* The fact that recording proved inconsistent without some knowledge of the social context demonstrated, on the one hand, that 'definitions' of dance are meaningless if they fail to take into account indigenous categories and the context; and on the other hand the necessity for anthropologists in the field to take more note of movement and physical expression than most monographs give evidence of. Since this was one of the 'practical aims' of the conference it was unfortunate that the point should have become most obvious in a paper which few members attended.

The system, then, has three major uses for anthropologists: it is a useful shorthand for what might otherwise take pages of writing; it provides a concrete picture of the constraints of the body against which selected movement in different contexts can be observed and analysed; and it generates questions, suggesting useful distinctions and criteria for distinctions and so stimulating analysis and leading to the formulation of problems.

**Editorial note:* Regrettably, this paper was not revised in time for publication.

ANATOMICAL CLASSIFICATION AND THE SEMIOTICS OF THE BODY

Roy F. Ellen

It is now generally accepted that in order to understand symbolic and more general semiotic systems it is necessary to go beyond them and investigate the classificatory structures in which they are set. In some cases the relationship between symbol and category has been worked out in detail, as exemplified in the now considerable literature on classificatory anomalies (Douglas 1966, 1973:113-67). In others the relationship is less well understood and systematic analysis is still in its infancy. In this paper I discuss the classificatory basis and some aspects of the semiotic use of that common source of potent symbols - the human body. Parts I to IV examine in general the classification of body-parts, while parts V to VIII are an attempt to relate such classifications to some consideration of their cognitive and cultural uses.

I

ANALYTIC AND SYNTHETIC CLASSIFICATIONS

In comparison with other fields in ethnoscience, the subject of anatomical classification has been neglected. Sturtevant (1964) makes no mention of it, and the exhaustive bibliography compiled by Conklin and his associates (1972) provides only a handful of items of direct relevance (Franklin 1963, Friedrich 1969, Lenormand 1950, Mahr 1950, Stark 1969, Werner and Begishe 1970). In fact, with a few interesting exceptions, the situation is much as Marsh and Laughlin described it almost two decades ago (1956:56), useful data being either scarce or non-existent. Considering the distinctiveness of such classifications when compared with those which have been subject to more extensive examination, and their relevance to the study of body symbolism, this is rather surprising. Papers by Lewis (1974) and Pekey (1975) are recent exceptions to the general lack of interest.

Those classificatory systems which have been subjected to the most intensive investigation are largely ethnobiological, and can be characterized as being composed of units arranged by a process of *aggregation* into units with still greater extension: categories are constructed and elements sorted into categories (Cole, Gay, Glick and Sharp 1971:59-91). Examples include systems based on a variety of classificatory principles: index, paradigms, trees, keys, typologies and taxonomies (Conklin 1964:39-40), and combinations of two or more of such. Studies that ostensibly concern the analysis of classificatory systems in general have tended to refer predominantly, if not exclusively, to classifications of this kind, particularly taxonomies (Perchonak and Werner 1969:229-37, Werner and Fenton 1973:543, 561). However, some classifications are not adequately described in such terms, either entirely or in part. Instead, the classificatory process commonly proceeds in an opposite fashion, through the *separation* of parts of a natural totality, or the de-composition into constituent parts of a system at a given level. This sort of operation, which Stark (1969:1) has termed 'partiality', is evident in human and non-human anatomical classification, plant morphology, topography and the classification of parts of cultural artifacts. Thus, relevant segregates might be 'man-arm-hand' rather than, say, 'man-mammal-vertebrate'.

This fundamental distinction between two classificatory procedures has not gone unrecognized, but its implications have been followed up with only the most desultory empirical research. Conklin (1962:131-2) phrased the difference as one between *part-of* (part-whole) classifications and those based on *kind-of* (class inclusion) relations, and following Nagel (1961:381-3) appreciated the kinds of ambiguities to which the latter can give rise (cf. Frake 1961, Berlin, quoted in Werner and Begishe 1970:250-5, Werner and Fenton 1973:561-2). A comparable distinction is made in structuralist theory between *structures* and *aggregates*, the former conferring on the whole overall properties distinct from the properties of the elements, and the latter being composed of elements that are independent of the complexes into which they enter (Piaget 1971a:6-10, cf. 1971b:169). All these dualistic distinctions indicate certain qualities of the classifications to which they refer. However, none of these terms conveys adequately and unambiguously the senses required. Consequently, rather than choose any one pair, I propose to term them *analytic* and *synthetic* classifications respectively, placing emphasis on process: on the one hand *analysis* (separation), on the other *synthesis* (aggregation). Among its several advantages, the emphasis on process avoids the misconception that the two types are quite independent. In fact, as is the case for the body, both may be present in the same arrangement. To summarize, the following pairs of terms emphasize different aspects of the same basic distinction:

part of	:	kind-of
structure	:	aggregate
concrete	:	abstract
analytic	:	synthetic

While it is inappropriate to explore these distinctions in any detail here, even less to discuss the range of other semantic relations which can be seen as derivatives (Casagrade and Hale 1967), it is useful to make a few points and map out some of the most significant coordinates of each column:

1. Conventionally, synthetic classifications are a means of distinguishing degrees of similarity and difference in natural species and other elements, the point of departure being plural clearly-bounded natural units. Analytic classifications are associated with the categorizing of items which are dependently (typically spatially) related, where the point of departure is a whole single natural element.
2. Synthetic classification proceeds through a process of aggregation (grouping together) of lower-order elements on the basis of the distinctive features of each. Analytic classification proceeds through a process of separation of parts of a single unit primarily determined by their spatially dependent relationships.
3. In synthetic classification the basic units are concrete entities, while higher order categories have no such existence and are artificial constructs. This, in part, accounts for the regularity and intellectual clarity which such classifications often possess: order can be imposed upon them. The lower-order units of analytic classifications exist only in relation to the whole of which they are part and usually have no existence by themselves; or their existence, apart from the wholes is of a different order altogether. However, unlike their synthetic counterparts, all levels and elements in analytic hierarchies frequently have a concrete existence.
4. The depth of synthetic classifications tend to be relatively shallow, but with considerable horizontal expansion. In analytic classifications, both vertical and horizontal expansion depends on concrete possibilities. Overall dimensions, therefore, tend to be more variable.
5. In synthetic classifications there appear to be certain restrictions on the number of unitary lexemes strung together in terminologically meaningful chains, whereas in analytic classifications it is possible to string together longer chains (Berlin in Werner and Begishe 1970:250).
6. In synthetic classifications, contrast between elements at

any one level is both essential and emphasized, though it may give rise to ambiguities. In analytic classifications, contrast at any one level exists but is often poor; in part due to the ambiguous status of link items between principal components, and in part due to the fact that 'levels' themselves are inadequately defined. Contrast operates more after the fashion of 'trees'.

7. Spatial terminology is generally absent in the constitution of most synthetic domains, other than in those specifically concerned with spatial phenomena. In analytic classifications, spatially-dependent terminology is important in the constitution of all domains.

These points, which relate primarily to folk-classifications, go some way to indicate the crucial importance of the constraints of material structure on the perception of forms classified analytically. Specifically, anatomical classification is structured through the geometry of the body: the constraints are somatic. A 'hand' does not have a separate existence from the 'body', whereas a 'cackatua' can easily and in a very real sense have a separate existence from the artificial category 'bird', or some differently defined higher order taxon. Moreover, a 'cackatua' can be placed in a slightly different category depending on various cultural and contextual criteria. For the Nuaulu of Seram in eastern Indonesia, the category *manwa* is not the equivalent of the English 'birds' or the Linnaean *Aves*, since the criteria for its definition are not the same, allowing for the inclusion of bats. Similarly, depending on the context, the cassowary is classified by the Nuaulu either as *manwa* or *peni*, a sacred category of game animals (Ellen 1972, 1975). The extent to which the cassowary belongs to the category 'bird' and its approximate glosses in other languages, is dependent upon the construction of the category and the distinctive features of the cassowary emphasized. The body, however, has its own concrete existence, not just a mental one, and the hand, according to all anatomical logics, is unambiguously classified as part of it.

The body provides the brain with a ready-made classificatory framework for its parts, though the operational image of the body may indeed be modified through the processes of cerebral organization and socialization. To this extent we might expect 'natural' analytic classifications to be less malleable than artificial ones, which are, after all, themselves cultural constructs. Some observational measure of this is to be found in Stark's consideration of the Nida-Conklin hypothesis (Conklin 1962:132, see also Frake 1961:121) - that there is a positive correlation between the lexical elaboration of a classification and its cultural importance. Stark found that this hypothesis was not borne out in her analysis of Quechua body-parts (Stark 1969:12), although the hypothesis would appear to hold for different levels and segregates within the domain (Franklin 1963:57). Nevertheless, both natural and artificial analytic classifications contrast with the fundamental mental construction of synthetic ones, however much the latter may be fashioned from distinctive features of species which exist in a concrete world, and however much the classificatory devices owe to the bricolage of cultural existence in nature. A cackatua remains a 'bird' just as long as the definition fits those properties of the cackatua which the classifier wishes to express. A hand remains a part of the 'body' since any change in the definition of the body is incapable of severing (metaphorically or physically) the association of hand and body. This may explain why isolated artisitc representations of body-parts sometimes appear odd and generate a certain unease until they become accepted in their new context. The difficulty arises in part from the unusual environment, but also from the fact of their being part of an unrepresented whole. The perception and classification of the physical relations between body parts is something that seems to occur very early in the development of a child[1] and although synthetic classifications also appear early, they tend to undergo rapid and more substantial cultural modification. Such a *gestalt* conception of the human body as an integrated whole has important implications for the study of its semiotic use.

THE RELATIONSHIP BETWEEN DIFFERENT CLASSIFICATORY PROCESSES

Despite the differences between the two types of classification briefly outlined, the relationship between them is not always clear at an operational level; both being observed with respect to the same group of elements and being transformable into each other (e.g. Stark 1969:14, footnote 6, Werner and Fenton 1973:561-2). Once constructed, synthetic structures can easily become reified as Lévi-Strauss has elegantly demonstrated for the category 'totemism' (1962). Such a practice is the source of many of the apparent anomalies in synthetic classifications. In Piagetian terms, an aggregate may appear to become a structure; the abstract is concretized. Even the very use in synthetic classification of the term 'kind-of relation' assumes that abstract categories have some 'concrete' existence, but this kind of reification is particularly apparent when classifications are written down, as in biological taxonomy.

A more relevant consideration for this paper is that the opposite may also occur. Thus an analytic break-down of the body into its constituent parts can be systematically subjected to recomposition. An important example of this is the process of establishing symmetry, so fundamental a feature of the body as both classificatory and symbolic system, whether this is in terms of entire body-halves or lower-level segregates (Needham 1973). Such a process of the analysis of a whole into its constituent parts and the subsequent grouping of like parts together represents the transformation of an analytic into a synthetic classification, from parts of a concrete structure to differently ordered aggregates of those parts. This transformation can be summed up diagrammatically in a simple version of Lévi-Strauss's totemic operator. Figure 1 illustrates such a transformation for two sets of binary discriminations, which also represent separate logical operations: the aggregation of elements into either 'feet' or 'hands' and into either 'left' or 'right'. The former operation necessitates only morphological discrimination of the items; the second, while it can strictly be achieved simply on a morphological basis, usually involves some recognition of body symmetry, the relationship of the parts to the whole, and their associated cultural connotations[2]. Thus, while the classification of body-parts is primarily (and of necessity) analytic, it becomes secondarily synthetic through reclassification of the parts so obtained.

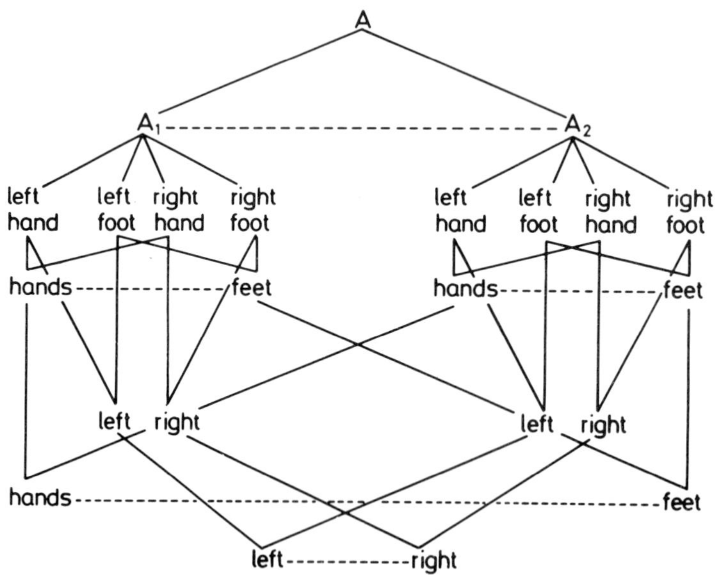

Evidence of synthetic operations in the classification of body-parts, however, goes beyond the simple symmetry just described, based on mirror images of body halves. It often involves groupings of a higher order, either as labelled categories ('limbs', 'blood-vessels', and so on), or as covert groupings with recognition of symmetry through the use of the same or similar terms for morphologically different parts at different levels of specificity, such as at the level of the entire limb or body half. There is generally a recognition of similarity between the upper and lower parts of the body, reflected in identifiable structural and terminological correspondences between upper and lower limbs. For the Kewa of the New Guinea highlands, we get the following terminological equations: 'ankle' = 'elbow', 'heel' = 'elbow' (different term), 'sole' = 'palm', 'toes' = 'fingers' and 'toenails' = 'fingernails' (Franklin 1963:61-2). Similar terminological parallelism based on the physical (and functional) resemblance of extremeties is on record for other languages (Mahr 1960:29-32, Stark 1969:8), and indeed appears repeatedly in many anatomical classifications. Additionally, there are instances where the labels for particular anatomical items are dependent on their superficial similarity, but where the resemblance is slight and not part of any general symmetrical pattern. For example, among the Aleut 'uvula' = 'little finger', 'little toe' (Marsh and Laughlin 1956:51).

All classifications of body-parts involve such operations, whether they are clearly related to the constraints imposed by visual perception of the totality, or whether they are not so constrained, as where true taxonomic relations are part of an overall analytic structure (Werner and Begishe 1970:253-5). Some analytic classifications of body parts *pre-suppose* an initial synthetic operation. Thus the statement 'blood-vessels are a *part of* the body', pre-supposes an operation in which morphologically similar items are identified as *kinds of* blood-vessels. Where the members are directly subordinate to the body, such synthetic classifications can be simple (Franklin 1963:57); alternatively they may be more complex, as in the division of the category 'blood-vessel' into 'veins', 'arteries' and 'capillaries'. Further operations may modify the analytically derived category: thus we have 'pubic hair', 'axillary hair', and so on (Franklin 1963:57). The extent to which classifications of body parts arrived at by analytic procedures may be so modified, varies from one language to another, and often such elements in one language remain unmodified elsewhere. Among the Nuaulu there is no term as abstract as the English 'hair': instead, there are separate terms for body hair (*hunua*) and hair on the head (*hu'au*) (cf Mahr 1960:8-9). Here there is no systematic modification of the kind mentioned.

Since anatomical classification proceeds not simply on the basis of a single body (the self or otherwise) but through observations of and interactions with a range of other bodies, the interrelationships can understandably be even more complicated than so far described. Thus, from the grouping of like parts of a single body, it is a logical next step to move onto the grouping of similar parts on different or plural bodies. Here the classification of hands of a single or plural body into one class is a procedure of the same kind as grouping cackatuas and cassowaries into the category 'bird': the same synthetic mechanism is at work However, while particular varieties of the species *Casuarius* may differ so much from a type specimen or the taxonomic definition of the species that they be considered in some way anomalous and uncertain members of the species, this does not happen to the same extent where human bodies are concerned. Genetic variation among *H. sapiens sapiens* may often give rise to symbolic usages and classificatory difficulties, but this is generally restricted to traits such as skin colour and hair form. In these cases ambiguity does not arise from the classification of various parts of a single whole body, but rather stems from the assignation of a series of individual bodies into classes, the boundaries of which are to a greater or lesser extent (but always to some extent) defined artificially and socially. In this sense, taxa used to classify human beings, such as races (Harris 1970), groups based on intelligence or even sex (Omerod in Douglas 1973:115-7), are comparable to those employed to classify other natural species.

On the other hand, in analytical classifications the ambiguity arises from the uncertain limits of otherwise unproblematic body-parts. Thus areas of transition, usually joints of various kinds are often seen as links between principal body-parts. In his analysis of Kewa anatomical classification, Franklin states that 'neck' and 'stomach area' are perceived as links between the three main divisions of the body, such that the connected series of parts may be written as 'HEAD-neck-UPPER TRUNK-stomach area-BELOW WAIST' (1963:55-7). The fact that elements at a single level in a classification can have different statuses has important implications for understanding the classification of the body and other analytic structures.

The two types of classification discussed are closely interrelated, and while the unique features of body-part classification stem from the primary analytic operation, it is possible to identify readily numerous examples of secondary synthetic operations which have much in common with taxonomies, including the property of reification of abstract categories such that their parts can be re-classified *as if* the categories were concrete totalities. If we now refer back to Figure 1, we see that it constitutes a model of conceptual apparatus in which, to paraphrase Lévi-Strauss, multiplicity can be transformed into unity, identity into diversity, and *vice versa* (Lévi-Strauss 1966:151-3). But what is important as far as a theory of the evolution of classification systems is concerned, is that the perception and classification of the physical relations between body parts appears to occur early in the development of a child (Fantz 1961), such that the establishment of analytic part-whole relations precedes class inclusion, both logically and in terms of socialization.

III

VARIATION IN ANATOMICAL CLASSIFICATION

While attributing to the body some kind of primacy, in emphasizing its *gestalt* properties and the effect of material constraints (as the assertion of pre-eminence of analytic procedures implies), it is, however, necessary to underscore and map the significance of any inter-cultural or sub-cultural variation that may exist. This is in order to avoid both the more crass material reductionism and emphasize the actual dimensions of the variation that in fact occurs.

Cultural variation in the classification of body-parts has already been documented, and it occurs in terms of the structure of classification systems, their lexical elaboration, and the semantics of terminology employed. Variation of one kind is frequently related to variation of another kind. For example, differential lexical elaboration at various levels has certain implications for overall structure. In fact, lexical elaboration is often at the root of more readily observable structural variation. Simple and useful indices of variation are understandably difficult to find; but some attempt has been made to calculate the total number of anatomical discriminations made in a particular culture as some reflection of complexity. For example, Marsh and Laughlin (1956:58-76) have distinguished terms for 330 such items among the Aleut Eskimo. The reasons for such differential elaboration can sometimes be explained in terms of the Nida-Conklin hypothesis, that the elaboration of terminologies (or portions of terminologies) is directly proportional to their cultural importance (but see section II above). Such importance will depend on ritual contexts (oracles, mummification, embalment, cannibalism, head-hunting, dismemberment), economic contexts (knowledge of animal anatomy through hunting, husbandry and butchery practices) and therapeutic practices (knowledge gained through autopsy, dissection, obstetrics, acupuncture, surgery, bodily manipulation and massage). Some workers have suggested some quite specific correlative hypotheses in this area. Thus, Marsh and Laughlin suggest that there is a general correlation between therapeutic means that focus on non-corporeal practices (e.g. shamanism) and the underdevelopment of systems of anatomical classification. They attribute Aleut knowledge in this sphere to their attempt at a rational practice of medicine (1956:39). Even the most superficial studies of this kind indicate that certain anatomical units are not universally singled out for labelling, while other units unfamiliar in terms of our own classifications are prominent in those of other cultures. Thus, the Nuaulu terminologically emphasize anatomical points and units which are not so recognized in English folk-classifications, such as the vertex, the highest point of the cranium (*huhui*), and the centre of the chest or xiphisternum (*hikai*). I have already pointed out the differentiation with regard to hair.

Variation can be equally as striking at the level of higher-order units. This can be illustrated with reference to three arrangements for the classification of the lower limb, although as it happens, it could just as well be the upper limb (cf Mahr 1960:31). The first arrangement is the situation where the term for the entire limb is used polysemously to refer also to the vertical portion of the limb in contrast to the horizontal part, as in English usage ('leg' = entire lower limb, but 'leg' contrasts with foot'). The second arrangement represents a situation which may be even more common, whereby no term approximately equivalent to the English 'foot' or the French *pied* exists, and where instead this part of the lower limb is included as a verbally unspecific part of a larger unit referring to the entire lower part of the limb from the knee downwards. Thus in Nuaulu usage *airiatua* = entire lower limb, and *atori* = lower limb from knee downwards and foot. The latter element is in a rather ambiguous position of contrast with individual items in the upper part of the limb. The final arrangement represents the situation found in some other Malayo-Polynesian languages, where the term for the entire limb is used polysemously to refer to both the limb as a whole and the horizontal position, which may or may not be in contrast with other elements (e.g. Malay *kaki*). Whether this represents a situation of true polysemy or whether it is simply that the equivalent of foot is not lexically indicated, must be carefully worked out for each case.

At the semantic level variation may occur, for example, in the extent to which terms are descriptive rather than specific. The fact that many folk anatomical classifications contain more descriptive than specific terms is not necessarily of importance. This is true of scientific anatomical terminology, although operationally each term is treated as though it were a specific one, either through ignorance or convenience (Marsh and Laughlin 1956: 50-1). It is also true that often expressions may be specifically invented by linguistic informants, but, as Marsh and Laughlin point out, from the point of view of synchronically studying their anatomical knowledge this does not make any difference, since their ability to make neologisms may merely indicate their capacity for recognition and the flexibility of their classifying systems. The difference between specificity and simple description is often a question of frequency of use: descriptive terms rapidly assume a specific status if they are used frequently (e.g. ear-lobes, eye-ball).

IV

OVERLAP BETWEEN CORPOREAL AND NON-CORPOREAL TERMINOLOGIES

The terminology of anatomical description clearly occurs in contexts other than in reference to the human body, and this is an extremely common phenomenon. The recognition of this fact is by no means novel (*ibid* 38). Correspondences of this kind are well documented with respect to non-human animals, plants and topography, but they are equally common with respect to the form of human artifacts (e.g. *ibid* 38).

In all languages terms used for parts of the human body are also used for non-human animals. The similarity in both the structure and terminology of the classifications is obviously related in many cases to morphological similarity. However, the extent to which we may assume that the classifications for other animals are derived from the human model, both terminologically and in terms of the morphological and functional relationships between parts is more awkward to demonstrate empirically (in either logical, operational or evolutionary terms). Nevertheless, the very fact that human beings perceive and think anthropocentrically in relation to the non-human universe, together with the demonstrable elaboration of human anatomical classification compared with that of other animals, suggests that, generally speaking, the human body is the primary model in both an evolutionary and logico-operational sense. Abstract references to anatomical terms invariably assume reference to the suggested human body part, while the same terms may be used for even the most remotely human forms in zoological terminology, as in the *pseudopodium* of the amoeba, the *cilia* of other protozoa and the 'foot' of the lowly chiton. (It might be noted that this kind of expression of cross-specific identity between body-parts and the relations between them generates the problem of the definition of the unit. The use of an anatomical term in more than one species, or even individual, transforms it from an analytic unit to a synthetic category.)

This is not to deny the existence of feedback relationships, which may take several forms. Some anatomical terminology appears to be derived from the cultural function of the parts, as in 'palm' = place of stone flaking (*ibid* 50). Perhaps more important, a knowledge of animal anatomy and physiology both in folk and western sciences, leads to a more sophisticated understanding of the structure and function of the human body: comparative zoology is an aid to human biology (*ibid* 40-1, Mahr 1960: 2). Thus, while in the last instance the human body is the reference structure for the classification of animal parts, and most terms are *expressive* transfers, there is an on-going dialectical relationship between what is known of animal forms and what is known of man.

Particular attention has been paid to the animal domain, since here the classificatory correspondences are so much more direct, but much the same remarks can be made for other domains. Consider, for example, the wide range of anatomical terms used in the English language to express topography, from phrases as simple as 'the foot of the hill' to more colourful expressions of the order, 'Cambridge Gulf is the arsehole of the world and Wyndham is four hundred miles up it'. That such terms are essentially corporeal ones used in a non-corporeal context would, I think, be an uncontroversial conclusion. Exgressive transfers of this kind are found widely and in a variety of domains (Stark 1969:9). Thus Fox, in an essay on Rotinese metaphorical language, provides anatomical-botanical equations of the order skin=bark, teats=sprouts, buds; coconut shell=human skull, and suggests a definite tendency towards ingression (1971:240). Similar examples are not difficult to find. Thus Stark records the equation knee=hill for the Quechua, stressing the fact that it is the usage for knee which is in fact derivative in this case (Stark 1969: 9). The situation may be more complicated than this in that a term may have a primary corporeal referent, a secondary non-corporeal one, and an apparently tertiary referent which seems to refer to a further body-part, or an allusion to a part. This is the case for the term *para* in the Tarascan language of Southwestern Mexico (Friedrich 1969:4):

para

animal, human back	A	B outside, exterior of roof
penis shaft, back as in 'behind one's back'	C	B

More common than this kind of practice is the use of suffixes in complex and compound lexemes for human body-parts to involve general topographical terminology, as in Quechua *maki-pampa* (hand-field) = palm of hand (Stark 1969:9). In fact, it is a likely and testable hypothesis that while single lexemes for body-parts are transferred egressively to the topographical domain, suffixes in complex and compound lexemes seem to be ingressive transfers. Much the same can be said for the relationship between anatomical terminology and parts of cultural artifacts.

From the remarks made above it will be clear that there are terminological and classificatory correspondences between a number of domains, all of which can be characterized by their concern with spatial relationships. It is also clear that the relationships, the nature of transferability, and feedback, are complex, such that in logical, operational and evolutionary terms the classificatory structures are, of necessity, closely interrelated. The problem is whether the human body is the primary referential set, other sets of referents simply representing metaphor-like extensions on the basis of underlying similarity; or whether all referential subsets are coordinate, all being equally characterized by the underlying abstract definition referring to spatial qualities (exterior/interior, above/below, and so on). In the latter 'theory of coordinate referential subsets' only a minority of recognized and labelled body-parts actually correspond to irreducible forms, the majority being morphologically regular derived nouns that are products of a root plus a complex suffix and nominalizing suffix (Friedrich 1969:4, 6-7). Werner and Begishe stress the importance of spatial addenda in defining foot parts, but here they are qualifiers at a binomial or trinomial level (1970:249). Such data suggest that none of the body-parts are primitive, but are secondary, usually labelled on the basis of their features as products of a root plus a spatial suffix, leading to the conclusion that the body is not the basic figure of orientation. However, as Friedrich points out, it is with the domain of the body that the degree of specificity and of overlap between referents is unquestionably the greatest. Moreover, terminologically non-corporeal elements occur in the suffixes of binomial composite lexemes, and not unitary ones (Werner and Begishe 1970:248). Friedrich (Friedrich 1969:9) illustrates this point effectively in relation to Tarascan:

> ..... 18 suffixes and 36 other body parts are covered by various extensions and specifications... 195 maximally productive roots which can 'take' all or nearly all the complex suffixes, form far more stems with reference to the body than to any other type of denotation... children apparently learn body-part meanings in a fairly complete way before they learn the other complete sets... the speaker's own body presumably enjoys some sort of real psychological primacy... 14 of 18 more or less corporeal suffixes for various parts of the genitalia is a symbolic condensation of this primacy...

For Friedrich, then, body-parts constitute a sort of *primus inter pares*, a physical model for the classificatory universe, although the lexemes are clearly re-interpreted in terms of their basic common features, being at once physiognomic and geometrical. Typically, it is the body part which motivates most innovation involving the term (*ibid:*9), while spatial terminology is important in classification and denotation of anatomical terms (Werner and Begishe 1970:34). Thus socialization, logical and operational evidence of the anthropocentric and egocentric nature of classificatory behaviour, argues strongly for some priority of the anatomical domain in both operational and evolutionary terms.

V

THE BODY AS A MEANS OF COMMUNICATION

If, as Mauss has remarked, Man's first and most natural instrument is his body, his first and most natural technical object, and at the same time technical means (Mauss 1973:75), it is also his first and most natural classifier and source of symbols, means of verbal and nonverbal communication through its effusions, movements and spatial orientation. The body has always been the subject of investigation and considered an important means of metaphorical expression and symbolic communication, transmitting 'information for and from the social systen of which it is part' (Douglas 1971:387; see also MacRae 1975). Such information may be transferred directly through the physical form of parts, their movement, the solid, liquid or gaseous substances they void, or by transformation and modification through mutilation, ageing, growth, decomposition, concealment, and adjustment. Indeed, the organized experiences associated with the perception and assignation of meaning to one's own body and the idea that a person's body is a psychological object that intrudes into social relationships and cannot be escaped, is an inevitable accompaniment of awareness.

The idea of the crucial pre-eminence of the body is stressed in all disciplines concerned with social communication: in philosophy (Spicker 1970), and in psychology in connexion with studies of art and perception and the work of the psycho-analytic school. For Jung, 'the symbols of the self arise in the depths of the body and they express its materiality every bit as much as the structure of the perceiving consciousness' (Jung 1958:138). In anthropology it has, in more recent years, been associated with the work of Mary Douglas, for whom it is a form for the expression of particular patterns of social relationship (Douglas 1970: xiii). Quite apart from its contemplation in these disciplines and in the arts, it has also been the subject of other studies encompassing a wide range of concerns. The use of the body, corporeal parts and terms in a general metaphorical or metanymic way, as well as for the ultimate life-symbols, has made body symbolism a central theme in the academic discussion of the subject. This is due partly to the ever-present temptation of reducing the social and the mental to some kind of physical reality, and partly to its 'autologous' character, that it is the scientific study of ourselves (Firth 1973:204).

Firth has suggested that we can distinguish at least four kinds of body symbolism (Firth 1973:226-8): communication through bodily action, the body as a set of abstract constituents, the treatment of social units in bodily terms, and the use of bodies or their parts actually or iconically. Here I shall primarily be concerned with the third of these categories. Furthermore I shall not directly concern myself with the transformation of body symbols into items associated with the body (e.g. clothes), which in themselves represent a rich area of research. Rather, I concern myself with the visually perceived and classified parts of the body form; not only the 'flesh, blood, breath and orifices, ingestions and excretions' (*ibid* 59), but also its organs, masses and protuberances.

VI

THE METAPHORICAL USE OF BODY PART RELATIONS

There is no shortage of ethnographic literature demonstrating the metaphorical use of the relations between body-parts, but on the whole it has not received the same attention as that paid to 'master symbols'. Here I concentrate on the use of anatomical terms to illustrate social and, sometimes, physical relationships, and *vice versa*, particularly drawing on the rich material contained in Griaule's *Conversations with Ogotemmêli* (1965) and on some sources from eastern Indonesia. In these cases we are dealing with metaphorical uses of the kind mentioned in Part IV, although here the term 'symbol' might seem more appropriate, since it stands for a complex non-material set of ideas. In looking at usages of this kind we might recognize a continuum from relational correspondences between different types of material phenomena, where one set of relations can be used to express the other, to complex relational correspondences that are predominantly social and non-material and where individual metaphors or relations between them convey a much greater cognitive load of a more abstract and less concrete kind.

Let us first look at the simple correspondences between body-parts and artifacts. Metaphorical correspondences of this kind usually involve some generalized recognition of material similarity. Such usages are found widely in the central Moluccas of

Indonesia (Jansen 1933). Among the Nuaulu, payments for lifting clan-specific dietary taboos on affines involve, for example, the following correspondences:

10 *kondai* (women's hair pins)	=	ribs
kain timor (hand woven cloth)	=	skin
long machete	=	spine
2 rings	=	eyes
5 bracelets	=	bones of arms and legs
1 *hoho* (large brass bell)	=	heart
1 large red porcelain plate	=	head
1 necklace	=	intestines

A similar set of correspondences exists for homicide payments. In the context of making ritual payments, but not necessarily in any other, the relationship of, say, rings to machete is the same as the relationship of spine to eyes. Although there is a concrete relationship system for body parts, this is not so for artifacts, other than what is the result of cultural convention. Consequently, the relational correspondences between the various artifacts is secondary and derivative, unlike the situation where there are two series of material items each with clearly defined and unalterable relationships; one is concrete both in terms and relations, while in the other, though the items are concrete enough, the relations are quite abstract. In Figure 2a, the abstract relations in the series B are defined totally in terms of series A. Because $A_1 = B_1$, $A_2 = B_2$ and A_1 is in a material relationship to A_2, therefore B_1 is in a corresponding relationship with B_2.

Figure 2

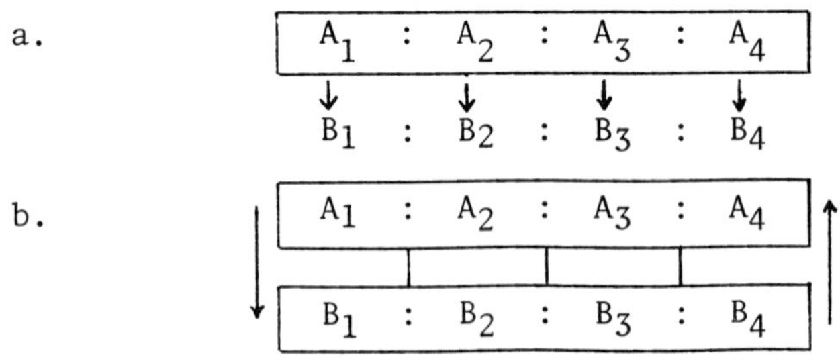

In the example mentioned in part IV, both series are composed of concrete items and concrete relations. This situation is illustrated in Figure 2b. $A_1 : A_2 :: B_1 : B_2$, and so on. This is well exemplified in the Dogon series relating parts of a granary to parts of a woman, the parts of the house compound with parts of the body and the village explained in terms of an extended anatomical metaphor (Griaule 1965:94-7):

(i) The Dogon granary is interpreted as being:

> like a woman, lying on her back (representing the sun) with her arms and legs raised and supporting the roof (representing the sky). The two legs were on the north side, and the door at the sixth step marked the sexual parts.... The granary and all it contained was therefore a picture of the world-system of the new order, and the way in which this system worked was represented by the functioning of the internal organs (p. 37).

Here we see the connection between anatomical classification, the metaphorical use of the relations of body-parts to each other, symbolically connected with grain, the most important Dogon food and principle of life-breath.

(ii) With regard to the house form, Griaule equates the vestibule with the male and the large central room with the female; the outside door represents the male sexual organ. The store rooms on either side are equatable with her arms and the communicating door (p. 96). A similar theme is further explored in the third example from Griaule.

> (iii) But a house of this sort is only one feature of the village. 'The village', said Ogotemmeli, 'should extend from north to south like the body of a man lying on his back.... The head is the council house, built on the chief square, which is the symbol of the primal field (p. 96).... To the east and west are houses for menstruating women; they are round like wombs and represent the hands of the village. The large family houses are its chest and belly; the communal altars at the south of the village are its feet.... The stones on which the fruit of the *Lannea acida* is crushed, placed in the centre of the village, represents its female sexual parts. Beside them should be set the foundation altar, which is its male sex organ; but out of respect for the women this altar is erected outside the walls.
> (p. 97, see also Fig. 8, p. 95.)

In these cases, socially important ideas are expressed by relating human anatomical nomenclature to parts of culturally created artifacts. In other instances social relations are not simply implied by two corresponding series of concrete sets of relationships, but are rather expressed *directly* in terms of the anatomical metaphor. A simple example of this is demonstrated in correspondences drawn by the Nuaulu between digits and relationships between primary kin:

a. hand:-	thumb	:	first finger	:	second finger	:	third finger	:	fourth finger
	mother	:	first son	:	second son	:	third son	:	daughter

b. foot:-	big toe :	first toe :	second toe	: third toe :	fourth toe
	father :	first son :	second son	: third son :	daughter

Thus as first finger is to thumb so eldest son is to mother, as fourth toe is to big toe so daughter is to father. It is significant that toes and fingers are used metaphorically in similar ways. As we have seen, they are anatomically and logically similar; as fingers are to the hand so toes are to the foot; there are five fingers and five toes and structurally there are physical resemblances between them: skin surface patterns, joints, nails, bone structure.

The same kind of situation has been demonstrated in more complex ways for other cultures. For example, the Dogon use anatomical relations (limb parts, joints and fingers) to express marriage relations.

In the same category as the Dogon examples must be assigned those instances where local political groups or larger entities are expressed as body parts of a single whole, and related metaphorical usages, such as those evident in terms such as the 'body politic'. Such usages stem directly from the characteristics of the body as a system of part-whole relations, of analytic classification; its properties are such that it is an excellent means by which to express the relation of the individual to the group. If the body is society, then its parts represent the parts of society[3] (Douglas 1971:387, Lévi-Strauss 1966:168-9, Perey 1975). Illustrative of these uses are Jansen's description of the Ambonese conception of local groups (*uli*) as bodies, and individuals and groups as specified parts of such bodies. Here the group that corresponds to the head usually has some kind of political hegemony and historical primacy, other villages both being historically derived from it and defining their relationship to it in anatomical terms. As with the Dogon, anatomical symbolism in general is an important means by which the Nuaulu describe the relations between the elements of their cosmos and social organization (c.f. Jansen 1933:457). A similar kind of situation is referred to by Steiner for parts of Polynesia, where 'parts of the body formed a fixed hierarchy which had some analogy with the rank system' (Steiner 1956:45-6). However, his ethnographic sources are unclear.

Given metaphorical correspondences between two different series, where these may be concrete, abstract, social or combinations of the three, the next logical step is to the expression of each through the mediation or addition of a third. For example, certain Nuaulu familial relations which may be expressed in terms of the relations between digits, are themselves expressions of items and relations in a third series, this time composed of other body-parts. This means that it is possible to construct the following anatomical equations: head = thumb, first finger = right hand, second finger = left hand, third finger = right foot, fourth finger = left foot. Although I would not wish to push this form of classificatory cross-referencing too far, its logical consistency in terms of some dominant Nuaulu dualisms can be seen in series read horizontally from the paired items given, such that, for example, eldest son = first finger = right hand = male.

These patterns are found in involved forms in Dogon ideas, involving cross-linkages between different but parallel material and social series. Here body-parts of different orders (organs, limbs, digits) may both mediate abstract and social relations, and be mediated themselves by other sets of correspondences, such that there is a complex set of cross-references between things so apparently diverse as granary parts, living beings, constellations, drums, language, techniques, numerals and so on. These associations are illustrated in Figure 3, indicating (a) some of the sets in terms of their relation to actual body geometry and (b) an abstracted list of some formal correspondences in sets. Things become more abstract still in, say, a consideration of the metaphorical mathematics of finger symbolism, where the sum of the numerals for each finger (or, in the case of the third and fourth, two fingers) is always 9; a figure equivalent to the head and signifying chieftainship (Griaule 1965:54).

All the sets of correspondences discussed are concerned with spatial relationships, since this is the idiom of expression. In some cases the space is abstract, in most it is social. All the abstract relationships are in essence the description of one set of spatial relationships in terms of another. Since at the level of classification there is a close relationship between conceptions of non-corporeal space and parts of the body (see Part IV), this is by no means unexpected. However, while at the level of the model we can say that items and relations in such series have an independent existence, it is possible, even probable, that metaphorical usage leads to some modification, usually in the more flexible series of a pair.

Figure 3

(a)

	Right female ancestors even numbers		Left male ancestors odd numbers		
upper	4	arm	arm	3	upper
lower	8	forearm	forearm	7	
upper	2	thigh	thigh	1	lower
lower	6	leg	leg		

(b)

	1	2	3	4	5	6	7	8
laterality	right	left	right	left	right	left	right	left
body-half	below	below	above	above	below	below	above	above
limbs	upper	upper	upper	upper	lower	lower	lower	lower
organ	stomach	gizzard	heart	little liver	spleen	intestine	great liver	bladder
fingers	second	fore	thumb	third	little	thumb	fore	second
sex	male	female	male	female	male	female	male	female

Such changes make the metaphorical relationships involved seem even more appropriate; as for example in the rearrangement of village components to mimic anatomical relations. So the formative processes are two-way: there is cognitive feed-back. The body is both symbol source and symbolized, a thing classified and a classifier of things, signifier and signified. In the same way as there is a feed-back relationship between the classification of body-parts and the use of the body to classify other things, so there is a corresponding relationship between the symbolism of body parts and the use of the body to symbolize other things. There is an indissoluble and dialectical link between the two.

VII

CLASSIFICATION OF BODY PARTS AND THE ANALYSIS OF SYMBOLS

In this paper I am concerned with the body as a communicative structure, as a source of a wide range of semiotic usages other than the specifically symbolic. In many ways the separate analysis of what are sometimes arbitrarily seen as symbols is a false point of departure. In Sapir's terms, condensation symbols should be seen in the context of a large collection of related referential ones, which cognitively operate at a simpler level. The Dogon case is a fine illustration of the arbitrariness of the distinction between sign and symbol and between the levels of classification and symbolism. Here, while body relations are being used as simple metaphorical devices to 'objectify', 'concretize', and discuss social relationships, they frequently transcend the level of the mnemonic and become part of a complex symbolic system in which the relations of body parts are an important mediating feature, conveying information about social relations and ultimately representing the most important material recognition of social reproduction. This process can be represented as a notional evolutionary and logical continuum between a *simple* pole representing the classification of the concrete and with a single semantic component, and a *complex* pole representing the classification of the abstract with a multiple semantic component (Figure 4). In order to control and handle the latter end of the continuum there is a constant attempt to explain the abstract in concrete terms, to contain and order the non-material in terms of the material. The properties of symbolic systems are those of classificatory systems. An analysis of body metaphors and symbols conducted in the context of overarching ideas of classification allows for a more systematic and therefore comprehensive approach than the narrow atomistic concentration on particular symbols. It also demonstrates the extent to which classification is both a logically necessary condition for symbol selection and usage and in a dialectical relationship with it, such that one cannot be satisfactorily investigated without the other. Consequently, not only is classification an integral part of the study of symbolic and metaphorical systems, but symbols should be analyzed dependently and from the same point of departure.

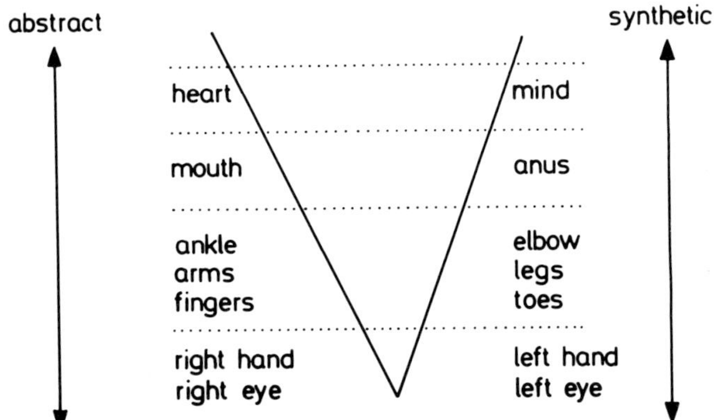

In their different ways, both Lévi-Strauss and Victor Turner have stressed the importance of looking at the structure of a series of symbols as well as individual ones in symbolic systems in general. Yet while the body has been the recognized source of potent individual symbols, and symbols *derived* from the body have been treated as elements in structural analyses of general symbolic statements and rituals, less attention has been paid to relations between body symbols in terms of their concrete relationship to other body parts and symbols. I am suggesting that the body can be seen as a structured set of symbols or possible symbols and items available for symbolic communication. In the same way that Turner has insisted that to understand a symbol it is necessary to examine its occurrence in a series of different contexts, or as one in a special cluster or gestalt of symbols, to investigate its 'positional Meaning' (Turner, 1966:295), so it is necessary to see the relations of symbols to the classificatory context of the part to the whole and the part to other parts. Thus we are concerned with 'the structural relations and transformations of symbols internally within the system', with the patterned totality of symbolic relations (Piaget 1971a:8-9).

Body symbols have traditionally been looked at individually, with concentration on single items - the mouth, anus, penis and so on. There is a danger in analyzing items of this kind as isolated items at an empirical level, simply on account of their attribution as single potent dominant or master symbols representing key physiological processes with fundamental symbolic qualities. While such items indeed carry a heavy semantic load which must be investigated systematically, such empirical, functional, physiological approaches tend to ignore the basic context, perception and classification of the body[4]. There has been a tendency to stress the function of individual parts rather than the relationships between them, either anatomically or functionally. The importance of the relational element in the definition of body-parts is expressed in much artistic representation of the human body, where different parts are not necessarily indicated by either their form or relative size, but in terms of their relation to other parts. A line represents a penis not because it is drawn between two other lines representing legs. This is still the case when it comes to body symbolism, although in the general area of classification the emphasis has shifted (Lévi-Strauss 1966:150-1). To understand the symbolic value of the penis, it is necessary to know something of its anatomical relations (as mediated by a particular form of classification) to the rest of the body, and to complementary organs of the opposite sex. In the same way that to understand an animal or plant symbol it is important to see it in relation to other categories, so also for body symbols; as eaglehawk is to crow, so penis is to vulva (different sex), or vulva to mouth (same sex).

While the analysis of classificatory context is now accepted as a necessary part of the analysis of symbols derived from different natural species, the same approach, considerations and rigour have not been applied to body symbols. This is as true for symbolic systems associated with analytical classifications as for synthetic ones. However, in the case of the former the validity of the structuralist approach is obvious, since the whole is a concrete entity and the relations of its parts are concrete and materially constrained. Simple one-to-one symbolic associations or contrasts give rise to specific problems if they alone define the co-ordinates of investigation; these are thrown into relief and countered by a more structured and comprehensive approach to the subject.

The importance of relations between different symbols drawn from a single domain is shown at its simplest in the notion of metaphoric and symbolic complementarity, in binary oppositions. The representation of the penis as a symbol implies the existence of something which is non-penis, generally or specifically. At the general level, penis may contrast with non-penis: in the anatomical domain it is penis and not anus, mouth, ear and so on. This is evident at a simple semantic and lexical level. Selection of any one body symbol, implies that something remains unselected. At the level of the semantics of symbolism, at a more abstract level of anatomical classification, it may be more specific, as in the contrast between penis and vulva. Another example of specific symbolic complementarity is in the contrast between right and left hand. To use one hand as a symbol frequently implies the existence of its complement. Here the complementarity is strikingly more morphological, whereas in the penis-vulva example morphological similarity is less important than functional and behavioural associations. Thus the symbolic use of the body operates on the complementarity of its parts, arrived at through the process of classification, which is at the same time both cerebral and social. Here again synthetic and analytic classifications are identical, for the right hand can only exist in the horizontal dimension in relation to the left, and in the vertical dimension in relation to arm, finger and so on. In the same way a taxonomic species only exists because it both contrasts horizontally with other species and vertically with higher and lower order groupings, varieties, phyla and so on. But having said this, it should also be remembered that the emphasis of particular pairs is at the expense of de-emphasizing other things. To contrast penis with vulva as symbols of masculinity and femininity is to emphasize sexual dissimilarities at the expense of sexual similarities, which are, after all, far greater.

The next step from simple complementarity is the recognition of the morphological and/or functional similarity of otherwise different body-parts: labia-lips, nose-penis, buttocks-cheeks, genitals-face, (e.g. Hallpike 1969:257). As Freud has shown, the secondary items in pairs may come to represent their primary 'equivalent' in certain circumstances. Such ideas of body part transposition are of great importance in symbolism (1954:387 and passim; see also Hallpike 1969:257).

While in both simple complementarity and more complex forms of symbolic association discrete organs and anatomical parts are usually specified, it is important to recognize that their complementarity or similarity does not rest solely on morphological or functional criteria, but is set in the broad context of body symmetry. The complementarity of left and right hands is structured, in part, in terms of their localization on opposite body halves. The transpositions cited above are structured in terms of the opposition of the upper and lower halves of the body. Other geometric dimensions may also be relevant. Mouth and anus are not simply opposed in terms of the contrast upper:lower, but also in terms of the contrast front:back. This is part of the *anatomical* classificatory context in which such distinctions are made. Additionally and quite crucially, the opposition is also functional, as in ingestion:exgestion, food:faeces.

It is at this level of abstraction that anomaly begins to become important, but it only does so because synthetic classification replaces analytic. The relationship of mouth to body is *unambiguous*; but if the mouth is isolated from the body and a class of mouth-like entities constructed (see Part I) it immediately becomes ambiguous and (most significantly) a potent symbol, having class similarities with other orifices such as vulvas and anuses. It is at this classificatory level (resynthesis) that Freudian notions of symbolism appear to operate. Anomaly is not an inherent property of analytic classifications because items occupy field-dependent spatial positions that cannot be claimed by others. There are boundary problems between items, but here they are due not so much to the ambiguous nature of the content as that of defining the material limits of the spatial area occupied and designated by a lexical item. As I have already mentioned, joints are of particular interest in this respect because they represent the links between parts, which can be areas of classificatory uncertainty. Equally, however, they may become reified and become a relational system in their own right, as in the case of the Dogon, where they may be seen as 'the chief thing in the body of the Man' (Griaule 1965:51, 140). The classificatory advantage of joints is that, in anatomical terms, they tend to be points rather than areas, places rather than spaces, which can be specifically located. In the Dogon example the joints represent different ancestors, and are connected with different colours associated with the celestial granary of Dogon cosmology or those of corresponding seeds[5].

The relationship between classification and semiotics is two-way. Of equal importance to the considerations so far discussed are what symbolic and metaphoric use of the human form can tell us about its classification. For example, the dualism of right and left and their pre-eminence in cosmology and ritual tell us something of the fundamental importance of symmetry, asymmetry, and contrast as means of ordering information (Needham 1973), while schematic representations of the body (either 'representational' or symbolic) tell us something of its key distinctive features. The '+' which Jung saw as being related to the human form (Firth 1973:59) is an indication that relationships between parts are often more important than individual features. In this case the key relations are head:trunk, left arm:right arm, and the relationship between the horizontal and vertical axes represented by these elements. In material figurines, the proportions can be the all important feature: the contrast between breast and buttock, head and trunk. Iconic representations such as these often enable us to arrive at the critical distinctions in body classification.

Classificatory complementarities are useful elements in symbolic systems, since they can at the same time code large quantities of information and yet do so in a simple and accessible form. But because of this it is important not to be misled by their apparent simplicity. Those derived from anatomy may involve minimal cultural modification. The contrast or association of pairs (arms:legs, fingers:toes) which often occur as figures of speech are apparently simple. Their origin in terms of the classification of body parts is obviously related to the parallelism of extremities already discussed (Parts II and VI). Semantically, however, they may acquire more complicated attributes. Other simple complementarities may involve the identification of functional similarities (eyes and ears). Still others may concern ideational speculation on Man and involve categories that are only in the most abstract sense anatomical (heart and mind, body and soul), at which point they merge with the host of abstract dualities associated with cosmological systems. Thus anatomical complementarities can be ranked on a scale of increasing abstractness. In Figure 4 I have divided this scale into four basic categories, but the allocations must to some extent be arbitrary. The lowermost category contains pairs of identical organs, or identical in everything except in terms of contraposition or lateral reversal. Category II contains all the obvious unambiguous anatomical parallelisms I have mentioned, which can be classified in anatomical terms alone. Category III contains morphological parallelisms of a more abstract kind, where they entail a high abstract content and reference to functional and behavioural considerations; morphological similarity may be of relative minor importance. Category IV contrasts are almost wholly defined in abstract terms with a superficial allusion to anatomical materiality. While the symbolic potency of pairs in the lower categories may be high, pairs higher up the scale are maintained almost solely because of their symbolic component.

VIII

THE MATERIALITY OF THE BODY

The materiality of objects, and the analytic process derived from it is an important determining and organizing factor in classification. The importance of material similarity of different parts of the same body has already been mentioned. However, the materiality of body parts is significant also in the extent to which it allows for unambiguous cross-cultural identification, as between, say, the words for 'penis'. Although there may be some semantic confusion stemming from differences in boundaries (such as whether it is the penis specifically or the entire male genitalia which are implied), because of their material referents and the analytical classification of which they are part, cross-cultural identity and consequently explanation of such items is a relatively straightforward business. This is in marked contrast to some symbols not derived from human anatomical classification: the more morphologically vague psycho-analytic notions, and symbols derived from entire representatives of natural species and their groups.

Since Durkheim, the stress has been on the cultural and social character of bodily symbols (Durkheim 1915:17) and the role of the underlying cerebral organization and physiology has been underestimated or neglected. Such emphasis is usually part of an attempt to demonstrate just how much of our perception of even the most natural and physical of things are reducible to our social organization of that reality (Berger and Luckmann 1967). The work of Mary Douglas has been essentially concerned with the way in which the 'social body' constrains the way the physical body is perceived. 'Physical nature', she writes, 'is masticated and driven through the cognitive meshes to satisfy social demands for clarity which compete with logical demands for consistency' (Douglas 1973:113). More specifically, the body as a symbol is often no more than society reified and sanctified in ideology, mediating the social structure by itself and its images (Douglas 1971:387, Ellen 1972:236-7).

It is difficult to deny the importance of cultural ideas in affecting the classification of body parts. For example, culturally-attributed functions of an organ may fundamentally affect the most 'primary' of classifications, as in mediaeval medicine (Singer 1957). But clearly, while this may be generally accepted, it would be wrong to dismiss altogether material influences on perception and classification and submit entirely to the vicissitudes of vacuous sociologism. While not seeing it quite in these terms, Firth has pointed to the 'great variety of circumstances in which physical structure and powers of the human body are perceived in direct relation to performance, irrespective of the character of the social body' (Firth 1973:307). My contention is simply that the human body forms the concrete basis of a characteristic classificatory apparatus for parts ordered analytically, which acts as a primary constraint in structuring 'cognitive meshes', and that this is not obliterated by the process of cultural re-organization, since the very mechanism of the process is itself partially constrained by it. It is only by assuming a core of materially constrained behaviour in perception and action that cross-cultural interpretation and recognition makes any sense. While a pot may be lifted in a variety of ways, there are a certain number of key actions involved which are everywhere the same and indicate cross-culturally the nature of the action. Similarly, while the body may be verbally classified in a number of ways, there are always key elements in both verbal and non-verbal classifications which are derived from the structure and function of the body.

In one sense it is true that body classification is not arbitrary: cognitive structures are, indeed, extensions of biological structures, but they are at the same time discovered through the process of cultural and social interaction. The process of classification may be visualized as operating at two co-existing and interpenetrating levels, represented by the analytic and the synthetic. At the first level, structures may be regarded as 'deep', dominant ones being determined by a combination of material constraints and cerebral apparatus, and influence of the cultural is superficial only in terms of providing lexical labels for culturally defined material parts. At the second level, culture is dominant, and the body, for example, is defined largely in cultural terms, although cultural categories are, if many stages removed, tied in various ways to material and cognitive constraints. The relationships between these two levels are subtle and complicated, since operational and observable reality will generally be composed of both in varying proportions, with the cultural component being self-amplifying. While the body, as with all nature, may be redefined and reified in cultural terms, it is not in the first instance of perception defined as such: the classification itself takes on a concretised form. In this it differs from the perception and classification of natural species, both in terms of socialization and the evolution of classificatory systems. Thus, while with Durkheim and Mauss (1901-2), I would recognize the intrusion of the social on the material, unlike them I argue the necessity to account for the dialectical relationship between the cerebral, the material, and the social, such that classificatory structure and function are virtually inseparable. I am suggesting that we should neither ignore that famous but uninvited guest of Lévi-Strauss (Mendelson 1967), not ignore his chair.

NOTES

1. For Inhelder and Piaget (1964) the earlier process takes the form of seriation. David Reason has suggested to me that in more general terms this might indicate the emergence of topological body space before Euclidean (oriented) space.
2. This does not preclude, however, confusion following from inadequate or erroneous information, sometimes sustained or determined by cultural factors; witness, for example, the cloaca-theory of psycho-analysis, in which the anus is confused with the vulva. The history of anatomy is full of similar if not so obvious confusions (Singer 1957:66-7, 78).
3. The same kind of notion is found in K.A. Scherner's ideas in his *Das Leben des Traumes* of the representation of the body in dreams as a house, and the parts of the house representing specific organs. Irrespective of the criticisms of Freud (1954) and others of his psychology, the use of the metaphorical relation of house-parts to body-parts illustrates the ubiquitousness of such metaphors.
4. In another paper I have drawn attention to the opposite tendency in the study of anthropomorphic symbols - the neglect of depth-studies of particular items. There too I was stressing, but for different reasons, the necessity of rooting explanation in the analysis of classification and the structure of categories (Ellen 1972).
5. There may be some relationship here between the classificatory and symbolic connotations of joints and the attention given them in primitive art.

REFERENCES

Berger, P.L. and Luckman, T. (1967). *The social construction of reality*. London: Allen Lane, The Penguin Press.

Casagrade, J.B. and Hale, K.L. (1967). Semantic relationships in Papago folk definitions. In Hymes, D. (ed.) *Studies in southwestern linguistics*, pp. 165-193. The Hague.

Cole, M., Gay, J., Glick, J.A. and Sharp, D.W. (1971). *The cultural context of learning and thinking*. Methuen: London.

Conklin, H.C. (1962). Lexicographical treatment of folk taxonomies. *Intern.J.Am.Linguistics 28*, 119-141. In *Problems in lexicography 4* (eds.) F.W. Householder and S. Saporta (Publication 21). Bloomington: Indiana University Research Centre in Anthropology, Folklore and Linguistics.

Conklin, H.C. (1964). Ethnogenealogical method. In Ward H. Goodenough (ed.) *Explorations in cultural anthropology*. New York: McGraw-Hill.

Conklin, H.C. (1972). *Folk classification: a topically arranged bibliography of contemporary and background references through 1971.* New Haven: Yale University, Department of Anthropology.

Douglas, Mary (1966). *Purity and danger.* London: Routledge and Kegan Paul.

Douglas, Mary (1970). *Natural symbols: explorations in cosmology.* Barrie and Rockliffe: London.

Douglas, Mary (1971). Do dogs laugh? A cross-cultural approach to body symbolism. *J.Psychosom.Res.,* 15: 387-90.

Douglas, Mary (Ed.) (1973). *Rules and meanings: the anthropology of everyday knowledge.* London: Penguin education.

Durkheim, E. (1915). *The elementary forms of the religious life.* (Translated from the French by J.W. Swain) London: George Allen and Unwin.

Durkheim, E. and Mauss, M. (1901-2). De quelques formes primitives de classification: contribution à l'étude des représentations collectives. *Annee Sociologique VI.*

Ellen, Roy F. (1972). The marsupial in Nuaulu ritual behaviour. *Man (N.S.) 7,* 223-238.

Ellen, Roy F. (1975). Variable constructs in Nuaulu zoological classification. *Social Science Information.* 14 (3/4): 201-28.

Fantz, R.L. (1961). The origin of form perception. *Sci.Amer.* 204, 66-72.

Firth, R.F. (1973). *Symbols public and private.* London: Allen and Unwin.

Fox, J. (1971). Sister's child as plant. In R. Needham (ed.) *Rethinking kinship and marriage.* Association of Social Anthropologists Monograph No. 11. London: Tavistock.

Frake, C.O. (1961). The diagnosis of disease among the Subanun of Mindanao. *Amer.Anthrop.* 63:113-32.

Franklin, K.J. (1963). Kewa ethnolinguistic concepts of body parts. *Southwestern J. of Anthropology* 19(1): 54-63.

Freud, S. (1954). *The interpretation of dreams.* (Trans. James Strachey) London: Allen and Unwin.

Friedrich, P. (1969). Metaphor-like relations between referential subsets. *Lingua* 24: (1): 1-10.

Griaule, M. (1965). *Conversations with Ogotemmêli: an introduction to Dogon religious ideas* (with an introduction by Germaine Dieterlen). London: Oxford University Press.

Hallpike, C.R. (1969). Social hair. *Man (N.S.)* 4(2): 256-67.

Harris, M. (1970). Referential ambiguity in the calculus of Brazilian racial identity. *Southwestern Journal of Anthropology* 26, 1-14.

Inhelder, B. and Piaget, J. (1964). *The early growth of logic in the child.* (Trans. E.A. Lunzer and D. Papert) London: Routledge and Kegan Paul.

Jansen, H.J. (1933). Inheemsche groepen-systemen in de Ambonsche Molukkan. *Adatrechtbundels 36*, 445-59.

Jung, C.G. (1958). *Psyche and symbol*, (ed.) Violet S. de Laszlo. New York: Doubleday.

Lenormand, M.H. (1950). Connaissance du corps et de la personne chez le Melanesien de Lifou (Iles Loyalty). *Journal de la Société des Océanistes* 6: 33-65/ Paris.

Lévi-Strauss, C. (1962). *Totemism.* (Translated from the French by Rodney Needham). London: Merlin Press.

Lévi-Strauss, C. (1966). *The savage mind.* London: Weidenfeld and Nicolson.

Lewis, Gilbert (1974). Gnau anatomy and vocabulary for illnesses. *Oceania* 45 (1): 50-78.

MacRae, D.G. (1975). The body and social metaphor. In J. Benthall and T. Polhemus (eds.) *The body as a medium of expression.*

Mahr, August C. (1960). Anatomical terminology of the eighteenth century Delaware Indians: a study in semantics. *Anthropological Linguistics* 2 (5) 1-65.

Marsh, G.H. and Laughlin, W.S. (1956). Human anatomical knowledge among the Aleutian islanders. *Southwestern Journal of Anthropology.* 12: 38-78.

Mauss, Marcel (1973). Techniques of the body (Translated by Ben Brewster). *Economy and Society* 2(1): 70-88.

Mendelson, M. (1967). The 'uninvited guest': ancilla to Lévi-Strauss on totemism and primitive thought. In E.R. Leach (Ed.) *The structural study of myth and totemism* (Ass. social Anthrop. Monogr. 5). London: Tavistock.

Nagel, E. (1961). *The structure of science: problems in the logic of scientific explanation.* New York: Harcourt.

Needham, R. (ed). (1973). *Right and left: essays on dual symbolic classification.* Chicago and London: The University of Chicago Press.

Perchonak, N. and O. Werner (1969). Navaho systems of classification: some implications for ethnoscience. *Ethnology* 8(3) 229-43.

Perey, A. (1975). Body and World in Oksapmin kin terms. *Oceania* 45(3): 235-6.

Piaget, J. (1971a). *Structuralism* (translated and edited by Chaninah Maschler). London: Routledge and Kegan Paul.

Piaget, J. (1971b). *Biology and knowledge: an essay on the relations between organic regulations and cognitive processes* (trans. Beatrix Walsh) Edinburgh: University Press.

Price-Williams, D.R. (1962). Abstract and concrete modes of classification in a primitive society. *Brit.J.Educ.Psych.* 32: 50-61.

Scherner, K.A. (1861). *Das Leben des Traumes-Entdeckungen auf dem*

Gebiet de Seele. Berlin: Schindler.
Singer, C. (1957). *A short history of anatomy and physiology from the Greeks to Harvey*. New York: Dover.
Spicker, S.F. (Ed.) (1970). *The philosophy of the body*. Chicago: Quadrangle Books.
Stark, Louisa R. (1969). The lexical structure of Quechua body-parts. *Anthropological Linguistics* 11(1): 1-15.
Steiner, Franz (1956). *Taboo*. London: Cohen and West.
Sturtevant, William C. (1964). Studies in ethnoscience. *American Anthropologist* 66(3): 99-131.
Turner, V. (1966). The syntax of symbolism in an African religion. In *Philosoph.Trans.Royal Soc. B*. No. 772. Vol. 251, 295-303.
Werner, O. and Begishe, K.Y. (1970). A lexemic typology of Navajo anatomical terms I: the foot. *Int.J.Amer.Linguistics*. 36: 247-65.
Werner, O. and Fenton, J. (1973). Method and theory in ethnoscience or ethnoepistemology. In R. Naroll and R. Cohen (eds.) *A handbook of method in cultural anthropology*. New York and London: Columbia University Press.

THE BODY AS A SOCIAL SYMBOL AMONG THE ROM

Anne Sutherland

> The social body constrains the way the physical body is perceived. The physical experience of the body, always modified by the social categories through which it is known, sustains a particular view of society. There is a continual exchange of meanings between the two kinds of bodily experience so that each reinforces the categories of the other (Douglas 1970: 65).

The idea of this essay is to demonstrate with a case study, the way that the body can provide a metaphor of social organization. Both Turner (1966) and Douglas (1970) have discussed the use of the body and bodily emissions as nonverbal categories in society. Following the interest sparked by Lévi-Strauss in dual organization and the work of Robert Hertz on the right hand (1960), a great deal has now been written on left and right hand symbolism and the relationship of such a dual model, based on a natural proclivity of the body, to other spheres of social behaviour (See, for example, Needham 1973).

Among the Rom, the separation of the body at the waist into two opposed halves and the perceived necessity to keep these two halves separate is the basis for Romany conceptions of purity and pollution. Purity is attained by avoiding polluting substances and areas, that is, the lower half of the body and bodily emissions emanating from there. Pollution is the failure to keep separate the upper and lower halves of the body. An unusual feature of Rom body symbolism, therefore, is the perceptual absence of left/right divisions of the body and the pervasiveness of upper/lower body symbolism. I have compared elsewhere the upper/lower and inside/outside boundaries among American Rom and English Gypsies (Sutherland 1976).

This conception of pollution is related to other concepts of social importance, namely, auspiciousness (*baXt*) and inauspiciousness (*prikaza*), health (*sastimos*) and illness (*naswalemos*).

Purity, health and luck, are inextricably linked. To enjoy any one, it is necessary to have all three. This conception of pollution and purity is also the basis for expressing the important social divisions or statuses of Rom society: 1) the crucial inside/outside boundary between Rom and non-gypsy, 2) internal political divisions, and 3) age statuses. These divisions are likewise influenced by the concepts of auspiciousness/inauspiciousness and health/illness.

The nature of the relationship between these concepts, boundaries, political divisions and statuses, provides a model for a number of everyday rules that may be seen as the performance of the deeper structure. The very performance of such an implicit structure consequently affects the conceptualization of the body itself. But the performance of body washing and ritual related to body separation, and the obvious practical difficulties of keeping separate two halves of a body which are always linked, also reflects metaphorically the very real problems that the Rom face in maintaining a moral boundary between themselves and non-gypsies, a boundary that must be crossed again and again through their economic relations with non-gypsies.

This boundary is handled with a set of apparent contradictions. Non-gypsies, for example, consistently think of Gypsies as 'dirty'. The Gypsies often support this image of themselves not only by talking about the 'dirty Gypsies down the road' but also through their actions: ignoring rubbish regulations, trashing up surrounding areas, chucking beer cans on the highways, letting children run around with a grubby appearance, scratching exaggeratedly for fleas in public places, etc. A 'dirty' image is often useful for keeping outsiders away. The Gypsies, however, think of the non-gypsies as 'dirty' because they mingle indiscriminately what should be kept separate, they lack 'shame' and morality in their sexual behaviour, and they are the source of venereal disease. This certainty of the Rom that they are superior in morality and cleanliness also maintains the boundary between inside and outside. They control very carefully their contacts with non-gypsies because of the dangers of pollution, and try to limit them as much as possible to the necessary economic pursuits. Patterns of lying and deceit have the same contradictory pattern. The Rom create for themselves a reputation as unreliable persons while at the same time they can feel superior by 'knowing' when another Gypsy is telling the truth or not, and by consistently fooling the non-gypsies and demonstrating how gullible and dull-witted they are.

The boundaries between purity and pollution, upper and lower halves of the body, inside and outside, are perceived in the same way by the Rom and present the same contradictions. They are constantly crossing them and they are constantly reinforcing the opposition. The aim of this paper is to show how the body, the society, and relations with outsiders are seen in terms of a set of boundaries forever being crossed.

WHO ARE THE ROM?

Not all Gypsies are Rom. There may be as many as 500,000 Gypsies in North America and by far the largest group of Gypsies are the vlaX-speaking Rom. Other groups of Gypsies include the Romanitshal, Sinti, Manouche, XoraXai, Zingari, Afrikaya, and Boyash. These groups are all socially and linguistically distinct from each other and rarely intermarry. A detailed analysis of Rom social organization is given in Sutherland 1975.

Linguistic studies of *Romanes*, a Sanskrit-based language, indicate that the Rom left India about 1000 years ago and travelled west through Persia and into Western Europe in the fourteenth century. The main immigration of Rom to North and South America occurred at the end of the nineteenth century. These Rom still speak inflected *Romanes* as their mother tongue. All Rom belong to one of four 'nations' or tribes, called *natsiyi* (sing. *natsia*): the Machwaya, Lowara, Kalderasha and Churara. Each *natsia* has its own character, which is expressed mainly in dialect differences and some variations in custom and appearance. The difference between the various *natsiyi* are insignificant compared with the differences between Rom and other Gypsies. The *natsiyi* are primarily status categories (from high to lower status - Machwaya, Lowara, Kalderasha, Churara). This ranking is important as a tern of reference in conflict situations between Rom. *Natsiyi* are not groups with any form of organization, nor do members of a *natsia* ever meet with each other formally.

Each *natsia* is theoretically composed of a number of restricted cognatic descent groups called *vitsi* (sing. *vitsa*), though in practice it seems that the Machwaya have very few *vitsi* and rarely refer to them, whereas the Kalderash have numerous *vitsi* which are very important for identifying an individual. Membership in a *vitsa* is cognatic by right, that is, any person has the right to belong to either his father's or his mother's *vitsa*, but because of patrilocal residence at marriage an individual usually chooses to affiliate with his father's *vitsa*. Like the *natsia*, each *vitsa* has a relative status and a name. The status of the *vitsa* is partially determined by the *natsia* to which it belongs and partially by its own reputation, size, power and wealth.

Each *vitsa* is composed of a number of *familiyi* (sing. *familia*), which are extended families usually spanning four generations, led by the eldest living progenitor of the family who may be male (*phuro*) or female (*phuri*). Members of a *familia* feel a very strong loyalty to each other. They live together or at least nearby, and they co-operate economically, often pooling income under the jurisdiction of the eldest member, who has considerable control over family members.

Thus, the formal organization of the Rom is composed of units of descending size: *natsia*, *vitsa*, and *familia*, but only the *familia* is always a discernable, functioning group. Cross-cutting this formal organization are two important groupings: the *kumpania* (pl. *kumpaniyi*) and the household (*tsera*). The *kumpania* is a residential group, composed of a number of households, and it can be recruited from different *natsia* or *vitsa*. Many *kumpaniyi* include persons from several *vitsi*, though frequently one *vitsa* or *natsia* predominates.

The *kumpania* defines a territory within which its members have the right to make a living. Members of a *kumpania* usually co-operate with each other as partners (*wortacha*) when exploiting the economic resources of the area. Some of the most common of the Rom's many ways of obtaining a living are fortune-telling, body and fender work, tarring roofs and driveways, repairing copper boilers, buying and selling used cars or trailers, and welfare. All these methods require that the Rom a) make their living from non-Gypsies (*gaje*) without accepting regular employment from them: earning money from another Rom is unacceptable; b) remain mobile while earning a living; c) do not undertake any job which is *marime* (polluted).

In addition to being an economic territory, the *kumpania* is the public group that is responsible for moral, social and political decisions on behaviour under the authority of the *kris romani*, the legal and judicial system of the Rom. The *familia*, and even the *vitsa*, cannot be solely responsible for regulating the behaviour of individuals, as these are units of kin who are expected to support each other. The *kumpania* contains a cross-section of kin groups and is the main functioning moral and political group.

POLLUTION CONCEPTS

Marime, meaning polluted, defiled or unclean, is at the same time the most crucial and the most difficult concept to understand. The origin of the term is unknown, thought it has been traced back at least to Eastern European Rom and probably stems from the Sanskrit root *Mrks* (to defile) from which the Indian word, *mala* (dirt or defilement) comes. Not all Rom in the world use the word *marime* or *mahrime*, which is known by all North American Rom, but all Rom have some term to indicate a defiled condition. Welsh Gypsies, for example, used *Maxado*, meaning stained, soiled, or defiled (Sampson 1926).

Marime has several meanings which fall into two basic categories. Firstly, it is used to indicate uncleanness or impurity of a physical as well as ritual or moral nature. But it always indicates a state of pollution rather than simply 'dirty' in a temporary and easily rectifiable way, which would be called *melalo*. For example, hands that are covered in grease from working on a car are *melalo*, whereas hands that have touched the genitals are *marime*.

The second meaning of *marime* is what the Rom translate as 'rejected'. The two meanings, impurity and rejection, overlap, since to become physically or morally impure could mean 'rejection' from the group. Non-gypsies are considered *marime* in a broad sense, that is, 'rejected' or outside the group's moral boundary, primarily because they fail to maintain correct conditions of body cleanness and therefore threaten the purity of the Rom who have contact with them. However, once a non-gypsy is known to observe proper cleanliness, this threat is diminished. In the few cases when non-gypsy women marry a Rom and live with the group, their cleanliness status must be verified, and often they become so conscious of this problem that they uphold the rules to an unbearable degree. In the community I studied, five per cent of the men had married non-gypsy women. Marriage of a Gypsy woman to a non-gypsy is rarer, but difficult to calculate exactly since such an action makes the woman instantly unclean and she must leave the group permanently (see Sutherland 1975).

Though these are the two primary meanings of *marime*, it is a broader concept than either body pollution or rejection from the Rom as a social group. The term is applied in a number of contexts as a defining mechanism for a whole series of social boundaries, and this gives it great importance for the Rom. It is also in this sense that *marime* is a flexible concept: it defines boundaries for new and changing situations and brings a moral order to bear in such situations. *Marime* is the moral language within which otherwise ambiguous social situations are given meaning.

The most important boundary is that between Rom and non-gypsy. Close relations with *gaje*, which would include sharing food and personal physical contact (particularly sexual contact), are in certain circumstances considered *marime*. To share food prepared by *gaje* is more likely to be considered *marime* than for *gaje* to share food prepared by the Rom. Sexual contact with *gaje* is always *marime* for a woman but not always for a man. In both cases, public knowledge and gossip increase the danger of an action being considered *marime*.

Non-gypsies are also considered to be the source of many diseases, particularly venereal disease, so that not only their persons but public places such as public toilets, hospitals, buses, schools, offices, jail and their homes, are all less clean than the homes of Rom and open outdoor spaces such as parks and woods. When they must be in *gaje* places the Rom generally avoid touching as many polluted surfaces as possible, and prolonged occupation of a *gajo* place, such as a regular job in an office or a stint in jail, means certain pollution. Many *gaje* places can be altered to make them more clean. A house formerly occupied by *gaje* can be painted and cleaned with bleach; food in a restaurant can at least be eaten with the hands rather than utensils; drinks can be taken from paper cups; paper towels can be used to turn on taps and open doors in toilets, and so on.

A second important boundary marked by pollution categories is that between the various sub-categories of Rom. It is very important for a Rom to know the *vitsa* and *natsia* of other Rom in order to understand their relative social positions both in terms of degree of kinship relatedness to each other, as well as their relative positions in terms of morality and purity. Though not as clearly demarked as the Rom-*gajo* boundary, in most situations the identity of a person's *natsia* and *vitsa* can be determined and used as a guideline of his social and moral standing in relation to the group.

Each *vitsa* as a whole has a purity status relative to other *vitsi*, a status that is determined by the actions of its members over time, so that a reputation or 'name' is built up around each *vitsa*. This reputation is based partly on the economic position of the members and partly on questions of purity and morality. There is therefore a kind of hierarchy of *vitsi*, and although it is fairly fluid and changeable as reputations and fortunes change, it is nevertheless a hierarchy that is expressed in the language of purity and impurity. A whole *vitsa* may be ostracized because of the *marime* actions of one or more of its members, or a *vitsa* may enjoy great prestige because its members are known for their moral purity, prosperity, good health and general good fortune.

A third important boundary marked by pollution categories is that between various categories of age. The Rom have terms for four age categories:

1.	*sha* (m) *shey* (f)	child
2.	*sha baro* (m) *shey bari* (f)	post-puberty juvenile
3.	*rom* (m) *romni* (f)	married adult
4.	*phuro* (m) *phuri* (f)	aged person (for women, post-menopause)

Each of these age categories has an accompanying general

purity or pollution status which is associated with the concept of shame (*lashav*), which is very like the Mediterranean idea of shame (see Campbell 1964) in that it is associated primarily with sexual morality, but not a concept restricted solely to women. A child, after the pollution of birth, at more or less six weeks, is considered *wuzho* or basically pure in body and action. Children can enjoy freedom from most social restraints and are not expected to understand or demonstrate 'shame' in their actions. Physical contacts defiling to adults are not necessarily defiling to children, who need not take many of the precautions that adults must take to ensure cleanliness in their daily lives. Children, for example, may eat food handled and prepared by non-gypsies, although post-puberty juveniles would reject it.

At puberty, boys and girls are introduced to the idea of personal shame. Thereafter both their bodies and their actions will be judged in terms of control: control of their own polluting secretions (e.g. menstrual blood, semen), and control of 'shameful' and polluting actions (sexual contact). Although modified by marriage and childbirth, these controls last through married life until old age. In old age, because of the supposed discontinuance of sexual relations, many regulations are relaxed. The aged are venerated and respected, as well as being politically powerful, because they enjoy *wuzho* status. Purity once again may be regarded as a language, in this case for expressing the secular authority of the elderly.

Purity and pollution categories are therefore important for defining various social boundaries at various levels of social organization. These categories are not the only factors involved in determining such boundaries. They are always accompanied by other categories based on economic or political status, fortune and health, etc., but they are very frequently the language chosen to express other such determinants. Without a reputation for morality and cleanliness, no man or woman will become a respected leader and a powerful influence among Rom. Conversely, those who suffer low status also suffer a reputation of dubious purity. The worst degradation that can befall an individual is to be declared *marime* by group decision, which entails a denial of commensality and close physical contact by fellow Rom and forces the individual to live in the world of the *gaje*. This brings me to a final concept, *Romania*, which sums up the relationship between all concepts, boundaries, and other factors mentioned so far.

Romania is the state of being a Rom; it includes the conditions and rules for identity as a Rom; it indicates the social behaviour that each individual Rom must accept, at least in principle, to be considered a Rom; it implies a purity status that is accepted by other Rom. *Romania* is the essence of their collective representations; it is the social order par excellence. It is the collective values, ideals, beliefs, actions, and behaviour of the Rom. In English, the Rom translate *Romania* in the only way possible to indicate the intensity of the concept: they call it their 'religion'.

Beliefs and actions that seriously deny, infringe or degrade *Romania* are *marime*. A person who disregards *Romania* in some important way is 'rejected' from the group. He becomes *marime*, defiled, polluted. If he recants and takes the correct procedures for reinstatement into the group, he can resume normal relations and renew his condition of purity. These include rectifying the behaviour, a public trial (*kris*), and possibly some punishment in order to be reinstated. The most usual punishment in serious cases is a set period of *marime* status when no Rom will eat with the *marime* person or his family, both of whom are socially ostracized.

THE BODY

The Rom divide the body at the waist, *maskar*, a term that also means the spatial middle. The upper half of the body is *wuzho* and must be kept that way. The lower half is *marime* and must be preventing from polluting the upper half. The hands (*was*) move between the two halves and have an indeterminate pollution status which depends on what they have touched. Shame (*lashav*) is associated with the lower half of the body but not with the upper half. The genital-anal area of the body is particularly shameful, and all excretions from this area are highly polluting. Excretions from the upper half of the body are neither polluting nor shameful. Menstruation, for example, is a polluting and shameful act, and menstrual blood is a polluting substance, whereas spittle and spitting are not. Spittle is, in fact, a clean substance that may be used to clean cuts or scratches, to ease childbirth pains, or to counter the evil eye. Blowing is also clean and sometimes curative, and *johai* (ghost's vomit) is the most effective cure of all.

Menstrual blood, urine, faeces, and semen, are highly polluting substances: contact with them can cause illness, and they must be controlled and contained as much as possible. I have no information on the control of semen, as this is exclusively a concern of males and never discussed with females, but I can illustrate the nature of these controls with female data. Menstruation, for example, is surrounded by a number of rules to control the ill effects of the bodily function. At menstruation a girl is introduced to shame and a dramatic change of status. Before menstruation she is a child (*shey*), can wear short skirts ('American clothes') and is not a pollution danger to men. Once a girl menstruates she must observe the washing, dressing, cooking, eating and behavioural rules of adult women, partly for her own protection and partly for the protection of men. The most immediately noticeable change is that she must begin wearing 'Gypsy clothes', separate top and long skirt, and can no longer use the clothes of pre-menstrual girls, but can now borrow the clothes of girls who have menstruated. Her clothes must be washed separately from men and from children, and then must be separated further according to 'top' clothes and 'bottom' clothes. If her hands have touched menstrual blood, she must not handle food before carefully washing them first. She must show respect to men by not passing in front of them, stepping over their clothes, or allowing her skirts to touch them. Finally, she begins to fast on Fridays.

A girl who menstruates must also begin to sleep separately from young girls because 'it might rub off on them', and during her menses she should sleep facing the wall in a corner with her legs crossed or clamped together. She can now join groups of post-puberty girls and enter into their discussions fully. On one occasion when two girls met for the first time, they exchanged information on their relationship to people in the *kumpania* and then inquired if the other had menstruated yet. This was necessary information for forming a friendship and knowing whether they could touch each other or not.

Many of the prohibitions for menstruating girls vary according to the *natsia*, *vitsa*, and household. Sleeping arrangements are not the same in each household; fasting varies from family to family; and the participation of women in cooking may depend partly on the size of the family and the ratio of boys and girls.

According to several women, modern paraphernalia for menstruation has released them from the most odious rules. Previously no cooking was allowed, and this is still the case for pregnant women, but recently careful cleanliness and avoidance of the touch of menstrual blood with the hands means that women can now cook and serve food during menstruation. However, there is still a certain reluctance to eat food cooked by a menstruating woman. One girl refused to eat food cooked by an unrelated woman and politely claimed she was dieting because she felt unsure of the woman's habits of cleanliness and was afraid she might have touched menstrual blood once. The girl, a Machwanka, stated that she would only eat food cooked by her own family because she could be sure they were clean; however, this was not exactly true, because she did eat at public rituals.

There are some further points to remember:

(1) Not all blood is polluting. Blood from the veins is clean, and red as a colour is associated with auspiciousness and happiness.

(2) Although urine and faeces and places associated with such

functions are always *marime*, there is more ambiguity towards sexual intercourse and the fluids associated with it. The experience of sexual intercourse, ideally at marriage, brings about the application of a whole series of rules about washing and public behaviour associated with the sense of shame. This does not necessarily mean that the private act of sexual intercourse is inhibited or avoided as something 'dirty'. Shame is the *public* expression of correct behaviour, and although there is some overlapping, it does not dictate the expression of private matters. One woman, for example, stated defiantly that although she would not step over her husband's clothes in public, in private she ignored such a precaution against polluting her husband.

Both separation, however, is only a general ideal, practised primarily in public rather than in private. No one can adhere to such an ideal at all times in its strictest sense, nor is it considered necessary to do so, but if certain illnesses occur, infringements of this ideal may be considered related to the onset of illness.

BODY CARE

Washing and bathing are necessary to have a clean body, but washing indiscriminately is worse than no washing at all, because the first principle to be observed is the correct separation of the upper and lower halves of the body. Both soap and towels are allocated either for use on the upper part of the body or the lower and these must not be allowed to mix (See Lee 1971: 29-30). The head, hair, and face are most important to be kept clean, and items used there, such as razors, soap, combs, towels, the *diklo* (head scarf), hats, etc., must not be allowed to become polluted. Miller reports the case of a Machwano who felt he would develop skin rash if a razor that had fallen on the floor was used on his face (Miller 1975: 42 footnote). The worst threat of pollution a woman can utter against a man is to throw her skirts over his head, thereby requiring him to exonerate himself in a public trial to remove the *marime* stigma, which can only be done by establishing that the act never occurred. The hair is also carefully controlled. Hair should not be combed on Friday (the Devil's day), when in mourning, or when ill with measles. Women should not cut their hair, except a small piece for luck on Fridays, and it should always be bound up and, for married women, covered with a headscarf. Young women do not always take such precautions, but older women instruct them to do so.

For ritual occasions, or when praying at the household shrine, washing is more carefully done. On Friday, a day of fasting, Julie explained that she always cleaned the house, washed her face and combed her hair, carefully donning a clean headscarf, abstained from smoking or drinking, and prayed for the good health of her family at the family shrine.

In mourning, such cleanliness standards are reversed. Relatives of the deceased must abstain from washing, shaving, combing hair or changing clothes for three days. This defiled condition of the body is accompanied by wailing, tearing out of hair, and beating the breast by the side of the coffin. Public flagellation and defilement of the upper body after death indicate a transition period accompanied by the expression of extreme grief. After the third day, both cleanliness and a more happy outward appearance are conventional.

All things that come into intimate contact with the body are subject to the principle of separation into *wuzho* and *marime* categories. The ideal of *wuzho-marime* separation is expressed by four basic divisions:

(a) *inside-outside*

There is a division between items that enter the body (food) and items that touch the body externally (clothes). Cooking and eating utensils, as well as the washing of related items (e.g. dishes, dish and table cloth, pans, etc.) must be separate from items contacting the body.

(b) *top-bottom*

Like the washing of the body, clothes used above the waist must be washed separately from clothes contacting the lower body. An important exception to this basic division was the apron (*katrinsa*) which is used to handle food (pans, mainly) and is always insulated from the body by several layers of clothes; therefore, it is regarded as a 'clean' item and is often washed with food items. The apron of a well-known ancestress has some ritual and medical efficacy. It is carefully kept long after death and bits of it are used to make the *bujo* (amulet) for young children.

(c) *male-female*

items used by men should ideally be washed separately from items used by women.

(d) *adult-child*

This division was somewhat ambiguous in practice, consistent with the generally ambiguous cleanliness status of a child (see Miller 1975: 43). Children's clothes are always washed separately from food items, but they were not consistently divided by the top-bottom or male-female categories. Most clothes for young girls were in one piece (e.g. dresses) anyway, but in some households, pre-puberty girls were not allowed to wear the clothes of their post-puberty sisters. There did not seem to be as great concern about the washing of children's clothes as there was for adults.

CLOTHES

The cleanest and purest clothes are new ones. Since they have never touched the body, they cannot have become polluted through contact with *marime* substances, or through improper separation in washing. For ritual occasions when purity is most important, new clothes are essential. At the Easter *slava*, which is a ritual partly associated with promoting good health in the *kumpania*, *every* Rom was dressed in new clothes for the occasion, including new underwear, socks, shoes, etc. Clothes worn by *gaje* are almost certainly impure through incorrect separation, so that most Rom, even when quite poor, will not wear clothes worn by *gaje* unless they know the user to be of high cleanliness and moral standards. Even then they usually limit themselves to upper body clothes, such as sweaters, or they take these items, unravel them, wash them carefully and crochet them into rugs and blankets. This process makes the items acceptable for use.

When washing clothes, separate tubs and wash loads are organized along the top-bottom and male-female divisions. Thus for clothes, the strictest households had four separate washings for a) male top, b) male bottom c) female top and d) female bottom. Face cloth (top), body towels (bottom), pillow cases (top) and sheets (bottom) were also put into the appropriate one of the four loads. Again in the households that maintained the strictest interpretation of pollution rules, four tubs were required for the four separate washes, but some women for convenience used one washing machine only, and a few of the less fastidious went to laundromats where *gaje* washed. In all households that I knew, another basin was kept specifically for food-related items (dishes, tablecloths, dishcloths, etc) which must be kept particularly pure, and yet another was used for the floor cloths, mops etc. which are *marime*. A dish cloth that falls to the floor is not used again for dishes or surfaces where food is prepared or served.

THE HOUSE

The house and its furnishings are also ordered by the divisions into *wuzho-marime* areas or items. Again, because the *gaje* do not observe this division, any previously occupied house must be scrubbed, painted and re-organized, or else the occupants will be susceptible to diseases attributed to contact with *gaje*. When a Rom family occupies a house that has had *gaje*, they usually set about knocking out walls to create one big room apart from the kitchen, which is a private family area, and the bathroom, which is a *marime* area. Then walls are painted or panelled and the floors are disinfected with astringents. Incense may be burned throughout the house to purify it as well as remove the 'smell' of the *gaje*. Apart from beds, sofas and coffee tables, very little furniture is considered essential. Those who can afford it buy new items, but if this is not possible used items that can be washed and purified are preferred. Lee (1971: 35-36) mentions that some homes have a special chair for *gaje*, and I found that many homes allotted special dishes for *gaje*.

The bathroom is a *marime* area, and it is considered impolite to mention its use or to be too obvious about going there. Floors are also *marime* areas, and *wuzho* items that fall are either discarded or must be washed carefully (e.g. soaking overnight in bleach). The floor often has food scraps, cigarette butts, and other discarded items, but it is swept at least every morning. To have rubbish scattered around on the floor is not considered *marime*, but only *melalo*, since it is an appropriate place for such things. Even so, a tidy house is an indication of moral and ritual purity and is considered to be conducive to good health. Besides, *Mamioro*, the spirit (*muli*) who is responsible for many serious illnesses, visits homes that are dirty as well as *marime* places, such as garbage dumps.

FOOD

It is very important to the Rom that they eat only food which is pure and clean. Food is associated not only with the upper half of the body, but also enters the body and has the double association of upper/inside as opposed to lower/outside. Consequently, there are many regulations regarding the preparation and handling of food. Eating is imbued with great social significance: to share food with a person is a statement of respect, friendship and acceptance of his cleanliness; to refuse to share food is to malign and insult a person, to deny him a social status as a pure (*wuzho*) person. The most serious punishment the Rom as a group can levy against any individual is to refuse him commensality. All rituals at which the Rom express important unifying social values involve the sharing of food. This includes life cycle rituals such as weddings, baptisms, and funeral feasts as well as other rituals such as various *slave* (Saint's day feasts), a *pakiv* (a feast in honour of an important person), and even the trials at which laws of social behaviour are resolved.

All food must be carefully prepared to avoid any *marime* contacts. Thus cooking as well as eating utensils and paraphernalia are always washed in a special basin with cloths reserved exclusively for these items. In many households, one soap is reserved for food-related items, and even the hands are washed only with one soap. Pregnant women, women in birth confinement (which varies from nine days to six weeks), and menstruating wommen, do not usually handle food, since they might contaminate it. Food prepared by *gaje* is known not to have been subject to correct separation from *marime* influences, and therefore is preferably avoided. This avoidance is not always possible, especially when travelling, and many Rom have for convenience taken to eating in restaurants though they still avoid *gaje* homes fairly strictly. When in restaurants, however, they try to avoid as many *marime* contacts as possible, and prefer establishments that serve wrapped take-away foods, paper cups, plastic forks, etc., or they may simply eat with their hands as is their custom at home. On one unforgettable occasion, when a group of Rom and I were forced to eat in a swanky restaurant in San Francisco, I was severely lectured on the terrible pollution dangers we were all exposed to and the diseases we were likely to come down with, while irate waiters and other customers stared in disgust at us because we ate everything with our hands.

Cleanliness in food preparation is particularly important on ritual occasions. This may be not only because the effectiveness of the ritual would be diminished, but also because at any ritual feast many Rom come together to eat, and there must be no cause for concern over food purity by any individual. To refuse to eat at a ritual is highly insulting. Even to fail to turn up at a ritual where one is expected is a slight on the reputation and cleanliness status of the host and can result in serious fights between rival groups. Therefore, for ritual occasions, special precautions are taken:

1. Fresh fruit and vegetables are bought in unpacked crates. Drinks are bought in cases or beer in kegs, and meat is bought on the hoof and is butchered by elderly Rom. Buying in bulk is not only convenient and economical for large gatherings, it is the best way to minimize handling of food by *gaje* and to ensure that the most fastidious care is taken in its preparation.

2. All cups, plates, napkins, and table cloths on a feast table must be pure; therefore, disposable items are always used, so that each person has unused table implements. The availability of paper and plastic products has greatly simplified the preparation of a feast table.

CONCLUSIONS

The Romany conception of purity and pollution, based on an ideal of separating the upper and lower halves of the body and including in certain instances male/female, adult/child and food/non-food separation is a metaphor for expressing various levels of social behaviour, relations with outsiders, internal political divisions, age statuses, ritual, arrangement and furnishing of the household, clothing and washing, cooking and eating, and the most personal activities and functions of the body.

These ideals of purity are, however, very difficult to maintain both for the body as well as for social boundaries. The Rom live within another culture, and although they strive to keep themselves separate from that culture, the boundary between themselves and outsiders is never static. Purity/pollution concepts and practices may ensure that enduring personal relationships with outsiders are nearly impossible, but economically the Rom are dependent on outside sources of income; therefore economic and political relationships with *gaje* must be more flexible. However, although economically the Rom are flexible, definitions of inclusion and exclusion from the group (*Romania*) remain constant or inflexible. The metaphor they use to discuss and determine inclusion and exclusion from the group is the purity metaphor, which centres on body divisions and extends to items that touch the body, food that enters the body, and the life cycle changes of the body. The metaphor extends also to internal social ranking, the status of an individual, his *vitsa* and his *natsia*. But it is at the widest social level - in relations between Rom and *gaje* - that the contradictions are most apparent. On the one hand, the Rom are aware that they live within a society that despises them and erects its own boundaries against them. To improve their situation in this society they often need to counter these prejudices and boundaries. On the other hand, they see themselves as morally superior to outsiders. They covet their separateness and are aware that outsider's prejudices and misconceptions help to keep them separate. Thus the most important boundary, that between Rom and *gaje*, is one that is forever being crossed for economic and political reasons. It is appropriate, then, that they would choose a metaphor which approximates this experience, that they would divide the body in such a way that the boundary (at the waist) is equally a difficult one to maintain. The physical experience of their bodies is similar to the widest social experience in an awareness of crossing boundaries. When these boundaries are crossed, the controls are most manifest.

REFERENCES

Campbell, J.K. (1964). *Honour, Family and Patronage*. Oxford: Clarendon Press.

Douglas, Mary (1970). *Natural Symbols, explorations in Cosmology*. London: Cresset Press.

Hertz, Robert (1960). *Death and the Right Hand*. (Trans. by R. & C. Needham) Cohen & West.

Lee, Ronald (1971). *Goddam Gypsy, an autobiographical novel*.

Montreal: Tundra Books.
Miller, Carol (1975). American Rom and the Ideology of Defilement. In Rehfisch (ed.) *Gypsies, Tinkers and Other Travellers*. Academic Press: London.
Needham, Rodney (ed.) (1973). *Right and Left: essays on Dual Symbolic Classification*. Chicago: University of Chicago Press.
Sampson, John (1926). *The Dialect of the Gypsies in Wales*. Oxford: Clarendon Press (Reprinted 1968).
Sutherland, Anne (1975). *Gypsies, The Hidden Americans*. London: Tavistock.
Sutherland, Anne (1976). *Gypsy Women and Body Symbolism*.
Turner, Victor (1966). Colour Classification in Ndembu Ritual. In ASA3 *Anthropological Approaches to the Study of Religion*. London: Tavistock.

SYMBOLS AND SOMATIC STRUCTURES

Lorna McDougall

As components of classification, collective representations have from the outset had a certain mentalistic quality. They are considered powerful expressions of abstract values and important social bonding mechanisms.

Because we take abstraction as characteristic of classificatory processes, we are inclined to see symbols as arbitrary, though compelling, conditioners of perception. While some levels of classification serve to reflect social and linguistic order, it seems likely that others represent principles of organization that can be directly perceived. The human body, like other organisms, is highly structured, and it would be surprising if our perceptual processes were entirely arbitrary.

The assumption that our perceptual world is ordered by our use of language raises the problem of relativity, and the difficulty of calculating classificatory common denominators; it arises from the traditional mind/body dichotomy, the discretion of mental and physical events, which allows categorization to be disconnected from the state and structure of the perceiving organism. This being so, we hesitate to consider the relation of symbols either to the body or that part of the body which mediates between the mental and the physical: the brain. The assumption has been that since its structure, like that of the body, does not vary significantly across cultures, it is an element that we need not question.

Attempts to return to the baseline of somatic structure and function in evaluating the form of certain common symbols have, however, indicated that attention to the body might repay effort. Victor Turner's ascription of colour symbolism to psychobiological determinants comes immediately to mind (Turner 1967: 91), as does Needham's search for the 'primary factors of human experience' (Needham 1974), and the trend towards a more comprehensive approach to behavioural and environmental interaction exemplified in ecological anthropology. These lines of study suggest the interdependence of man and nature, and provide a basis for taking into consideration the specifics of the 'natural' human body and collective representations and practices in the varied settings of time and space. If this be granted, a number of interesting questions can be raised relative to the consequences of collective representations at the somatic level and the reverse.

The issue to which I want to draw attention in this context concerns certain persistent pan-cultural symbols. In our efforts to attain high levels of description in idiographic studies, this kind of problem has been somewhat neglected. But it deserves attention, and the partial answer that I present here for certain symbolic forms across cultures has as much to do with bodies as with minds.

Abstract geometrical designs may not seem obvious candidates in the search for somatic influences on symbolic form. Jung, however, recognized certain geometrical figures, including the mandala or circle, as archetypes, which are by definition reduced to invariants resulting from commonalities of the human body and its structure. I shall examine this position in the light of experimental evidence that relates to the generation of a class of image to which some archetypes belong: phosphenes.

PHOSPHENES

Phosphenes are defined as luminous endogenous patterns experienced with the eyes closed (Knoll *et al.* 1963:222). They are, among other things, what we see when we 'see stars' on emerging from darkness into light, or vice versa. Phosphenes are known by different names in different fields: to psychiatrists as visual hallucinations, to opthalmologists as entopic phenomena, whilst psychologists recognize them by the name we are using here. The reader may begin by experimenting with his own (harmless) production of phosphenes by the simple method of placing the fingers on closed eyelids and exerting differing degrees of pressure. Weak pressure produces phosphenes of curvilinear form, discs, concentric circles or arcs. These tend to appear at the side of the visual field opposite the point at which the pressure is applied; strong pressure exerted in the same way produces a 'checkerboard effect' (Oster 1970:84).

Phosphenes have a venerable history. They are mentioned by Deschales in 1690 (Helmholtz 1909:189-191). LeRoy experimented with them in 1755; Volta, between 1796 and 1800, Helmholtz (1911), and later, Klüver (1942), all contributed to their study, as well as many others (Gebhard 1952; Clausen 1955; Bliss and Clark 1962: 93).

In more recent times intensive experimentation has been carried on in Japan by K. Motokawa and his associates, and in Germany by M. Knoll and his colleagues. Knoll's enduring interest in visual phenomena is best known in his development of the electron microscope with E.R. Rushka. In the U.S., J. Clausen and J. Gebhard have carried out numerous experiments on phosphenes, while others have exhibited interest in the experimentation in the light of its possible practical applications (Clausen, 1955:1-26). Oster, for example, is interested in using phosphenes as vehicles of visual prosthetics, to provide impulses for the blind or visually handicapped (Oster 1970).

Many of the investigations of phosphenes have relied on producing them under laboratory conditions, often in conditions of sensory deprivation, using either electrical or chemical stimulation. Outside the laboratory there are numerous situations in which phosphenes occur spontaneously. We may pay little attention to them in many circumstances, for they are often fleeting. They are provoked by a state of fatigue and drowsiness (Ahlenstiel and Kauffmann 1953:503, 506); as a result of psychological (Bachem *loc. cit.*) or mechanical shock; during fasting and meditation (Bash *et al.* 1954:205; Ahlenstiel and Kauffmann 1952: 188) or chemical intoxication. They further manifest themselves as vehicles of illumination in mystical traditions, and in the phenomenon known as the 'prisoner's cinema', experienced by people incarcerated in dark dungeons; they may perhaps be related to phantoms and ghosts; they are the hazard recognized by the long-distance truck-driver peering for hours into a snowstorm, and by pilots flying alone at high altitudes in empty, hence visually cueless, skies. Toxins such as those associated with scarlet fever produce phosphenes (Oster 1970:83) as do hallucinogenic drugs (that is those which have by definition a selective affinity for the visual system): mescaline (Serko 1913), psilocybin (Weidmann and Cerletti 1960), and lysergic acid diethylamide (Knoll *et al.* 1963). Some have also been reported while the open cortex has been stimulated locally during brain surgery, using a frequency of 60 cyc/sec.

Experimentation with phosphenes includes investigations of generating sites and mechanisms (Clausen 1955; Knoll *et al.* 1963), effective types of stimulus (Höfer 1963), and the identification and lifetime of single phosphenes (Knoll *et al.* 1963). The influence of chemical stimulants on their shape (Knoll *et al.* 1963) has also been investigated.

Phosphenes tend, by virtue of their form, to fall into one of two broad categories, *representational*, and *geometrical* or *abstract*. It is with the latter that we are concerned here.

Phosphene shapes have been classified into fourteen distinct groups (Höfer 1963), within each of which there are variants. The classification is provisional, and in Figure 1 a fifteenth class

has been added:-

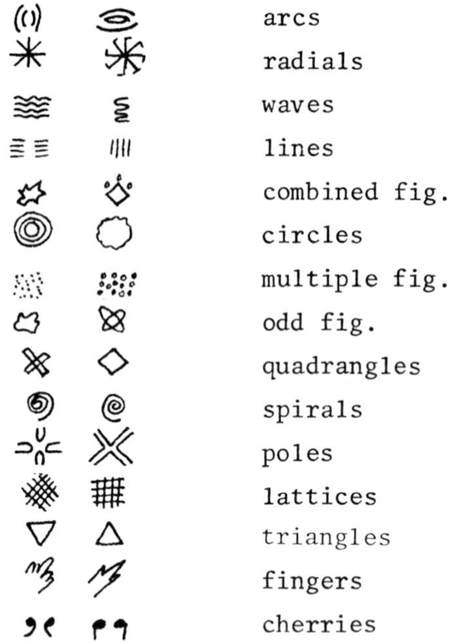

Fig. 1. *Fifteen types of phosphene, in order of frequency of occurrence*. (After Kellogg, Knoll and Kugler 1965:1129).

The provisional classification of phosphenes cited here is the result of experiments involving 373 subjects who were stimulated by electrical pulses in the electroencephalographic range 5-35 cyc/sec. The patterns generated vary with the frequency within a very narrow band-width range and remain stable for the individual over a period of time in some cases (Knoll and Kugler 1959: 1824; Knoll *et al.* 1962). Various factors are also known to alter the frequencies at which individual phosphenes are generated.

Phosphenes can be shown to differ according to the method of stimulation, electrical or chemical. Electrically induced phosphenes are of abstract or ornamental type patterns with only vibrational or repeating movements, whilst chemically-induced phenomena also include landscapes and living or man-made objects, such as flowers, animals, machines, and fast-moving, firework-type patterns (Knoll *et al.* 1963:219). With combined stimuli, electrical stimulation used at the peak of chemical stimulation does not enhance further production of phosphenes, although it makes much brighter those that do appear (*loc. cit.*). The probable reason for the distinction between the two modes of stimulation is the contrast between the broad spectrum of drug effects and the narrower electrical effects which are confined to a few brain regions of high electrical conductivity (*ibid.*: 220). Chemical stimulation thus induces subjective representational images as well as a greater variety of phosphenes than does electrical stimulation. The characteristic movements of chemically induced phosphenes include those categorized as *vibrational* or *repeating, high speed,* and *transitional.* Different drugs also produce different effects in relation to speed of image, colour, etc. (*ibid.*: 221). The movement is an interesting feature. It has been suggested that Op-Art is an attempt to capture this directly.

While most of the research has centered on eliciting the specific aspects of the visual system responsible for the production of phosphenes, upon which there are several views, Clausen (1955) records other somatic functions that are important influences in determining the frequency threshold for phosphene production. He reports that CO_2-O_2 balance in the blood is effective in altering the phosphene threshold. This balance is altered by hyperventilation. Motokawa also indicates that oxygen content is critical (Motokawa and Iwama 1949). Although research has been less intensive in the more general area of somatic states, there are reports of other influential endogenous factors, including adrenalin levels (Bogoslowski 1944). The relationship between phosphene threshold and blood pressure or vaso-motor reflexes is reported to be negative (Clausen 1955).

Both hyperventilation and adrenalin levels are known to affect states of mind, sometimes in gross and sometimes in more refined ways. Hyperventilation can cause blackout, an extreme effect, but more sophisticated breathing techniques such as those employed in Yogic traditions, alter functioning in more refined ways and, it is frequently asserted, assist in the development of the 'higher' states of consciousness. Phosphenes are probably not causes but effects of certain styles of somatic functioning. As Clausen points out, there is a possibility of using phosphene thresholds as indicators or measures of psycho-physiological functions. This suggestion has not been followed up.

Although there is no consensus about the site that generates phosphenes, there is general agreement that phosphene vision differs from normal vision. The most recent evidence suggests that phosphene designs reflect the orderliness of neuronal structure, and that they represent the brain looking at itself with an internal eye, so to speak. It is hypothesized that the different designs are generated at different loci along the visual pathway (Oster 1970:83-86). Most of the conditions which favour the spontaneous generation of phosphenes are, as we have seen, characterized by the lack of directed attention. It is as though, when the mind 'idles', it is scanning the inner landscape in an undirected, spontaneous fashion. It is clear that phosphenes are a function of the interaction of somatic factors with specific types of stimulus. There are accordingly two relevant features, the states of the somatic system involved, and the stimulus applied. While the variance in individual thresholds for specific phosphenes has been noted, it has not been further probed. Also the classification of phosphenes, although valuable in itself, tells us nothing about their relation to what we experience as 'meaning'. Laboratory conditions that facilitate investigation of the visual generating mechanisms are not in general suitable for intensive investigation of the more social and subjective elements of phosphene experience, although some subjective descriptions have, in fact, been recorded (Knoll and Kugler 1959).

Passing reference may be made to phosphenes as 'visual hallucinations'. Visual hallucination is derived from its Greek origin in a term meaning 'to wander in mind'. Whilst those who see phosphenes may be wandering in mind, as we have seen, they are not necessarily deluded. Ordinarily, we consider the hallucinating individual abnormal. He or she is seeing things which, to the best observation of others present, are not there. It is only in certain instances that such is in fact the case. A hallucinating subject may be involved not in seeing things which are not there, but only in seeing things which, generated as they are within the confines of the brain, are not available to be shared by companions.

We must now turn our attention to the connections between phosphenes as we recognize them experimentally and as they are manifest in both artistic and experiential form.

PHOSPHENES AS SYMBOLS

If we make the not unreasonable assumption that phosphenes are discovered in the body rather than built into it by cultural experience, it is not surprising that phosphenes commonly inform vocabularies of symbols occurring in 'primitive' art (Oster 1970: 83). They also inspire neolithic art (Kellogg *et al.* 1965:1130). It is not possible to reproduce here the impressive range of recurring symbols adduced by Kellogg to demonstrate her contention concerning the universality of phosphenes in art forms throughout the world (see Figure 2). The reader's attention is drawn to this evidence, contained in her well documented volume, *Analyzing Children's Art* (1969), and covering a great variety of different cultures. A brief look at one or two other examples will have to serve our purposes here. Phosphene designs appear to play an important role in the development of a child's expression in drawing or painting, where the abstract forms precede the representational, and begin at about 2 years (*loc. cit.*). Phosphene images

(a) (b)

Fig. 2. *Phosphene designs present in (a) Egyptian motifs* (after Kellogg 1969:213) *and (b) Australian motifs* (after Black 1964:21).

parallel these early geometric forms. Kellogg refers to these formal similarities as composing a 'natural' art (Kellogg 1969: 208-225), since children of similar ages in different cultures appear to draw a limited and definable range of designs which from the start resemble phosphene types. Knoll considers that many of the phosphenes belong to the category of archetypal images (Knoll 1958:124).

It has been observed that among the geometrical designs produced by visual hallucination, the figure of the mandala, in its more specialized form, the yantra, appears frequently. This is an occidental parallel to the eastern symbolic manifestations (Ahlenstiel and Kauffmann 1952:189; Bash, Ahlenstiel, and Kauffmann 1954), of which the physiological correlations are not, unfortunately, available. The large range of geometrical images, wheels, stars, checkerboards, and zigzags have been classified as 'pre-yantra' images (Ahlenstiel and Kauffmann 1952:189).

Eliade defines the mandala as a series of circles which may or may not be concentric, inscribed in a square (Eliade 1963:373). Jung describes the formal elements of archetypal mandala symbolism as follows:

1. Circular, spherical, or egg-shaped formation.
2. The circle is elaborated into a flower (rose, lotus) or a wheel.
3. A centre expressed by a sun, star, or cross, usually with four, eight, or twelve rays.
4. The circles, spheres, and cruciform figures are often represented in rotation (swastika).
5. The circle is represented by a snake coiled about a centre, either ring-shaped (uroboros) or spiral (Orphic egg).
6. Squaring of the circle, taking the form of a circle in a square or vice versa.
7. Castle, city, and courtyard (temenos) motifs, quadratic or circular.
8. Eye (pupil and iris).
9. Besides the tetradic figures (and multiples of four), there are also triadic and pentadic ones, though these are much

rarer. They should be regarded as 'disturbed' totality pictures, as we shall see below (Jung 1950:361-362).

The word mandala is the Sanskrit word meaning 'circle'; as Jung points out, it is the Indian term for the circles drawn in religious rituals (Jung 1950:355). The mandala diagram, in an elaborate and sometimes pictorial form, is traced on the ground or on paper, or it may be found etched in stone or metal, as in the great temples.

The Yantra is the purely geometrical mandala. It is used in both Hindu and Buddhist traditions as a support for meditation (see Figure 3A). The contemplation of the yantra, in conjunction with Yogic practices, is said to take the believer along the path leading to enlightenment. The interplay of the elements is understood to be an expression of all the cosmic manifestations, beginning and ending in the unity of the centre visualized as a dot.

Both the mandala and the yantra often show the typical quaternary character. Since these are cosmic diagrams, it follows that they manifest the general tendency to visualize or represent the world as divided into four quadrants. In desacralized terms the quadrants represent to modern man areas delimited by the axes joining the cardinal points. In some traditions the axes are at an angle of 45° to the horizontal axis, like the cross of St Andrew in Heraldry.

(a) (b)

Fig. 3. *(A) Linear Yantra* (after Zimmer 1947:plate 36) *(B) Papago mandala* (after Sides 1961:plate 42).

THE INTERPRETATION OF PHOSPHENES

Phosphenes are images associated with subjectively different experiences, having varying physiological and bioelectrical correlates. The symbols themselves are not mentalistic categories, but direct representations of experiential events involving sensory processes as one component, and they are objectively distinguishable. But can we then say that physiological or biochemical measures reflect similar types of experience? From research in psychology, we know that apparently similar physiological states may be experienced differently in terms of pleasure and pain. Setting and expectation are important factors here, as well as the question of methodology and accuracy of procedures to assess 'similarity'. The brain may well be like a computer in utilizing both hardware and software, and the interpretation of neurological events may be subject to programming, although the nervous system works in predetermined ways.

A detailed comparison of native interpretations of phosphenes has not yet been carried out. Among the Dogon we find the geometrical zig-zags, lattices, circles and lines composing phosphenes taken to be fundamental to the structure of the material universe, and the zig-zag, for example, interpreted as the alternation of life and death, day and night, themselves symbolic of the interplay of the forces of order and disorder in the cosmos. In very general terms, phosphene-like shapes in primitive art are associated with 'sacredness', and principles of order.

In this respect it is interesting to note that there are many rituals in which techniques are employed which are in principle capable of triggering phosphenes. It is not my purpose to argue that these constitute objectives of ritual or to inflate their importance. I am inclined to think that they are most likely to accompany desired alterations in perception brought about by altered physiology. The means of altering somatic functioning are legion, and only a certain amount is known of the systematic alterations of perception attendant upon each type. Barbara Lex (1975) has recently shown that a number of ritual practices, such as sweating, fasting, and abstention from salt and water, can be elucidated in terms of modern bio-chemistry, and she has indicated that some such practices may stimulate the production of endogenous, mind-altering substances. The evidence suggests functioning brings about an alteration of awareness and perception, at least at an empirical level. Myerhoff's data from the Huichol peyote ritual may be cited as a case in point (Myerhoff 1974): several techniques which alter body functioning are used to alter perception, and it seems that phosphenes of many kinds, including a flower-like mandala, are generated. It may be observed that the experience of the mandala is typically associated with the dissolving of boundaries between the Self and the objective world. Tucci remarks that this transition is facilitated by the correspondence of micro- and macro-cosm (Tucci 1961). In the process of generating the mandala, the brain appears to become aware of itself as both subject and object of perceiving consciousness. This is familiar to anthropologists as the 'body as model of the universe' paradigm, so often encountered in the study of cosmologies, and embodied in temples and houses.

SYMBOLS AND ARCHETYPES

Firth characterizes the modern view of symbols 'not as surrogates or evasions of reality, but as "higher forms of reality"' Firth (1973:166). Yet we often find that the 'higher reality' is social reality in disguise. 'What ritual performances do is to recall and present in symbol form the underlying order that is supposed to guide the members of the community in their social activities' (*ibid.*:167). This represents one of the most frequently found approaches in the analysis of symbols. It owes much to the influence of Durkheim's concept of collective representations, which directs our attention to the reflection of social structure in symbolic structure, in which respect it is extremely useful. It is interesting, however, to set Jung's delineation of archetypes in complementary relationship with the views of more sociologically-minded thinkers.

Jung claims that the concept of archetypes corresponds with 'the collective representations of Lévy-Bruhl, the categories of the imagination of Hubert and Mauss and the elementary or primordial thoughts of Adolf Bastian'. But Jung approaches his material from a position that differs considerably from that of Durkheim or contemporary adherents to his philosophy, even if he concurs with Durkheim and Mauss in attributing an affective component to the sources of the symbolic configurations (Durkheim and Mauss 1963:85).

Jung frequently draws attention to our typically Western misconstruction of symbolic meaning in terms of a category. He considers that while the archetype is the primordial image, the 'autochthonous mythical motif' is 'subject to progressive incrustation with rational forms' (Jung 1948:147). This occurs in a definite sequence, and our apprehension of symbols is heavily overlaid with such rationalization. According to Jung, the archetype degenerated to thought after Descartes, and subsequently Kant, had reduced the archetype to a limited number of categories of understanding (*ibid.*: 55-56). Through this veil of history we are erroneously led to view symbols as intellectual constructs. Jung concludes, 'Western consciousness is by no means the only kind of consciousness there is; it is historically conditioned and geographically limited, and representative of only one part of mankind' (*ibid.*: 55).

Jung is ever concerned to take account of what he perceives to be the intuitive basis of knowledge. In sociological terms, this basis is less significant than the configuration that symbols assume in serving to encode messages about social structure. Whether these two approaches are ultimately mutually exclusive seems, however, open to question.

Jung describes the most salient features of archetypal forms as follows: chaotic multiplicity and order; duality; the opposition of light and dark, upper and lower, right and left; the union of opposites in a third; the quaternity (a square or cross); rotation (circle, sphere); and finally the centering process and radical arrangement that usually forms some quaternary system Jung 1948:136).

These are the recurrent images to which the infinitely varied forms of his patient's fantasies could consistently be reduced. These archetypal forms represent 'well-defined themes and formal elements' (Jung 1954:203). They are the images of fantasy, and the term image is intended to express not only the form of activity taking place, but the typical situation in which the activity is released.

Jung goes further, however, and asserts that as an archetypal form, the mandala is not merely a motif in the 'psychology of primitives', but a pattern of instinctive behaviour and even 'an unconscious image of the instincts themselves' (Jung 1948:136). It appears that he interprets instinct as the arbiter of organized understanding in the broadest sense. He quotes Herbert of Cherbury:

> the natural instincts are found in every normal man, through which the internal conformity of things such as the cause, meaning, or purpose of things, the good, the bad, the beautiful, the pleasing, etc. are brought into conformity independently of discursive thought (*loc. cit.*).

Such instinctual patterns are not individually generated. Archetypes have a universal quality, and their origin must have coincided, Jung insists, 'at least with the origin of the species' (Jung, 1954:78). The universal quality they manifest is what connects them with the 'collective unconscious'; this in turn owes its existence and substance exclusively to (human) heredity. In its universal aspect the collective unconscious contrasts with the personal unconcsious, since the latter is composed of events of a uniquely individual nature, correctly classified not as archetypes but as complexes (Jung 1948:134).

The archetypes, first described by Dionysius Aeropagite and

found extensively in the Corpus Hermeticum of the Mediaeval Alchemists, are symbolic images and the collective unconscious, as repository of the archetypes, ultimately reveals 'the psychic expression of identity of brain structures irrespective of racial differences' (Jung, 1967:11), a position of considerable interest in view of the preceding evidence. At their root, many of the categories which inform the social symbolic system are of a sensory-intuitive sort and it may be that this is why analogies based upon them are so compelling.

It is appropriate to note that Huxley has already recognized that social symbolism need not represent exclusively the imposition of category on experience. In his discussion of the ritual of voodoo and the symbolism of the body, he asserts that 'images in fact seem to be connected with awareness of physical states which cannot be verbally defined and ritual can be seen as a way of resolving the mind-body dichotomy by acting out the forces which the body proposes to the mind' (Huxley 1966:427).

Jung raises another issue of importance in the interpretation of symbols, stemming from the possible *levels* of interpretation. Mandalas are not all alike, even though they share certain representational similarities. He comments that among other things, even in India, the centre is treated in very different ways in different representations, depending on the requirements of the ritual, the grade of initiation of the contemplator, and the sect to which he belongs (Jung 1950:356). In addition, mandalas can be more or less regular. Jung refers to 'disturbed mandalas' and mentions that the 'best' are Tibetan (*loc. cit.*). He expresses the opinion that 'no European mandalas known to me (I have a fairly large collection) achieve the conventionally and traditionally established harmony and perfection of the Eastern mandala' (Jung 1967:56), a value judgement that might bear further investigation.

The representation is therefore available to anyone confronted with it who has eyes to see, although the endogenous image may only be generated under specific physiological conditions. To the untrained, the yantra is nothing more than a symmetrical and rather pleasing design. To cognize its inner meaning, the contemplator must be aware of identity with the design, which he or she can achieve only in appropriate neurological circumstances. Even so, the perception of the same phosphenes under apparently identical neurological circumstances may not have the same meaning to two different persons, especially in different cultures. Again, although experimental evidence of conditions conducive to phosphene production makes certain kinds of symbol more accessible to study, the observation of a phosphene-generating person clearly need not simultaneously give to the experimenter direct understanding of the image or experience of which he is observing and measuring the physiological correlations.

Different levels of symbolic understanding are recognized in several cultures. For example, the Dogon initiate develops his symbolic understanding in response to personal growth. The novice is originally instructed in only the major meanings of the symbolic universe. Further understanding results from curiosity and experience, motivated by a high degree of interest (Griaule and Dieterlen 1965:51). The young initiate will learn by asking questions as they occur and seem relevant. And 'même au cours 'un systeme, il [l'instructeur] ne lui en donne pas la clef; le postulant doit la trouver' (*ibid.*: 52).

The initiate gradually learns about the structure of the universe; but, 'il s'agit également de la vivre, au sens le plus complet du mot, par le corps comme par l'esprit' (*ibid.*:53). In the process of initiation it is not only symbols that change: understanding of them becomes more complete

Evaluation

The direct relationship between body and imagistic structure represents an extension of the search alluded to above for the 'primary factors of human experience'. It emphasizes the importance of considering the body as a shaping factor in social representations rather than as the resultant mould. In this, it contrasts with several contemporary findings. Mary Douglas maintains that 'the social body constrains the way the physical body is perceived' (Douglas 1970:65). While this is undoubtedly true, it does not mean that society can condition the body in any way it pleases. It may be overstating the case to agree with Marcel Mauss that in relation to techniques of the body there can be 'no such thing as natural behaviour' (*ibid.*:65). There are, equally clearly, limits within which collective representations are constrained and, in addition, body is neither able to be dissociated from mind nor is mind a totally disembodied creative force.

The evidence reviewed in this paper offers a glimpse of the body as a source of symbols rather than as an entity to be shaped and formed by social systems and arbitrary devices. The end of natural science is to understand nature, and since nature's ways are to some extent open to direct intuitive sensuous knowledge it follows that 'to know nature well, the natural body is to be trusted, cherished, and made sensitive to its natural environment' (*loc. cit.*). Thus certain somatic states may, through their intrinsic structure, be capable of engendering common forms adopted as collective representations and elaborated in accordance with cultural styles.

REFERENCES

Ahlenstiel, H. and Kauffmann, R. (1952). Über die Mandalaform des 'Linearen Yantra'. *Schweizerische Zeitschrift für Psychologie, 11,* 188-197.

Ahlenstiel, H. and Kauffmann, R. (1953). Geometrische Gestalten in Optischen Halluzination. *Archiv für Psychiatrie und Nerven Krankheiten, 190,* 503-529.

Bash, K.W., Ahlenstiel, H. and Kauffmann, R. (1954). *Über Praeyantraformen und ein Lineares Yantra: Studien zur analytischen Psychologie C.G. Jung.* Zurich: Rascher Verlag.

Black, R. (1964). *Old and New in Australian Aboriginal Art.* Sydney: Angus and Robertson.

Bliss, E.L. and Clark, L.D., (1962). Visual Hallucinations. In West, J.L. (ed.), *Hallucinations.* New York and London: Grune and Stratton, 92-107.

Bogoslowski, A.I. (1944). The Problem of the Electrical Sensitivity of the Eye. *Prob.Physiol.Opt. 2,* 136-172.

Cerletti, A. (1959). Teonanactl and Psilocybin. *Deutsche medizinische Wochenschrift, 34,* 2317-2321.

Clausen, J. (1955). *Visual Sensations (Phosphenes) produced by AC Sine Wave Stimulation.* Copenhagen: Ejnar Munksgaard.

Douglas, M. (1970). *Natural Symbols: Explorations in Cosmology.* London: Barrie and Rockliffe.

Durkheim, E. and Mauss, M. (1963). *Primitive Classification.* Translated from the French by Rodney Needham. Chicago: The University Press.

Firth, R. (1973). *Symbols Public and Private.* Ithaca, New York: Cornell University Press.

Gebhard, J.W. (1952). Thresholds of the Human Eye for Electric Stimulation by Different Wave Forms. *Journal of Experimental Psychology, 44,* 132-140.

Griaule, M. and Dieterlen, G. (1965). *Le Renard Pâle.* Travaux et Mémoirs de L'Institut d'Ethnologie *72,* Paris: Institut d'Ethnologie.

Helmholtz, H.V. (1909-11). *Handbuch der Physiologie: Optik.* 3 Vols. Leipzig: Leopold Voss.

Höfer, O. (1963). Über die Abhängigkeit elektrisch induzierter subjektiver Muster (Phosphene) vom zeitlichen Verlauf der anregenden Wechselstrórne. *Elektromedizin, 8,* 72-78.

Huxley, F. (1966). The Ritual of Voodoo and the Symbolism of the Body. *Philosophical Transactions, The Royal Society of London,* series B, *251,* 423-427.

Jung, C.G. (1948). Instinct and the Collective Unconscious. In *Structure and Dynamics of the Psyche.* Collected Works, vol. 8, 2nd ed., 1969. New York: Pantheon Books.

Jung, C.G. (1950). Concerning Mandala Symbolism. In *The*

Archetypes and the Collective Unconscious. Collected Works, vol. 9(1). New York: Pantheon Books, 1959.

Jung, C.G. (1954). On the Nature of the Psyche. In *The Archtypes and the Collective Unconscious*. Collected Works, vol. 8. New York: Pantheon Books, 1969.

Jung, C.G. (1967) Commentary on the Secret of the Golden Flower. In *Alchemical Studies*. Collected Works, vol. 13, 5th ed. New York: Pantheon Books, 1969.

Kellogg, R., Knoll, M. and Kugler, J. (1965). Form Similarity between Phosphenes of Adults and Pre-school Children's Scribblings. *Nature*, 208, No. 5015, 1129-1130.

Kellogg, R. (1969). *Analyzing Children's Art*. Palo Alto: National Press Books.

Kluver, H. (1942). Mechanism of Hallucinations. Ch. X in *Studies in Personality*. New York: McGraw-Hill.

Kluver, H. (1958). Discussion on a Paper by W.P. Chapman. *Research Publications, Association for Research in Nervous and Mental Diseases*, 36, 258-277.

Knoll, M. and Kugler, J. (1959). Subjective Light-Pattern Spectroscopy in the EEG Range. *Nature, 184;* no. 4701, 1823-1824.

Knoll, M., Höfer, O., Lawder, S.D. and Lawder, U.M. (1962). Die Reproduzierbarkeit von elektrisch angeregten Lichterscheinungen (Phosphenes) bei zwei Versuchspersonen innerhalb von 6 Monaten. *Elektromedizin, 7*, 235-242.

Knoll, M., Kugler, J., Höfer, O. and Lawder, S.D. (1963). Effects of Chemical Stimulation of Electrically Induced Phosphenes on their Band-width Shape, Number and Intensity. *Confinia Neurologica, 23*, 201-226.

Lex, B. (1975). A Proposed Bio-Anthropological Approach Linking Ritual and Self-intoxication. Paper presented at the 74th Annual Meeting. AAA. San Francisco.

Motokawa, K. and Iwama, K. (1949). The Electric Excitability of the Human Eye as a Sensitive Indicator of Oxygen Deficiency. *Tohoku Journal of Experimental Medicine, 50*, 319-328.

Myerhoff, B. (1974). *Peyote Hunt: the Sacred Journey of the Huichol Indians*. Ithaca, New York: Cornell University Press.

Needham, R. (1974). *Right and Left: a Study in Religious Polarity*. Chicago: The University Press.

Oster, G. (1970). Phosphenes. *Scientific American, 222*, no. 2, 82-87.

Sides, D. (1961). (1963). *Decorative Art of the South West Indians*. New York: Dover Press.

Tucci, G. (1961). *The Theory and Practice of the Mandala*. Translated from the Italian by Alan Houghton Brodrick. London: Rider and Co.

Turner, V.W. (1967). *The Forest of Symbols: Aspects of Ndembu Ritual*. Ithaca, N.Y.: Cornell University Press.

Weidmann, H., Cerlett, A. and Taeschler, M. (1960). Studies on Psilocybin and Related Compounds. *Helvetia Physiologica und Pharmacologica Acta*, *18*, 43-49.
Zimmer, H. (1947). *Myths and Symbols in Indian Art and Civilization*. New York: Harper and Row.

NOTES ON CONTRIBUTORS

ARNOLD, KATHERINE LOUISE. Born 1948, London. Studied at Durham University, B.A., M.Phil.
Research Assistant on the Irish government's project on the Irish language 1971-1972. Training in child psychotherapy at the Tavistock Clinic, London, 1973-.
Author of 'The whore in Peru - the splitting of women into good and bad, and the isolation of the prostitute', in S. Lipshitz ed. (forthcoming).

BAILY, JOHN SEBASTIAN. Born 1943, England. Educated at Oxford University, B.A. (Psychology and Physiology), University of Sussex, D.Phil.
Research Fellow, Department of Social Anthropology, The Queen's University of Belfast, on grants from the Social Science Research Council, 1973-.
Author of *Krishna Govinda's Rudiments of tabla playing* (1974); 'Arm-body adaptation with passive arm movements', *Perception and Psychophysics* (1972); 'Recent changes in the *dutar* of Herat', *Asian Music* (1976).

BENESH, RUDOLF. Born 1916, died 1975. London. Portrait painter and Fellow of the Institute of Chartered Accountants.
Copyrighted the Benesh Movement Notation in 1955 and founded the Institute of Choreology in 1963.
Co-author of *An Introduction to Benesh Movement Notation* (1956 and 1969).

BENESH, JOAN (née Rothwell). Born 1920, Liverpool. Studied at the Studio School of Dance and Drama, and ballet under Lydia Sokolova.
Joined the Royal Opera Company in 1950 and the Sadlers Wells Ballet the following year. In 1957 transferred to the Royal Ballet school to teach ballet and choreology. Founded the Choreology training course in 1963 and retired as principal in July 1976.

Co-author of *An Introduction to Benesh Movement Notation* (1956 and 1969). *Reading Dance* (1977).

BLACKING, JOHN. Born 1928, England. Studied at Cambridge University, B.A., University of Witwatersrand, Ph.D. and D.Litt.
Lecturer in Social Anthropology and African Government, University of Witwatersrand, Johannesburg, 1959-1965; Professor and Head of Department, and Chairman of African Studies Programme, 1966-1969; Visiting Professor at Makerere University College, Kampala, 1965; Professor of Anthropology, Western Michigan University, 1970-1972; Professor of Social Anthropology, The Queen's University of Belfast, 1970-.
Author of *Black Background: the childhood of an African girl* (1964); *Venda children's songs: a study in ethnomusicological analysis* (1967); *Process and product in human society* (1969); *Man and Fellowman* (1974); *How musical is man?* (1973 and 1976).

EKMAN, PAUL. Born 1934, Washington D.C., U.S.A. Studied at the University of Chicago, New York University, and Adelphi University, B.A., M.A., Ph.D.
Post-Doctoral Research Fellow, Langley Porter Institute, 1961-63; Assistant Professor of Psychology, San Francisco State University 1963-1966; Career Scientist Development Award, Langley Porter Institute, 1966-1971; Professor of Medical Psychology, Department of Psychiatry, University of California Medical School (San Francisco) and Director, Human Interaction Laboratory, Langley Porter Neuropsychiatric Institute, 1971-.
Co-Author of *Emotion in the Human Face* (1972), and of *Unmasking the Face* (1975). Editor of *Darwin and Facial Expression* (1973).

ELLEN, ROY F. Born 1947, United Kingdom. Studied at London School of Economics, B.Sc., Ph.D.
Social Science Research Council Studentship, 1968-1972; temporary lecturer in anthropology, London School of Economics, 1972-73; Lecturer in Social Anthropology, University of Kent at Canterbury, 1973-.
Author of 'The marsupial in Nuaulu ritual behaviour', *Man* (1972); 'Variable constructs in Nuaulu zoological classification'; and 'Non-domesticated resources in Nuaulu ecological relations', *Social Science Information* (1975); 'The development of anthropology and colonial policy in the Netherlands, 1800-1960', Journal of the History of the Behavioural Sciences (1976).

HANNA, JUDITH LYNNE. Born 1936, USA. Educated at University of California at Berkeley and Los Angeles, B.A. (Political Science), Michigan State University, M.A. (Political

Science); Columbia University, M.A., Ph.D. (Anthropology). Lecturer in African dances, Michigan State University, 1965; Research Scientist, American University, 1965-68; Adjunct Associate Professor of Liberal Arts, Fordham University, 1970-73; Research Scientist for Program Development and member of School of Arts and Humanities graduate faculty, University of Texas at Dallas, 1976-.

Co-author of *Urban Dynamics in Black Africa,* 1971; articles on urban politics, university students, nonverbal communication, and dance.

HUXLEY, FRANCIS. Born 1923, England. Studied at Oxford University, M.A., B.Sc.

Assistant Curator, Liverpool Museum, 1953-55; Research Fellow at St. Catherine's College, Oxford, 1963-68; social anthropologist, Weyburn Hospital, Canada, 1968-70; Co-founder of Survival International; Director of Studies of the Philadelphia Association, 1974.

Author of *Affable Savages* (1956); *The invisible* (1966); *The Way of the Sacred* (1974); *The Raven and the Writing-desk* (1976). Co-author of *Tribes of the Amazon Basin in Brazil* (1972).

KUBIK, GERHARD. Born in Vienna. Studied ethnology at Vienna University; Ph.D. Vienna 1971.

Lecturer at the Institute of Ethnology and the Musicological Institute of the University of Vienna.

Author of *Música Traditional e aculturada dos !Kung' de Angola* (Lisboa, 1970); *Recording and classification of oral literature in Tanzania and some other parts of Africa* (Dar-es-Salaam 1971), and numerous papers.

LANGE, RODERYK. Born 1930, Poland. Studied at the Nicolaus Copernicus University Torún, Poland and at the Wroclaw State University, M.A., Ph.D. (Ethnology, Cultural Anthropology); at the Folkwang Hochschule, Essen, Germany (Movement analysis and notation).

Head of dance research department, Torún Ethnographic Museum, Poland, 1954-67; Lecturer on Dance anthropology, Nicolaus Copernicus University, 1965-67; Senior Lecturer, Laban Centre for Movement and Dance, England, 1967-72; Co-director, Centre for Dance Studies, Jersey, C.I., 1971-; part-time Lecturer on dance anthropology, Queen's University, Belfast, 1975- and University of London, Goldsmiths' College, 1976-.

Author of *The Nature of Dance* (1975).

LEWIS, GILBERT AGUILAR. Born 1938, England. Studied at University of Oxford and University College Hospital, London, B.A., B.M., B.Ch.(Oxon), M.R.C.P. (London); London School of Economics, Ph.D.

Medical posts, 1962-65; Junior Research Fellow, Medical Research Council, 1965-67; Junior, then Senior, Research Officer, London School of Economics, 1967-71; Assistant Lecturer in Social Anthropology, Cambridge, 1971-75; Fellow of St. John's College, Cambridge, 1972-; Lecturer in Social Anthropology, Cambridge, 1975-.

Author of *Knowledge of Illness in a Sepik Society* (1975).

LOUDON, JOSEPH BUIST. Born 1921, United Kingdom. Educated at Oxford University, B.A. (natural science), M.A., B.M., B.Ch. (medicine), and at London School of Economics, A.P.D.S.A. (social anthropology).

Medical appointments 1946-56; Assistant Lecturer in Social Anthropology, London School of Economics 1956-57; Member of Scientific Staff, Medical Research Council 1957-64; Lecturer in Anthropology 1964-66, Senior Lecturer since 1966, University College of Swansea, University of Wales.

Author of *White Farmers and Black Labour-tenants in Natal* (1970); 'Religious order and mental disorder' in M. Banton (ed.), *The Social Anthropology of Complex Societies* (ASA Monograph 4) (1965); 'Teasing and socialization on Tristan da Cunha' in A. Mayer (ed.), *Socialization: the approach from social anthropology* (ASA Monograph 8) (1970). Editor of *Social Anthropology and Medicine* (ASA Monograph 13) (1976).

McDOUGALL, LORNA. Born 1942, Scotland. Educated at Trinity College, Dublin, B.A. (mod), and Oxford University, B.Litt.

Assistant Professor of Anthropology, Pitzer College and the Claremont Graduate School, The Claremont Colleges, California, 1971-.

Author of 'Lévi-Strauss in Fairyland', in A. Bharati (ed.), *Religion and Ritual*, and 'The Quest of the Argonauts', in T. Williams (ed.) *Psychological Anthropology*.

POLUNIN, IVAN VLADIMIROVITCH. Born 1920, England. Educated Oxford University, B.Sc., M.A., D.M.

Reader in the Department of Social Medicine and Public Health, University of Singapore.

Author of *The Medical history of Malayan aborigines* (1953); 'Health and disease in contemporary primitive societies; in D. Brothwell and A.T. Sandison (eds), *Diseases in Antiquity;* 'The magical medical system of the Jah Heut tribe of Central West Malaysia', *Proc. VIIIth ICAES*, 1970; 'Visual and sound recording apparatus in ethnographic fieldwork', *Current Anthropology* (1970).

ROSENTHAL, LEORA NADINE. Born 1950, South Africa. Studied at the University of the Witwatersrand, B.A., B.A. (Hons.), M.A., Ph.D.

Lecturer, University of the Witwatersrand 1974-1975. Currently residing in New York.

Author of 'Contagion and Cataclysm: a theoretical approach to the study of ritual pollution beliefs', *African Studies* (1973).

ROUGET, GILBERT. Born 1916, France. Studied at the Sorbonne, Paris.

Research Director, Centre Nationale de la Recherche Scientifique; Head of the Department of Ethnomusicology of the Musée de l'Homme, Paris.

Author of numerous articles on ethnomusicology, mainly on African Music (1946-1976), and of *La musique, l'extase et la transe. Eléments pour une théorie générale des relations de la musique et de la possession* (in press).

SKULTANS, VIEDA. Born 1944, Riga. Studied in University College, London. B.A.Hons. (Philosophy), Postgraduate Diploma in Social Anthropology. Ph.D. (Social Anthropology) Wales.

Senior Research Assistant in Department of Sociology and Social Anthropology, University College of Swansea, 1968-1971; Lecturer in Mental Health, University of Bristol, 1971-.

Author of 'The Symbolic Significance of Menstruation and the Menopause', *Man*, December 1970; *Intimacy and Ritual: A Study of Spiritualism, Mediums and Groups* (1974); *Madness and Morals: Ideas on Insanity in the Nineteenth Century* (1975).

STRATHERN, ANDREW JAMIESON. Born 1939, England. Studied at Cambridge University, B.A., Ph.D.

Research Fellow, Trinity College, Cambridge 1965-8; Research Fellow, Department of Anthropology, Research School of Pacific Studies, Australian National University, Canberra 1969-72; Professor of Social Anthropology, University of Papua New Guinea, Port Moresby, 1973-6; Professor of Anthropology, University College, London, 1976-.

Author of *The rope of moka* (1971); *One father, one blood* (1972); *Self-decoration in Mount Hagen* (with Marilyn Strathern), 1971; *Melpa courting songs*, 1974.

SUTHERLAND, ANNE HARTLEY. Born 1945, Texas. Studied at University of Texas, B.A. and Oxford University, dipl., B.Litt., D.Phil.

Lecturer in Portuguese, University of Texas, 1965; Lecturer in Anthropology, Durham University, 1972-; Consultant Anthropologist RAI/BBC anthropology film project, 1976.

Author of *Gypsies, The Hidden Americans* (1975); *Face Values*, 1977.

SUBJECT INDEX

A

B

C

INDEX OF SOCIETIES

AUTHOR INDEX